A-Level Year 1 & AS

Physics

Revising for Physics exams is stressful, that's for sure — even just getting your notes sorted out can leave you needing a lie down. But help is at hand...

This brilliant CGP book explains **everything you'll need to learn**, all in a straightforward style that's easy to get your head around. We've also included **exam questions** to test how ready you are for the real thing.

There's even a free Online Edition you can read on your computer or tablet!

How to get your free Online Edition

Go to **cgpbooks.co.uk/extras** and enter this code...

1762 0304 5527 0419

This code only works for one person. If somebody else has used this book before you, they might have already claimed the Online Edition.

A-Level revision? It has to be CGP!

3303197043

Contents

Topic 7 — Particles

Topic 8 — Quantum Phenomena

Published by CGP

Editors:
Emily Garrett, David Maliphant, Frances Rooney and Sarah Williams.

Contributors:
Tony Alldridge, Jane Cartwright, Peter Cecil, Mark Edwards, Barbara Mascetti, John Myers and Andy Williams.

ISBN: 978 1 78294 294 8

With thanks to Rachael Marshall, Sam Pilgrim and Charlotte Whiteley for the proofreading.
With thanks to Jan Greenway for the copyright research.

Printed by Elanders Ltd, Newcastle upon Tyne.

Based on the classic CGP style created by Richard Parsons.

The Scientific Process

'How Science Works' is all about the scientific process — how we develop and test scientific ideas. It's what scientists do all day, every day (well, except at coffee time — never come between a scientist and their coffee).

Scientists Come Up with **Theories** — Then **Test Them**...

Science tries to explain **how** and **why** things happen — it **answers questions**. It's all about seeking and gaining **knowledge** about the world around us. Scientists do this by **asking** questions, **suggesting** answers and then **testing** their suggestions to see if they're correct — this is the **scientific process**.

1) **Ask** a question about **why** something happens or **how** something works. E.g. what is the nature of light?
2) **Suggest** an answer, or part of an answer, by forming a **theory** (a possible **explanation** of the observations) — e.g. light is a wave. (Scientists also sometimes form a **model** too — a **simplified picture** of what's physically going on.)
3) Make a **prediction** or **hypothesis** — a **specific testable statement**, based on the theory, about what will happen in a test situation. For example, if light is a wave, it will interfere and diffract when it travels through a small enough gap.
4) Carry out a **test** — to provide **evidence** that will support the prediction (or help to disprove it). E.g. shining light through a diffraction grating to show diffraction and interference (p.92-93).

The evidence supported Quentin's Theory of Flammable Burps.

A theory is only scientific if it can be tested.

...Then They **Tell** Everyone About Their **Results**...

The results are **published** — scientists need to let others know about their work. Scientists publish their results as reports (similar to the lab write-ups you do in school) written up in **scientific journals**. Scientific journals are just like normal magazines, only they contain **scientific reports** (called papers) instead of the latest celebrity gossip.

1) It's important that the **integrity** (trustworthiness) of the reports published in scientific journals is checked. Scientists (like anyone else) might be **dishonest** or **biased**, or an investigation might have made **invalid** conclusions (see page 13).
2) The report is sent out to **peers** — other scientists that are experts in the **same area**. They examine the data and results, and if they think that the conclusion is reasonable it's **published**. This makes sure that work published in scientific journals is of a **good standard**. This process is known as **peer review**.
3) But peer review **can't guarantee** the science is **correct** — other scientists still need to **reproduce** it.
4) Sometimes **mistakes** are made and bad work is published. Peer review **isn't perfect** but it's probably the best way for scientists to self-regulate their work and to publish **quality reports**.

...Then **Other Scientists** Will **Test** the Theory Too

Other scientists read the published theories and results, and try to **test the theory** themselves. This involves:

- Repeating the **exact same experiments**.
- Using the theory to make **new predictions** and then testing them with **new experiments**.

If the **Evidence** Supports a Theory, It's **Accepted — for Now**

1) If all the experiments in all the world provide good evidence to back it up, the theory is thought of as **scientific 'fact'** (for now).
2) But it will never become **totally indisputable** fact. Scientific **breakthroughs or advances** could provide new ways to question and test the theory, which could lead to **new evidence** that **conflicts** with the current evidence. Then the testing starts all over again...

And this, my friend, is the **tentative nature of scientific knowledge** — it's always **changing** and **evolving**.

The Scientific Process

So scientists need evidence to back up their theories. They get it by carrying out experiments or studies.

Evidence Comes From Controlled Lab Experiments...

1) Results from **controlled experiments** in **laboratories** are **great**.
2) A lab is the easiest place to **control variables** so that they're all **kept constant** (except for the one you're investigating).

Pages 6-13 are all about how to design and carry out a lab experiment.

...That You can Draw Meaningful Conclusions From

1) You always need to make your experiments as **controlled** as possible so you can be confident that any effects you see are linked to the variable you're changing.
2) If you do find a relationship, you need to be careful what you conclude. You need to decide whether the effect you're seeing is **caused** by changing a variable (this is known as a **causal relationship**), or whether the two are just **correlated**. There's more about drawing conclusions on page 13.

When planning and carrying out experiments, you need to make sure you're working ethically — you need to think about how the experiment will affect the environment as well as other people and animals.

Society Makes Decisions Based on Scientific Evidence

1) Lots of scientific work eventually leads to **important discoveries** or breakthroughs that could **benefit humankind**.
2) These results are **used by society** (that's you, me and everyone else) to **make decisions** — about the way we live, what we eat, what we drive, etc.
3) All sections of society use scientific evidence to make decisions, e.g. politicians use it to devise policies and individuals use science to make decisions about their own lives.

Other factors can **influence** decisions about science or the way science is used:

1) **Economic factors:**
Society has to consider the **cost** of implementing changes based on scientific conclusions — e.g. the cost of reducing the UK's carbon emissions to limit the human contribution to **climate change**.
Scientific research is often **expensive**. E.g. in areas such as astronomy, the Government has to **justify** spending money on a new telescope rather than pumping money into, say, the **NHS** or **schools**.
2) **Social factors:**
Decisions affect **people's lives** — e.g. when looking for a site to build a **nuclear power station**, you need to consider how it would affect the lives of the people in the **surrounding area**.
3) **Environmental factors:**
Many scientists suggest that building **wind farms** would be a **cheap** and **environmentally friendly** way to generate electricity in the future. But some people think that because **wind turbines** can **harm wildlife** such as birds and bats, other methods of generating electricity should be used.

Practice Questions

Q1 What is peer review?

Q2 Explain why a theory supported by a paper published in a scientific journal is not necessarily scientific 'fact'.

Q3 Explain why considering scientific evidence alone is not enough when making decisions in society.

Exam Question

Q1* A scientist has developed a proposal for a power station which generates electricity renewably from the energy of the tides, without producing greenhouse gases that can lead to climate change. The proposed location of the power station is off a stretch of scenic coastline that attracts a lot of tourism. She claims that her evidence suggests that if built in this location, the power station would produce sufficient electricity to entirely replace a nearby coal-fired power station that uses non-renewable fossil fuels and produces greenhouse gases. Discuss the economic, social and environmental factors that need to be considered by decision-makers considering her proposal. [6 marks]

* The quality of your extended response will be assessed in this question.

No. Borrowing your friend's homework to 'check' it is not peer review...

Hopefully these pages have given you a nice intro to the scientific process, e.g. what scientists do to provide you with 'facts'. You need to understand this, as you're expected to know how science works yourself — for the exam and for life.

Quantities and Units

Learning Physics is a lot like building a house — both involve drinking a lot of tea. Also, both have important foundations — if you skip this stuff everything else is likely to go a bit wrong. So, here goes brick-laying 101...

A Physical Quantity has both a Numerical Value and a Unit

1) Every time you measure something or calculate a quantity you need to give the **units.**
2) The **Système International** (S.I.) includes a set of **base units** for physical quantities from which lots of other units are derived. Here are the S.I. base units that you're likely to come across:

Quantity	S.I. base unit
mass	kilogram, kg
length	metre, m
time	second, s
current	ampere, A
temperature	kelvin, K
amount of a substance	mole, mol

Don't forget — the units on both sides of an equation must be the same.

Kilograms are a bit odd — they're the only S.I. unit with a scaling prefix (see the next page).

You're more likely to see temperatures given in °C.

3) Many more units can be derived from these base units — e.g. newtons, N, for force are defined by $kg\,ms^{-2}$. The newton is an **S.I. derived unit**.
4) The S.I. derived units you'll need will be covered throughout the book and you need to remember them.
5) You also need to have a rough idea of the size of each S.I. base unit and S.I. derived unit in this book, so that you can **estimate quantities** using them.

Remembering how S.I. derived units are defined will help you make sure the other quantities in your equations are in the right units.

You Can Check Your Units Mathematically

The units in any equation must always be the **same on both sides** — this is called **homogeneity of units**. You can use this rule to work out some of the simpler S.I. derived units, like speed:

They always checked the homogeneity of their outfits before leaving the house.

Example: Show that the S.I. derived unit for speed is ms^{-1}.

You know that speed = distance ÷ time

Distance is a length, so its S.I. base unit is the metre, m.

The base unit of time is the second, s.

To find the unit for speed, just put the units for distance and time into the equation for speed: $\mathbf{m \div s = ms^{-1}}$

You can also use this rule to help you to **check your working** if you have to combine or rearrange equations:

Example: For an object moving with uniform acceleration, $v = u + at$ and $s = \frac{1}{2}(u + v)t$. Combine these equations to get an equation for s in terms of u, v and a. Check the homogeneity of your answer.

There's more about these equations on pages 20-21.

Rearrange $v = u + at$ to get: $t = \frac{v - u}{a}$

The substitute this into $s = \frac{1}{2}(u + v)t \Rightarrow s = \frac{1}{2}(u + v)\left(\frac{v - u}{a}\right)$

Then simplify the equation: $\mathbf{s = \frac{1}{2a}(v^2 - u^2)}$

v = final velocity, u = initial velocity, a = acceleration, t = time and s = displacement

To check the units are the same on both sides, substitute the units for each quantity into the equation, then cancel down (you can ignore any numbers, e.g. the 2):

s is a length in metres, v and u are velocities in ms^{-1} and a is acceleration in ms^{-2}.

$$m = \frac{1}{ms^{-2}}((ms^{-1})^2 - (ms^{-1})^2) = \frac{1}{ms^{-2}}(m^2s^{-2} - m^2s^{-2})$$

There are only metres left on both sides of the equation, so the equation is homogeneous.

Quantities and Units

Prefixes Let You Scale Units

Physical quantities come in a **huge range** of sizes. Prefixes are scaling factors that let you write numbers across this range without having to put everything in standard form.

These are the most common prefixes you'll meet:

prefix	**pico (p)**	**nano (n)**	**micro (μ)**	**milli (m)**	**centi (c)**	**deci (d)**	**kilo (k)**	**mega (M)**	**giga (G)**	**tera (T)**
multiple of unit	1×10^{-12}	1×10^{-9}	1×10^{-6}	0.001 (1×10^{-3})	0.01 (1×10^{-2})	0.1 (1×10^{-1})	1000 (1×10^{3})	1×10^{6}	1×10^{9}	1×10^{12}

If you're a bit uncertain about moving between these scaling factors, then convert quantities into the standard unit before you do anything else with them:

Example 1: Convert 1869 picometres into nanometres.

First, convert the value to metres: 1869 pm = 1869×10^{-12} m

Then divide by 1×10^{-9} to convert to nanometres: $1869 \times 10^{-12} \div 1 \times 10^{-9}$ = **1.869 nm**

Or, you can convert between prefixes directly:

Example 2: Convert 0.247 megawatts into kilowatts.

1 MW = 1×10^{6} W and 1 kW = 1×10^{3} W

So the scaling factor to move between MW and kW is:

$(1 \times 10^{6}) \div (1 \times 10^{3}) = 1 \times 10^{3}$.

So 0.247 MW = $0.247 \times 1 \times 10^{3}$ = **247 kW**

It's really easy to get muddled up when you're converting between prefixes. The rule is, if you're moving to the right in the table above, your number should get smaller, and if you're moving to the left the number should get larger. If your answer doesn't match the rule, you've made a mistake.

Be careful with using these prefixes in the middle of calculations — they'll change the units of your final answer, and could get you in a mess. It's generally safest to do your calculations with everything in its S.I. base units or S.I. derived units, then convert the answer to include a sensible prefix when you're done.

Practice Questions

Q1 What is the S.I. unit of mass?

Q2 What is meant by an S.I. base unit and an S.I. derived unit?

Q3 What does the term homogeneity of units mean?

Q4 What is: a) 20 000 W in kilowatts, b) 2×10^{-6} W in milliwatts c) 1.23×10^{7} W in gigawatts?

Exam Question

Q1 The density, ρ, of a material gives its mass per unit volume. It is given by $\rho = m / V$, where m = mass and V = volume.

a) Express the units of density in terms of S.I. base units. [1 mark]

b) Calculate the density of a cube of mass 9.8 g, and side length 11 mm. Give your answer in the units stated in part a). [2 marks]

What's the S.I. base unit for boring...

Not the most exciting pair of pages these, I'll admit, but it's important that you have the basics down, or else you're leaving yourself open to simple little mistakes that'll cost you marks. So make sure you're familiar with all the S.I. units in the table, then try and write down all the prefixes and their scaling factors. If you don't get them all first time, keep trying until you can. Remember, you need to know the units for every other quantity you meet in this book, too.

Planning and Implementing

Science is all about getting good evidence to support (or disprove) your theories, so scientists need to be able to spot a badly designed experiment, interpret the results of an experiment or study, and design their own experiments too...

You Might have to **Design an Experiment** to Answer a **Question**

1) You might be asked to design a physics experiment to **investigate** something or answer a question.
2) It could be a **lab experiment** that you've seen before, a **new experiment** you aren't familiar with, or it could be something **applied**, like deciding which building material is best for a particular job.
3) Whatever you're asked, you'll be able to use the physics you know and the skills covered on the next few pages to figure out the best way to do the investigation.

A **Variable** is Anything that has the Potential to **Change** in an Experiment

1) First, you need to identify your **independent** and **dependent variables**:

The **independent** variable is the thing you **change**.
The **dependent** variable is the thing you **measure**.

Example 1: If you're investigating how changing the potential difference across a component affects the current through it, the **independent variable** is the **potential difference**, and the **dependent variable** is the **current**.

2) Apart from the independent and dependent variables, **all other variables** should stay the same during your experiment. If not, you can't tell whether or not the independent variable is responsible for any changes in your dependent variable, so your results won't be **valid** (p.12). This is known as **controlling variables**. It might be worth **measuring control variables** that are likely to change during your experiment to check that they really are under control.

Example 1 (continued): In the example above, you need to use the same **circuit components**, and to keep the **temperature** of the apparatus **constant** — e.g. by letting the circuit cool down between readings.

Example 2: If you're investigating the value of **acceleration due to gravity** by dropping an object and timing its fall, **draughts** in the room could really mess up your results. Picking an object that is more **resistant** to being blown about (like a ball-bearing) will help make your results more **precise** (p.12).

Select Appropriate **Apparatus** and **Techniques**

1) You need to think about what **units** your measurements of the independent and dependent variables are likely to be in before you begin (e.g. millimetres or metres, milliseconds or hours).
2) Think about the **range** you plan on taking measurements over too — e.g. if you're measuring the effect of increasing the force on a spring, you need to know whether you should increase the force in steps of 1 newton, 10 newtons or 100 newtons. Sometimes, you'll be able to **estimate** what effect changing your independent variable will have, or sometimes a **pilot experiment** might help.
3) Considering your measurements before you start will also help you choose the most appropriate **apparatus** and **techniques** for the experiment:

There's a whole range of apparatus and techniques that could come up in your exam. Make sure you know how to use all the ones you've come across in class.

Example:
- If you're measuring the length of a **spring** that you're applying a force to, you might need a **ruler**. If you're measuring the diameter of a **wire**, you'd be better off with a set of **callipers**.
  If the wire **doesn't stretch** easily, you may need to use a **very long wire** to get an extension that's big enough to measure, so you might need to use a **pulley** like in the Young modulus experiment on p.56.
- If you're measuring a **time interval**, you could use a **stopwatch**. If the time is **really short** (for example if you're investigating acceleration due to gravity), you might need something more sensitive, like **light gates**.

4) Whatever apparatus and techniques you use, make sure you use them **correctly**. E.g. if you're measuring a length, make sure your eye is level with the ruler when you take the measurement.
5) While you're planning, you should also think about the **risks** involved in your experiment and how to manage them — e.g. if you're investigating a material that might snap, wear safety goggles to protect your eyes.

Planning and Implementing

Figure Out how to *Record* your Data Before you *Start*

Before you get going, you'll need a **data table** to record your results in.

1) It should include space for your **independent variable** and your **dependent variable**. You should specify the **units** in the headers, not within the table itself.
2) The readings of the **independent variable** should be taken at **evenly spaced intervals**.
3) Your table will need enough room for repeated measurements. You should aim to **repeat** each measurement at least **three times**. Taking repeat measurements can reduce the effect of random errors in your results (see p.12) and makes spotting **anomalous** results, like this one, much easier. If there's no way to take repeat readings, then you should **increase** the **total number** of readings that you take.
4) There should be space in your table for any data processing you need to do, e.g. calculating an **average** from repeated measurements, or calculating speed from measurements of distance and time.
5) Most of the time, your data will be **quantitative** (i.e. you'll be recording numerical values). Occasionally, you may have to deal with **qualitative** data (data that can be observed but not measured with a numerical value). It's still best to record this kind of data in a table, to keep your results **organised**, but the layout may be a little **different**.

	Current / A			
P.d. / V	Trial 1	Trial 2	Trial 3	Average
1.00	0.052	0.047	0.050	0.050
1.50	0.079	0.075	0.077	0.077
2.00	0.303	0.098	0.097	...
2.50	0.129	0.125	0.130	...
3.00	0.149	0.151	0.145	...
...	...	...	...	...

You Could be Asked to *Evaluate* An *Experimental Design*

If you need to evaluate an experimental design, whether it's your own or someone else's, you need to think about these sorts of things:

- Does the experiment **actually test** what it sets out to test?
- Is the method **clear** enough for someone else to follow?
- Apart from the **independent** and **dependent variables**, is everything else going to be **properly controlled**?
- Are the **apparatus** and **techniques appropriate** for what's being measured? Will they be used correctly?
- Are enough **repeated measurements** going to be taken?
- Is the experiment going to be conducted **safely**, for those involved and those **nearby**?
- Are there any other **ethical considerations**? For example, does the experiment produce **harmful waste products**? Is the apparatus being used in a way that won't **damage it**?

Greta was paying the price for not planning her experiment properly.

Practice Questions

Q1 What is meant by the term independent variable? What is a dependent variable?

Q2 Why do you need to plan to control all of the other variables in an experiment?

Q3 What do you need to consider when selecting your apparatus?

Q4 Why should you take repeated measurements in an experiment?

Exam Question

Q1 A student is investigating the effect of the light level on the resistance of an LDR (light-dependent resistor). The student connects the LDR to a power supply, and measures the resistance of the LDR at various distances from a light source using a multimeter.

a) State the independent and dependent variables for this experiment. [1 mark]

b) State two variables that the student needs to control in order to ensure his results are valid. [2 marks]

The best-planned experiments of mice and men...

...often get top marks. The details of planning and carrying out an experiment will vary a lot depending on what you're investigating, but if all this stuff is wedged in your brain you shouldn't go far wrong, so make sure you've got it learned.

Analysing Results

Now it's time to look at the results of your experiment (or some results given in your exam) a bit more closely...

Do any **Calculations** You Need to **First**

1) Before you calculate anything, check for any **anomalous results.** If there's something in the results that's **clearly wrong**, then don't include it in your calculations — it'll just **muck everything up**. Be careful though, you should only exclude an anomalous result if you have **good reason** to think it's wrong, e.g. it looks like a decimal point is in the **wrong place**, or you suspect that one of the control variables **changed**. And you should talk about any anomalous results when you're evaluating the experiment (pages 12-13).
2) For most experiments, you'll at least need to calculate the mean (average) of some **repeated measurements**:

$$\text{mean (average) of a measurement} = \frac{\text{sum of your repeated measurements}}{\text{number of repeats taken}}$$

In class, you could use a spreadsheet to process your data (and plot graphs), but it's important that you know how to do it by hand for the exam.

3) Calculate any quantities that you're interested in that you haven't **directly measured** (e.g. pressure, speed).

You should try to give any values you calculate to the **same number of significant figures** as the data value with the **fewest significant figures** in your calculation. If you give your result to too many significant figures, you're saying your final result is more **precise** than it actually is (see p.12).

Present Your Results on a **Graph**

Discrete and continuous data is usually presented with a scatter graph (see below). Categoric data (data which can be sorted into categories, e.g. types of material — wood, brass, steel etc.) is usually presented with a bar or pie chart.

The type of graph you plot depends on the type of data you've recorded:

1) **Discrete data** — you get discrete data by **counting**. E.g. the number of weights added to the end of a spring would be discrete. You can't have 1.25 weights.
2) **Continuous data** — a continuous variable can have **any value** on a scale. For example, the extension of a spring or the current through a circuit. You can never measure the exact value of a continuous variable.

1) Usually, the **independent variable** goes on the ***x*-axis** and the **dependent variable** goes on the ***y*-axis**. Both axes should be **labelled** clearly, with the quantity and **units**. The **scales** used should be sensible (i.e. they should go up in sensible steps, and should spread the data out over the full graph rather than bunching it up in a corner).

If you need to use your graph to measure something, select axes that will let you do this easily (e.g. by measuring the gradient or the intercept, see the next page).

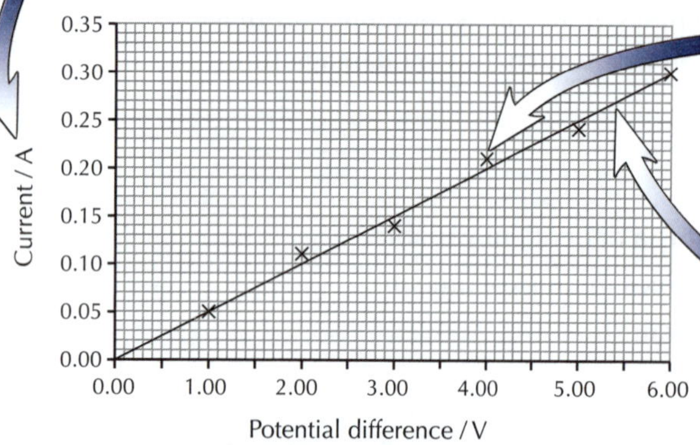

2) Plot your points using a **sharp pencil**, to make sure they're as **accurate** as possible.

3) Draw a **line of best fit** for your results. Around **half** the data points should be above the line, and half should be below it (you should ignore anomalous results). Depending on the data, the line might be **straight**, or **curved**.

Graphs can Show Different Kinds of **Correlation**

Remember, correlation does not necessarily mean cause.

The **correlation** describes the relationship between the variables. Data can show:

Positive correlation:

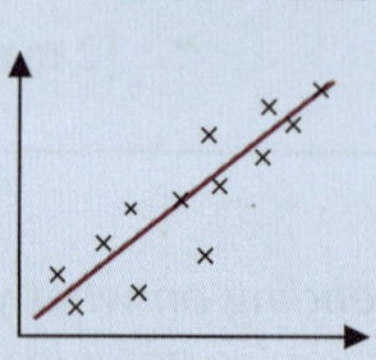

As one variable increases the other increases.

Negative correlation:

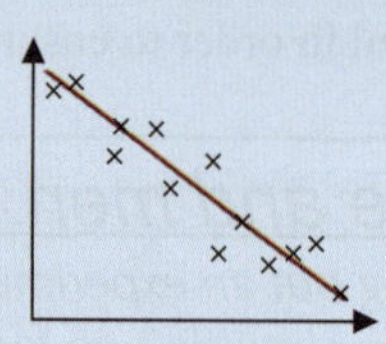

As one variable increases the other decreases.

No correlation:

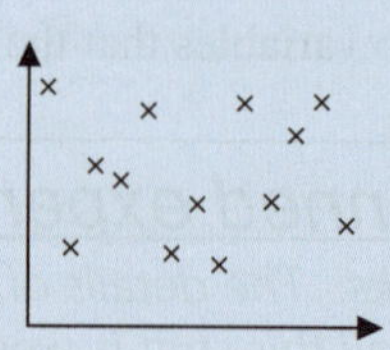

No relationship between the variables.

Analysing Results

You Might Need to Find a *Gradient* or *Intercept*

If the line of best fit is **straight**, then the graph is **linear**. This means a change in one always leads to a change in the other.
The **line of best fit** for a linear graph has the **equation**: $y = mx + c$

Where m is the **gradient** of the line and c is the **y-intercept**.

If the line of best fit goes through the origin (c is 0), you can say the variables are **directly proportional** to each other: $y \propto x$

$\propto$ just means 'is directly proportional to'.

Example: This graph shows displacement against time for a motorbike travelling west. Find the bike's velocity.

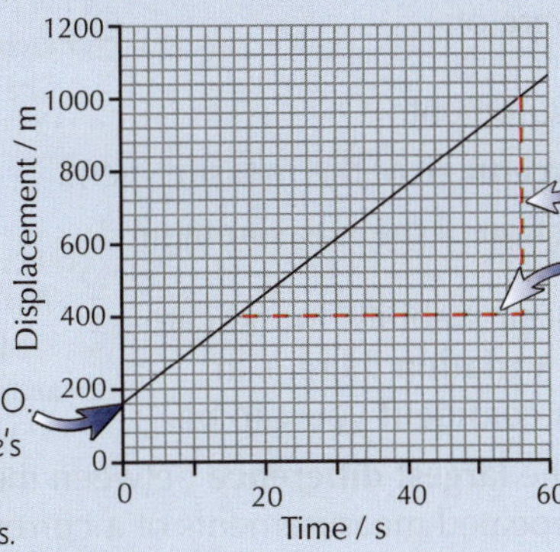

The y intercept is 160. This means the bike's displacement was 160 m at time 0 s.

For a displacement-time graph, the gradient gives the velocity (as velocity = displacement ÷ time).

$\Delta y = 1000 - 400 = 600$ m

$\Delta x = 56 - 16 = 40$ s

Δ means 'change in'.

velocity = $\frac{\text{displacement}}{\text{time}} = \frac{\Delta y}{\Delta x} = 600 \div 40 =$ **15 ms^{-1} west**

If a graph has a **curved** line of best fit, you can find the gradient of a given point on the line by drawing a **tangent** to the curve (see page 17). It's sometimes helpful to choose axes that turn a curved graph into a straight one:

Example:

For a given force, the graph of **pressure** applied against the **area** that the force is applied over looks like this:

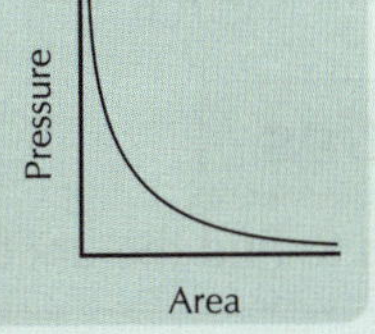

If you plot pressure against **1 ÷ area**, the graph looks like this:

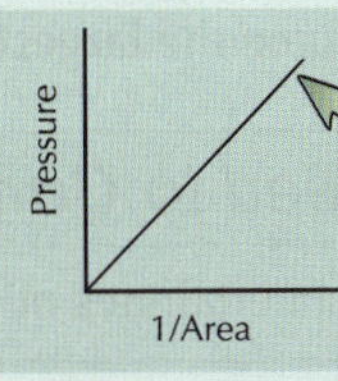

The **gradient** is:
pressure ÷ (1 ÷ area)
= pressure × area
= force applied (p.38)

pressure = force ÷ area.

Practice Questions

Q1 Describe what you should do with anomalous results when processing data.

Q2 How do you calculate an average of repeated results?

Q3 Sketch a graph showing a negative correlation.

Exam Question

Q1 An engineer is investigating the performance of a prototype car with a new kind of environmentally-friendly engine. The data below shows the speed of the car, going from stationary to 70 kilometres per hour. (In this question, you may use the formula: acceleration = change in speed ÷ time taken to change speed.)

Time / s	0.0	1.0	2.0	3.0	4.0	5.0	6.0	7.0
Speed / km per hour	0	3	8	22	40	54	66	70

a) Draw a graph showing speed against time for this data. [4 marks]

b) State the times, to the nearest second, between which the graph is linear. [1 mark]

c) Using the graph, calculate the maximum acceleration of the car. [4 marks]

My level of boredom is proportional to the time I've spent on this page...

This stuff can get a bit fiddly, especially measuring the gradient of a curved line, but for the most part it's not too bad, and you should have seen a lot of it before. So dust off your pencil sharpener, and get to work...

Measurements and Uncertainties

There are errors and uncertainties in every measurement. You need to be aware of them.

All Results have Some *Uncertainty*

The smallest measuring interval on an instrument gives you its resolution.

1) **Every** measurement you take has an **experimental uncertainty**.
2) The smallest uncertainty you can have in a measurement is the uncertainty due to the **resolution** of your equipment. It's ± **half** of one division on the measuring instrument used. E.g. using a thermometer with a scale where each division represents 2 °C, a measurement of 30 °C will at **best** be measured to be 30 ± **1 °C**. And that's without taking into account any other errors that might be in your measurement.
3) The ± sign gives you the **range** in which the **true** length (the one you'd really like to know) probably lies. 20 ± 0.5 cm tells you the true length is very likely to lie in the range of 19.5 to 20.5 cm. The maximum difference between your value and the true value (here 0.5 cm) is sometimes called the **margin of error**.
4) There are three measures of uncertainty you need to be aware of:

Percentage uncertainty can also be called percentage error.

Absolute uncertainty — the **total uncertainty** for a measurement.
Percentage uncertainty — the uncertainty given as a **percentage** of the measurement.
Fractional uncertainty — the uncertainty given as a **fraction** of the measurement.

Measuring larger values reduces the percentage and fractional uncertainty.

5) An uncertainty should also include a level of **confidence** or probability, to indicate how **likely** the true value is to lie in the interval. The **more variation** there is in your results, and the **fewer repeats** you have done, the **less confident** you can be.
6) The uncertainty on a **mean** (see p.8) of repeated results is the **largest difference** between the mean and any of the values used to calculate it. So if you take repeated measurements of a current and get values of 0.1 A, 0.4 A and 0.4 A, the mean current is 0.3 A, and the uncertainty on the mean is 0.3 – 0.1 = **± 0.2 A**. You can estimate the uncertainty on the mean as **half the range** — e.g. (0.4 – 0.1) ÷ 2 = **0.15 A**.
7) If no uncertainty is given for a value, the **assumed uncertainty** is **half the increment** of the **last** significant figure that the value is **given** to. E.g. 2.0 is given to 2 **significant figures**, so you would assume an uncertainty of 0.05. You should always assume the **largest** amount of uncertainty when doing an experiment.

Sometimes You Need to *Combine Uncertainties*

You have to combine the uncertainties of different measured values to find the uncertainty of a calculated result:

Adding or *Subtracting* Data — *ADD* the *Absolute Uncertainties*

Example: A wire is stretched from 4.3 ± 0.1 cm to 5.5 ± 0.1 cm. Calculate the extension of the wire.

1) First subtract the lengths without the uncertainty values: 5.5 – 4.3 = 1.2 cm
2) Then find the total uncertainty by adding the individual absolute uncertainties: 0.1 + 0.1 = 0.2 cm
So, the extension of the wire is **1.2 ± 0.2 cm**.

Multiplying or *Dividing* Data — *ADD* the *Percentage Uncertainties*

Example: A force of 15 N ± 3% is applied to a stationary object which has a mass of 6.0 ± 0.3 kg. Calculate the acceleration of the object and state the percentage uncertainty in this value.

1) First calculate the acceleration without uncertainty: $a = F \div m = 15 \div 6.0 = 2.5\ \text{ms}^{-2}$
2) Next, calculate the percentage uncertainty in the mass: % uncertainty in $m = \frac{0.3}{6.0} \times 100 = 5\%$
3) Add the percentage uncertainties in the force and mass values to find the total uncertainty in the acceleration: Total uncertainty = 3% + 5% = 8%
So, the acceleration = $\mathbf{2.5\ ms^{-2} \pm 8\%}$

Raising to a Power — *MULTIPLY* the *Percentage Uncertainty* by the *Power*

Example: The radius of a circle is r = 40 cm ± 2.5%. Find the percentage uncertainty in the circle's area (πr^2).

The radius is raised to the power of 2 to calculate the area. So, the percentage uncertainty is 2.5% × 2 = **5%**

Measurements and Uncertainties

Error Bars Show the Uncertainty of Individual Points

1) Most of the time, you work out the **uncertainty** in your **final** result using the uncertainty in **each measurement** you make.
2) When you're plotting a **graph**, you can show the uncertainty in **each measurement** by using **error bars** to show the **range** the point is likely to lie in.
3) You can have error bars for just one variable (see below), or **both** the dependent and the independent variable. Error bars on both variables give you an '**error box**' for each point.

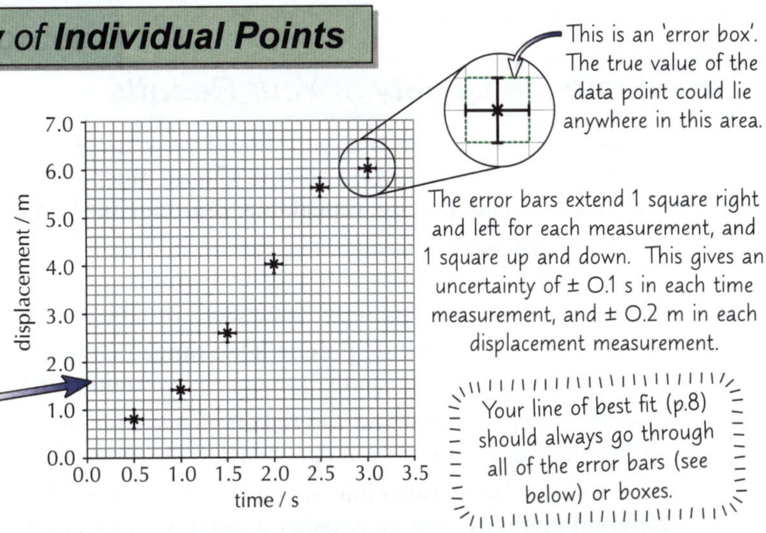

You Can Calculate the Uncertainty of Final Results from a Line of Best Fit

Normally when you draw a graph you'll want to find the **gradient** or **intercept** (p.9). For example, you can calculate k, the **force constant** of the object being stretched, from the **gradient** of the graph on the right — here it's about 20 000 Nm^{-1}. You can find the **uncertainty** in that value by using **worst lines**:

1) Draw lines of best fit which have the **maximum** and **minimum** possible slopes for the data and which should go through all of the **error bars** (see the pink and blue lines on the right). These are the **worst lines** for your data. If your data has errors in **both** the independent and dependent variables, then the worst lines should go through the **corners** of the **error boxes**.
2) Calculate the **gradients** of the worst lines. The blue line's gradient is about 21 000 Nm^{-1} and the pink line's gradient is about 19 000 Nm^{-1}.
3) The **uncertainty** in the gradient is given by half the **difference** between the **worst gradients** — here it's 1000 Nm^{-1}. So this is the uncertainty in the value of the force constant. For this object, the force constant is 20 000 ± 1000 Nm^{-1} (or 20 000 Nm^{-1} ± 5%).
4) Similarly, the uncertainty in the ***y*-intercept** is just half the **difference** between the **worst** intercepts (although there's no uncertainty here since the worst lines both go through the origin).

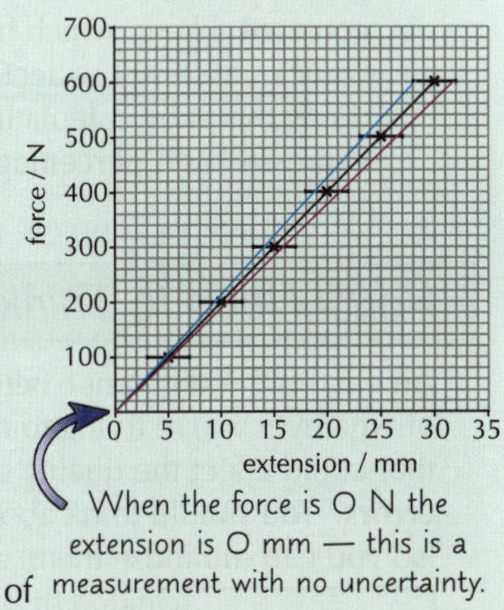

Practice Questions

Q1 What is meant by experimental uncertainty? How does it relate to an instrument's resolution?

Q2 What are the rules for combining uncertainties?

Q3 What are worst lines? How could you use them to find the uncertainty in the intercept of a graph?

Exam Question

Q1 A student is investigating the acceleration of a remote controlled car. The car has an initial velocity of 0.52 ± 0.02 ms^{-1} and accelerates to 0.94 ± 0.02 ms^{-1} over an interval of 2.5 ± 0.5 s.

a) Calculate the percentage uncertainty in the car's initial speed. [1 mark]

b) Calculate the percentage uncertainty in the car's final speed. [1 mark]

c) Calculate the car's average acceleration over this interval. Include the absolute uncertainty of the result in your answer. (acceleration = change in velocity ÷ time taken). [4 marks]

My percentage uncertainty about these pages is 0.99%...

Uncertainties are a bit of a pain, but they're really important. Learn the rules for combining uncertainties, and make sure you know how to draw those pesky error bars, and what to do with them when you've got them.

Evaluating and Concluding

Once you've drawn your graphs and analysed your results, you need to think about your conclusions.

Evaluate the Quality of Your Results

Quality control — also important in construction.

Before you draw any conclusions, you should think about the quality of the results — if the quality's not great you won't be able to have much confidence in your conclusion. Good results are **valid**, **precise**, **repeatable**, **reproducible** and **accurate**.

1) **Valid measurements** measure what they're supposed to be measuring. If you haven't **controlled** all the variables (p.6) then your **results** won't be valid, because you won't just be testing the effect of the independent variable.
2) You can say your results are **precise** if the **range** that your repeated data is spread over is small. The precision of a result is only influenced by **random errors** (see below).
3) Results are **repeatable** if **you** can **repeat** an experiment (using the **same** method and equipment) multiple times and get the **same results**. You can measure the repeatability of your results by looking at their **precision**.
4) For a result to be **reproducible**, it needs to be obtained by **different experimenters**, using **different equipment** and **different methods**. Testing whether a result is reproducible is a better test of its quality than testing whether it is repeatable, as it's less likely that the same **systematic errors** could have affected both methods (see below).
5) An **accurate result** is one that's really close to the **true value**. You often **can't know** the accuracy of a result, since you can never know the true value of what you're measuring. The exception to this is if you're measuring a **known constant**, like *g*, which has been **tested many times**, and is known to a good degree of certainty. In cases like this, you can assess how accurate your results are by **comparing** them to this known value. One way of doing this is by calculating the **percentage difference** — the difference between your result and the true value, expressed as a **percentage** of the **true value**.

You'll Need to Think About Random and Systematic Errors

An error is the difference between your **measured value** and the **true value** of whatever you're trying to measure. There are two different kinds of error that could affect the quality of your results — **random errors** and **systematic errors**. You should think about these when you're planning your experiment (so you can minimise them) and when you're evaluating your results.

Zero errors are caused by the apparatus failing to read zero when it should do, e.g. when no current is flowing through an ammeter.

Systematic Errors

1) **Systematic errors** (including **zero errors**) are the same every time you repeat the experiment (they shift all the values by the same amount). They may be caused by the **equipment** you're using or how it's **set up**, e.g. not lining up a ruler correctly when measuring the extension of a spring.
2) Systematic errors are really **hard to spot**, and they affect the **accuracy** of your results, but not the **precision** — so it can look like you've measured something **really well**, even if your results are actually **completely off**. Systematic errors might show up when you **compare results** with other people in your class who have done the **same experiment** if they're caused by a mistake you made, or when you compare your results with someone else who used a **different method** if there is something wrong with the procedure itself (i.e. when you check the reproducibility of your results).
3) It's always worth **checking your apparatus** at the start of an experiment, e.g. measure a few known masses to check that a mass balance is **calibrated** properly, to make systematic errors less likely.

Random Errors

1) **Random errors** vary — they're what make the results a bit different each time you repeat an experiment. For example, if you measured the length of a wire 20 times, the chances are you'd get a slightly different value each time, e.g. due to your head being in a slightly different position when reading the scale. It could be that you just can't keep controlled variables (p.6) exactly the same throughout the experiment.
2) Unlike systematic errors, you can at least generally tell when random errors are there, as they affect your precision.
3) Using apparatus with a **better resolution** (p.10) can reduce the size of random errors in **individual measurements**.
4) Doing more **repeats** can also reduce the effect of random errors on **calculated results** like the mean (p.8).

Evaluating and Concluding

Draw **Conclusions** that Your Results **Support**

Your conclusion can't be more precise than the margin given by the uncertainty in your results.

1) A conclusion **explains** what the data shows.
   You can only draw a valid conclusion if you have valid data **supporting** it.
2) Your conclusion should be limited to the **circumstances you've tested** it under — if you've been investigating how the current flowing through a resistor changes with the potential difference across it, and have only used potential differences between 0 and 6 V, you can't claim to know what would happen if you used a potential difference of 100 V, or if you used a different resistor.
3) You also need to think about how much you can **believe** your conclusion, by evaluating the quality of your results (see previous page). If you can't believe your results, you can't form a **strong conclusion**.

Think About how the Experiment Could be **Improved**

Having collected the data, is there anything you think should have been done **differently**?
Were there any **limitations** to your method?

1) If the results aren't **valid**, could you change the experiment to fix this, e.g. by changing the data you're collecting?
2) If the results aren't **accurate**, what could have caused this?
   If there are **systematic errors** in your results, what could you do to prevent them?
3) Are there any changes you could make to the **apparatus** or **procedure** that would make the results more **precise**?
   - There are some simple ways to reduce **random errors** or their effects (see p.12), including using the most **appropriate** equipment, (e.g. swapping a millimetre ruler for a micrometer to measure the diameter of a wire) and **increasing** the number of **repeats**.
   - You can also use a **computer** to collect data — e.g. using light gates to measure a time interval rather than a **stopwatch.** This makes results more **precise** by reducing **human error**.
4) Are there any other ways you could have **reduced the errors** in the measurements?

Practice Questions

Q1 What is a valid result?
Q2 What is the difference between a repeatable result and a reproducible result?
Q3 What is the difference between saying the results of an experiment are precise and saying that they are accurate?
Q4 Give two examples of possible sources of random error and one example of a possible source of systematic error in an experiment. Which kind of error won't affect the precision of the results?
Q5 What should you think about when you are trying to improve an experimental design?

Exam Question

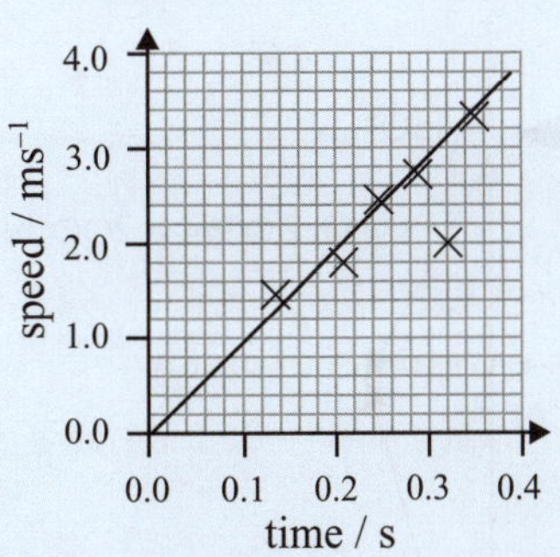

Q1 A student is investigating how the speed of a falling object is affected by how long it has been falling for. He drops an object from heights between 10 cm and 60 cm and measures its speed at the end of its fall, and the time the fall takes, using light gates. He plots a graph of the final speed of the object against the time it took to fall, as shown on the left.

a) Identify the anomalous result. [1 mark]

b) The student concludes that the speed of any falling object is always proportional to the time it has been falling for.
Explain whether or not the results support this conclusion. [2 marks]

In conclusion, Physics causes headaches...

Valid, precise, repeatable, reproducible and accurate... you'd think they all mean the same thing, but they really don't. Make sure you know the difference, and are careful about which one you use, or you'll be throwing marks away.

Scalars and Vectors

And now it's time to draw some lovely triangles. Please, don't all thank me at once...

Scalars Only Have Size, but Vectors Have Size and Direction

1) A **scalar** has **no direction** — it's **just an amount** of something, like the **mass** of a **sack of meaty dog food**.
2) A **vector** has magnitude (**size**) and **direction** — like the **speed and direction** of next door's **cat** running away.
3) **Force**, **velocity and momentum** are all **vectors** — you need to know **which way** they're going as well as **how big** they are. Here are some of the common scalars and vectors that you'll come across in your exams:

Scalars	Vectors
mass, time, energy temperature, length, speed	displacement, force, velocity, acceleration, momentum

4) Vectors are drawn as **arrows** (to show their direction) with their **size** written next to them (see Example 1 below). In the exam, you might see quantities written with **arrows** above them, e.g. $\vec{v}$, to show that they're vectors.

Sometimes vectors are printed in bold, e.g. **v**, but it's quite hard to handwrite in bold, so the arrow is used too.

You can Add Vectors to Find the Resultant

1) Adding two or more vectors is called finding the **resultant** of them.
2) You should always start by drawing a **diagram**. Draw the vectors '**tip to tail**'. If you're doing a **vector subtraction**, draw the vector you're subtracting with the same magnitude but pointing in the **opposite direction**.
3) If the vectors are at **right angles** to each other, then you can use **Pythagoras** and **trigonometry** to find the resultant.
4) If the vectors aren't at right angles, you may need to draw a **scale diagram**.

Example 1: Jemima goes for a walk. She walks 3.0 m north and 4.0 m east. She has walked 7.0 m but she isn't 7.0 m from her starting point. Find the magnitude and direction of her displacement.

Jemima's 'displacement' gives her position relative to her starting point, see p.16.

First, draw the vectors **tip-to-tail**. Then draw a line from the **tail** of the first vector to the **tip** of the last vector to give the **resultant**:

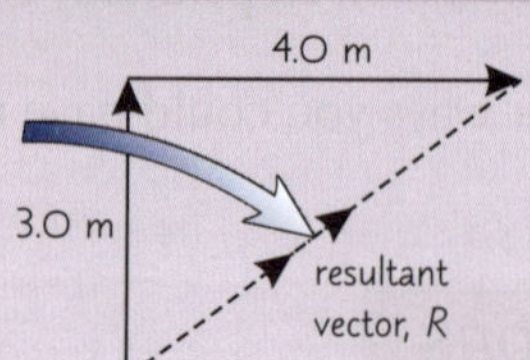

Because the vectors are at right angles, you get the **magnitude** of the resultant using Pythagoras:

$R^2 = 3.0^2 + 4.0^2 = 25.0$ So $R = \mathbf{5.0\ m}$

Now find the **bearing** of Jemima's new position from her original position.

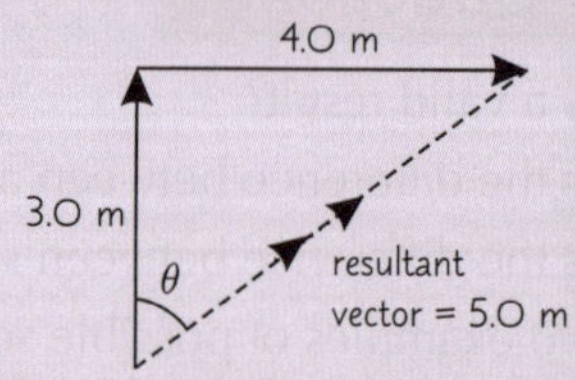

Jemima

You use the triangle again, but this time you need to use trigonometry. You know the opposite and the adjacent sides, so you can use:

$\tan\theta = 4.0 / 3.0$ So $\theta = \mathbf{053°}$ **(to 2 s.f.)**

A bearing is just an angle measured clockwise from the north line, represented by three digits, e.g. 10° = 010°.

Example 2: A van is accelerating north, with a resultant force of 510 N. A wind begins to blow on a bearing of 150°. It exerts a force of 200 N (to 2 s.f.) on the van. What is the new resultant force acting on the van?

The vectors **aren't** at right angles, so you need to do a scale drawing. Pick a sensible scale. Here, 1 cm = 100 N seems good.

Using a really sharp pencil, draw the initial resultant force on the van. As the van is going north, this should be a 5.1 cm long line going straight up.

The force of the wind acts on a bearing of 150°, so add this to your diagram. Using the same scale, this vector has a length of 2.0 cm.

Then you can draw on the new resultant force and measure its length. Measure the angle carefully to get the bearing.

The resultant force has a magnitude of 350 N (to 2 s.f.), acting on a bearing of 017° (to 2 s.f.).

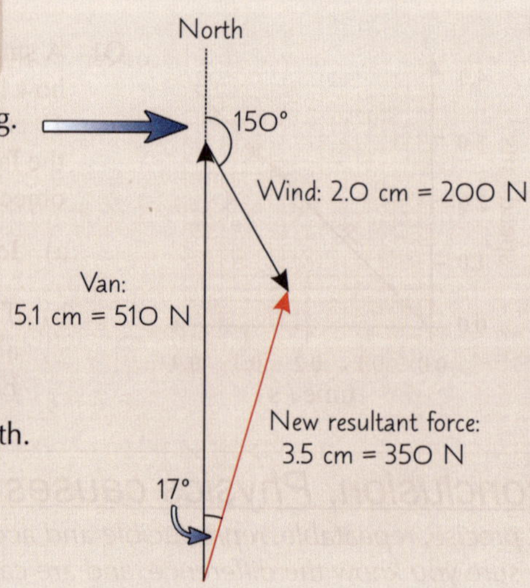

Scalars and Vectors

*It's Useful to Split a **Vector** into **Horizontal** and **Vertical Components***

This is the opposite of finding the resultant — it's called **resolving**. You start from the resultant vector and split it into two **components** at right angles to each other. You're basically **working backwards** from Example 1 on the last page.

Resolving a vector v into horizontal and vertical components:

You get the **horizontal** component v_h like this:

$$\cos\theta = v_h / v$$

$$v_h = v\cos\theta$$

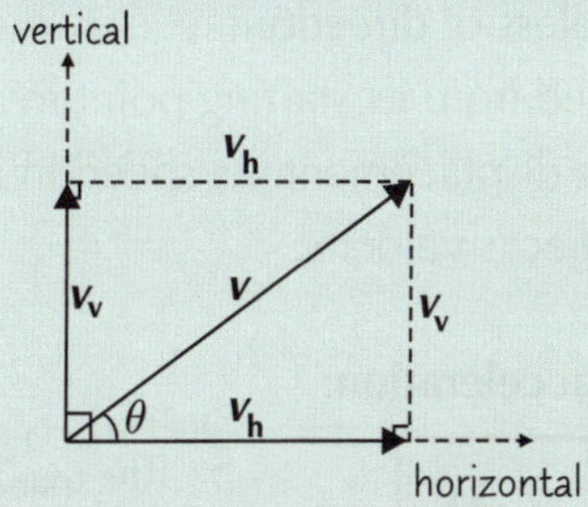

...and the **vertical** component v_v like this:

$$\sin\theta = v_v / v$$

$$v_v = v\sin\theta$$

Where θ is the angle from the horizontal.

Example: Charley's amazing floating home is travelling at a speed of 5.0 ms^{-1} at an angle of 60° (to 2 s.f.) up from the horizontal. Find the vertical and horizontal components.

The **horizontal** component v_h is:

$v_h = v\cos\theta = 5.0\cos 60° = \mathbf{2.5\ ms^{-1}}$

The **vertical** component v_v is:

$v_v = v\sin\theta = 5.0\sin 60° = \mathbf{4.3\ ms^{-1}}$ **(to 2 s.f.)**

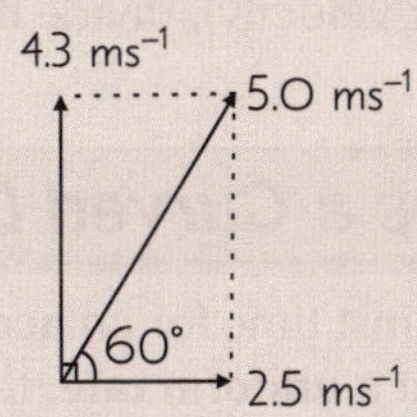

Charley's mobile home was the envy of all his friends.

You can also resolve vectors using **scale drawings**. It's easiest to do this on **squared paper**. Just draw the vector to **scale** in the **correct direction**, then draw a **horizontal** line and **vertical** line from the tip and tail of the vector to form a **right-angled triangle**. Then measure the **lengths** of the lines and use the **scale** to convert them — this gives you the **horizontal** and **vertical** components of the vector.

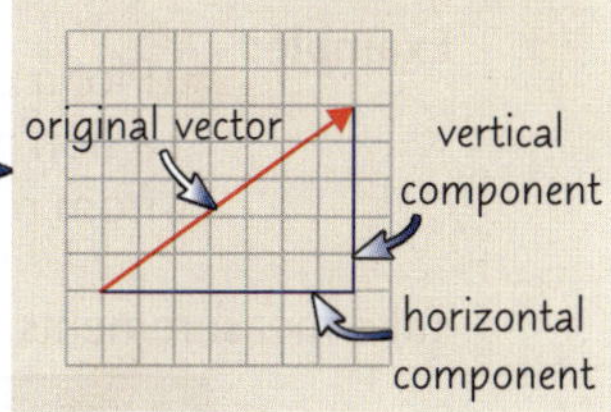

Resolving is dead useful because the two components of a vector **don't affect each other**. This means you can deal with the two directions **completely separately** (there's more on this on page 36).

Practice Questions

Q1 What is the difference between a vector and a scalar? Give three vector quantities and three scalar quantities.

Q2 Describe how to draw a vector, and how to use vector notation.

Q3 Describe how to use a scale diagram to: a) find a resultant vector
b) resolve a vector into components at right angles to each other.

Exam Questions

Q1 The wind applies a horizontal force of 20.0 N on a falling rock of weight 75 N.
Calculate the magnitude and direction of the resultant force. [2 marks]

Q2 A glider is travelling at a velocity of 20.0 ms^{-1} at an angle of 15.0° below the horizontal.
Calculate the horizontal and vertical components of the glider's velocity. [2 marks]

Q3 A remote controlled boat is placed in a river. The boat produces a driving speed of 1.54 ms^{-1} at an angle of 60.0° to the current (travelling with the current). The river is flowing at 0.20 ms^{-1}.
By resolving the vectors into their horizontal and vertical components, show that the resultant velocity of the boat is 1.6 ms^{-1} at an angle of 54° to the current. [4 marks]

I think I'm a scalar quantity, my Mum says I'm completely direction-less...

*Lots of different ways to solve vector problems on these pages, it must be your lucky day. Trigonometry comes up all over the shop in physics, so make sure you're completely okay with it. Remember: **S**in θ = **O**pposite ÷ **H**ypotenuse, **C**os θ = **A**djacent ÷ **H**ypotenuse, and **T**an θ = **O**pposite ÷ **A**djacent. Or SOH CAH TOA.*

Displacement-Time Graphs

Drawing graphs by hand — oh joy. You'd think examiners had never heard of the graphical calculator. Ah well, until they manage to drag themselves out of the dark ages, you'll just have to grit your teeth and get on with it.

Displacement, Velocity, and Acceleration are All Linked

Displacement, velocity and acceleration are all **vector** quantities (page 14), so the direction matters.

Speed — How fast something is moving, regardless of direction.
Displacement (s) — How far an object's travelled from its starting point in a given direction.
Velocity (v) — The rate of change of an object's displacement (its speed in a given direction).
Acceleration (a) — The rate of change of an object's velocity.

You need to know these formulas for **velocity** and **acceleration**:

$$v = \frac{\Delta s}{\Delta t} \quad \text{and} \quad a = \frac{\Delta v}{\Delta t}$$

The triangle symbol is the Greek 'delta', and it means 'the change in'.

The speed (or velocity) of an object at any given point in time is known as its **instantaneous** speed (or velocity). To find the **average** speed (or velocity), divide the **total** distance or displacement by the total time.

Acceleration Means a Curved Displacement-Time Graph

A graph of displacement against time for an **accelerating object** always produces a **curve**.
If the object is accelerating at a **uniform rate**, then the **rate of change** of the **gradient** will be constant.

Example: Plot a displacement-time graph for a lion who accelerates constantly from rest at 2 ms^{-2} for 5 seconds.

Take measurements to obtain a **table of values**...

t (s)	0	1	2	3	4	5
s (m)	0	1	4	9	16	25

...then plot the **graph**:

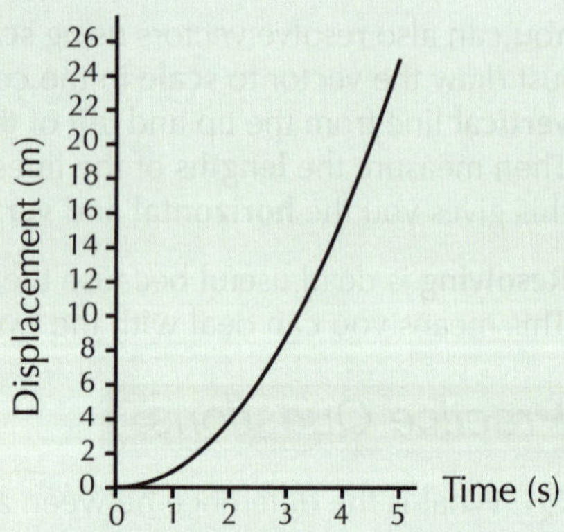

Different Accelerations Have Different Gradients

In the example above, if the lion has a **different acceleration** it'll change the **gradient** of the curve like this:

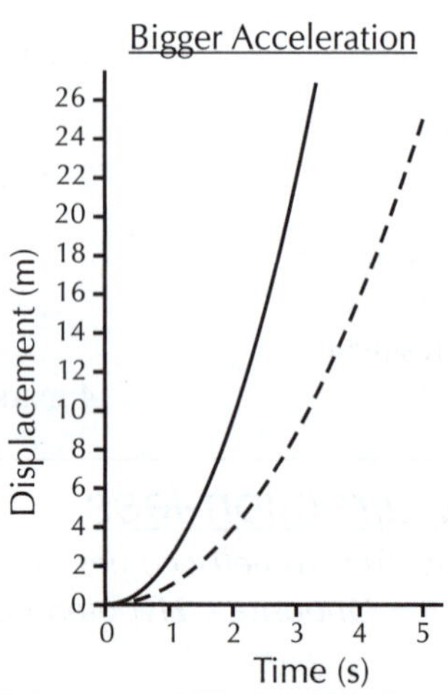

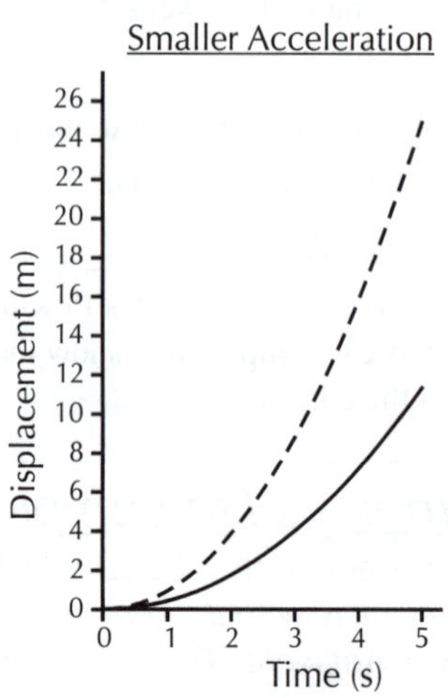

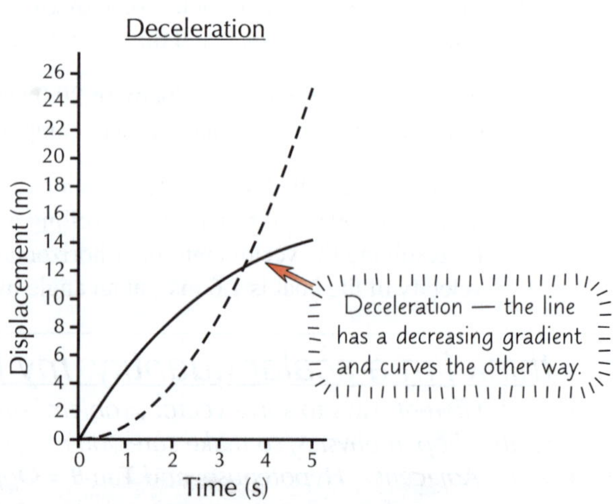

Displacement-Time Graphs

The Gradient of a Displacement-Time Graph Tells You the Velocity

When the velocity is constant, the graph's a **straight line**.
As you saw on the previous page, velocity is defined as...

$$\textbf{velocity} = \frac{\textbf{change in displacement}}{\textbf{change in time}}$$

On the graph, this is $\frac{\text{change in } y\ (\Delta y)}{\text{change in } x\ (\Delta x)}$, i.e. the gradient.

So to get the velocity from a displacement-time graph, just find the gradient.

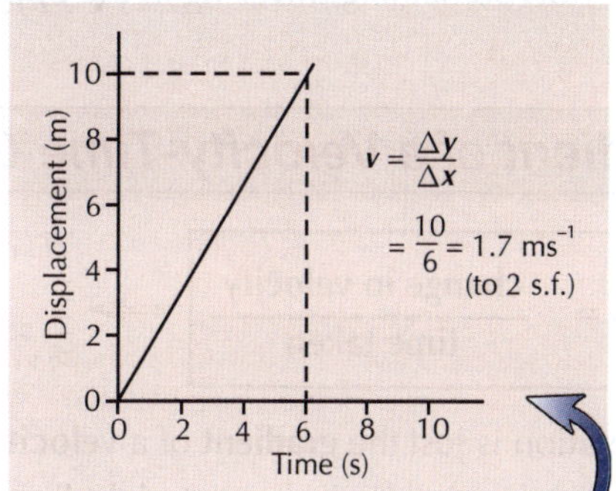

Acceleration is $\frac{\text{change in velocity } (\Delta v)}{\text{change in time } (\Delta t)}$, so it is the rate of change of this gradient. If the gradient is constant (straight line) then there is no acceleration, and if it's changing (curved line) then there's acceleration or deceleration.

It's the Same with Curved Graphs

If the gradient **isn't constant** (i.e. if it's a curved line), it means the object is **accelerating**.

To find the **instantaneous velocity** at a certain point you need to draw a **tangent** to the curve at that point and find its gradient.

To find the **average velocity** over a period of time, just divide the total change in displacement by the total change in time — it doesn't matter if the graph is curved or not.

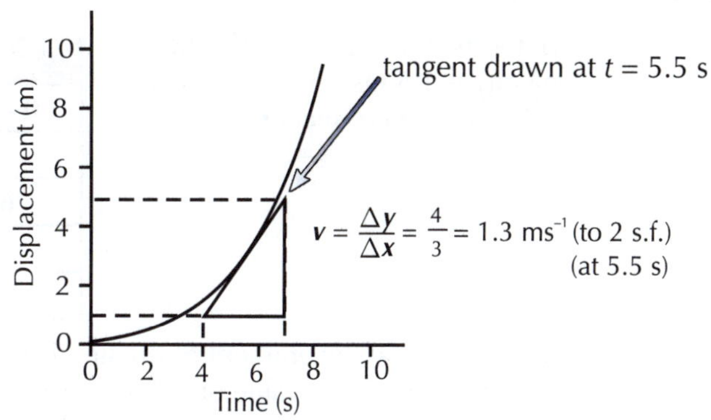

Practice Questions

Q1 What is given by the slope of a displacement-time graph?

Q2 Sketch a displacement-time graph to show: a) constant velocity, b) acceleration, c) deceleration

Exam Questions

Q1 Describe the motion of the cyclist as shown by the graph below. [4 marks]

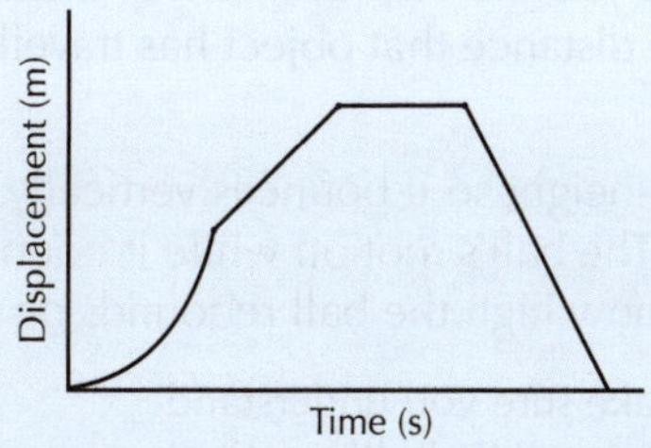

Q2 A baby crawls 5 m in 8 seconds at a constant velocity. She then rests for 5 seconds before crawling a further 3 m in 5 seconds. Finally, she makes her way back to her starting point in 10 seconds, travelling at a constant speed all the way.

a) Draw a displacement-time graph to show the baby's journey. [4 marks]

b) Calculate her average velocity at all the different stages of her journey. [2 marks]

Some curves are bigger than others...

Whether it's a straight line or a curve, the steeper it is, the greater the velocity. There's nothing difficult about these graphs — the problem is that it's easy to get them muddled up with velocity-time graphs (next page). Just think about the gradient — is it velocity or acceleration, is it changing (curve) or constant (straight line), is it 0 (horizontal line)...?

Velocity-Time and Acceleration-Time Graphs

Speed-time graphs and velocity-time graphs are pretty similar. The big difference is that velocity-time graphs can have a negative part to show something travelling in the opposite direction:

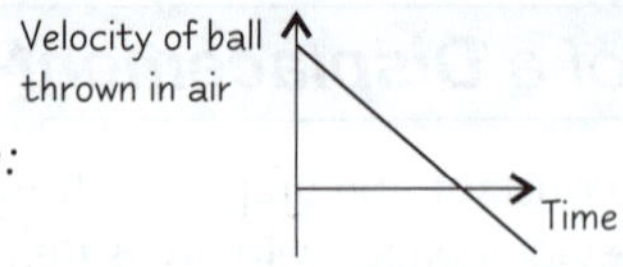

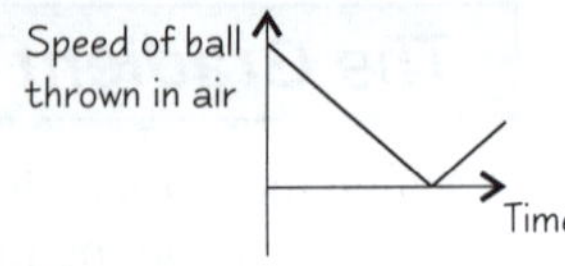

The Gradient *of a* Velocity-Time Graph *Tells You the* Acceleration

acceleration = $\frac{\text{change in velocity}}{\text{time taken}}$ → likewise for a speed-time graph

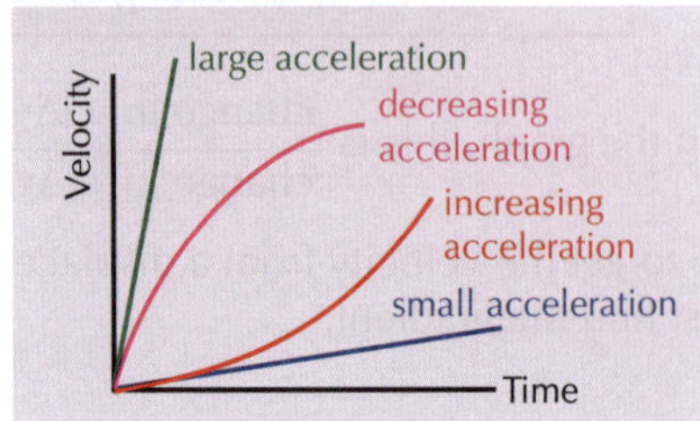

So the acceleration is just the **gradient** of a **velocity-time graph**.

1) **Uniform** acceleration is always a **straight line**. The **steeper** the **gradient**, the **greater** the **acceleration**.
2) A **curved** graph shows **changing** acceleration. **Increasing** gradient means **increasing acceleration**, and **decreasing** gradient means **decreasing acceleration** (or deceleration).

Example: A lion strolls along at 1.5 ms^{-1} for 4 s and then accelerates uniformly at a rate of 2.5 ms^{-2} for 4 s. Plot this information on a velocity-time graph.

Norman (the lion*)...

So, for the first four seconds, the velocity is 1.5 ms^{-1}, then it increases by **2.5 ms^{-1} every second**:

t (s)	*v* (ms^{-1})
0 – 4	1.5
5	4.0
6	6.5
7	9.0
8	11.5

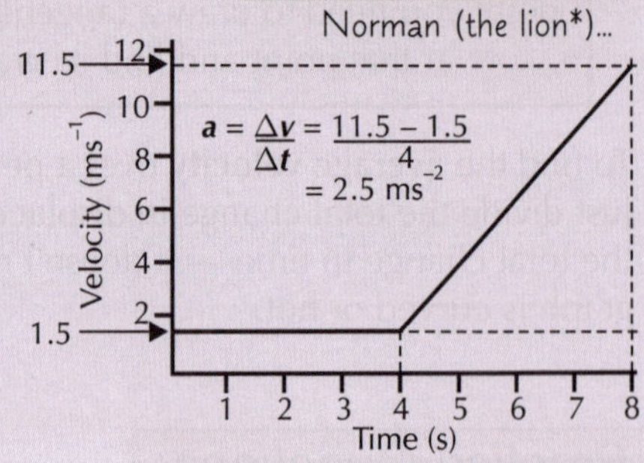

You can see that the **gradient of the line** is **constant** between 4 s and 8 s and has a value of 2.5 ms^{-2}, representing the **acceleration of the lion**.

*Yes, I know — I just like lions, OK...

Displacement = *Area Under* *Velocity-Time Graph*

You know that: **distance travelled = average speed × time**

The **area under a velocity-time graph** tells you the displacement of the object. The magnitude of this displacement is the distance that object has travelled.

Example: A ball is dropped from table-height so it bounces vertically. It bounces twice before someone catches it. The ball's motion while it bounces is shown on the *v-t* graph below. Calculate how high the ball rebounds on the first bounce.

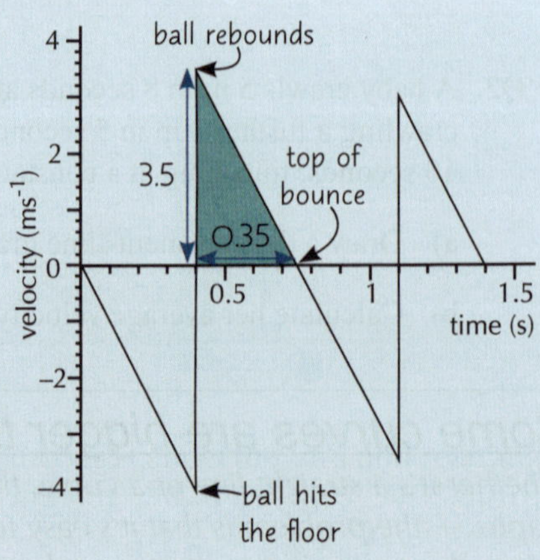

Before you try and calculate anything, make sure you understand what each part of the graph is telling you about the ball's motion.

1) When the ball is first dropped, the velocity of the ball is **negative** — so downwards is the negative direction.
2) The points where the **ball hits the floor** are shown by the **vertical straight lines** on the graph — the ball's **speed** remains roughly the **same**, but its **direction** (and **velocity**) changes the instant it hits the floor.
3) The points where the ball's velocity is **zero** show where the ball reaches the **top of a bounce** before starting to fall downwards.

The height of the first bounce is the **area under the graph** between the time the ball first rebounds from the floor and the time it reaches the top of the bounce.

displacement = area under graph
= (3.5 × 0.35) ÷ 2 = 0.6125 = **0.61 m (to 2 s.f.)**

Velocity-Time and Acceleration-Time Graphs

Acceleration-Time (a-t) Graphs are Useful Too

An **acceleration-time graph** shows how an object's **acceleration** changes over time.

1) The **height** of the graph gives the object's **acceleration** at that time.
2) The **area** under the graph gives the object's **change in velocity**.
3) A negative acceleration is a **deceleration**.
4) If **$a = 0$**, then the object is moving with **constant velocity**.

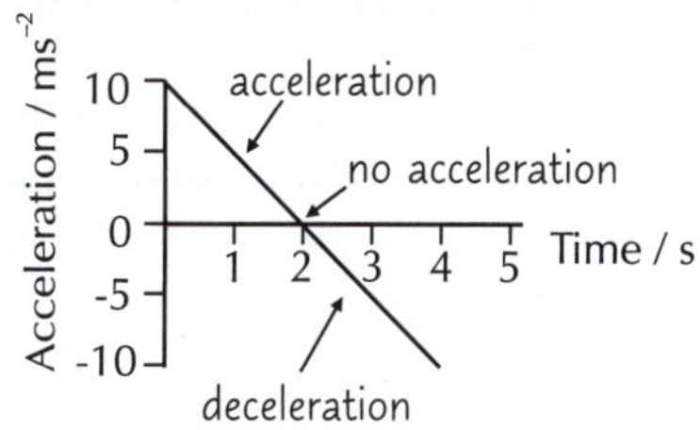

You Have to **Estimate** the Area Under a **Curved** Graph

If an object's acceleration **isn't constant**, you won't get a straight line *a-t* graph. You need to know how to **estimate** the area under a curved graph. If the graph is on **squared paper**, you can work out the value represented by the **area** of **one** square and multiply by the approximate **number of squares** under the curve. Another way is to split the area approximately into simple shapes, calculate the value of the **area** of each of them, and then **add** them all up.

Example: The acceleration of a car in a drag race is shown in this acceleration-time graph. Calculate its change in velocity.

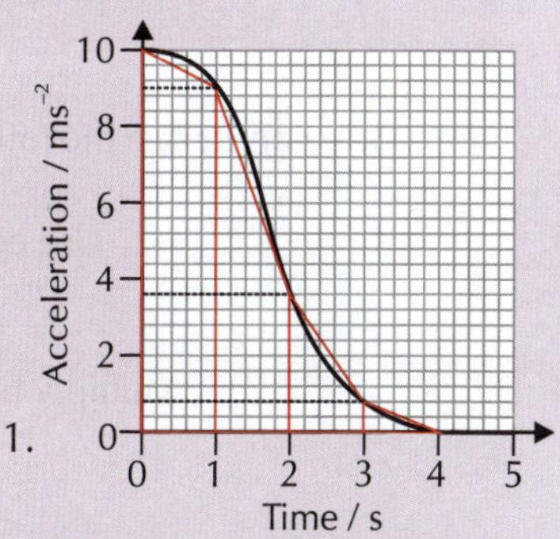

Change in velocity = area under graph

Split the area under the curve up into trapeziums and a triangle.

0-1 s — estimate the area using a trapezium. Area = $\frac{1}{2}(a + b) \times h$

a is the length of the first side, $a = 10$

b is the length of the second side, $b = 9$

h is the width of each strip, so $h = 1$. Area = $\frac{1}{2}(10 + 9) \times 1 = 9.5$ ms^{-1}

1-2 s — this can also be estimated with another trapezium. $a = 9$, $b = 3.6$, $h = 1$.

So area = $\frac{1}{2}(9 + 3.6) \times 1 = 6.3$ ms^{-1}

2-3 s — estimated with another trapezium. $a = 3.6$, $b = 0.8$, $h = 1$. So area = $\frac{1}{2}(3.6 + 0.8) \times 1 = 2.2$ ms^{-1}

3-4 s — this estimation uses a triangle. Area = $\frac{1}{2}$(base × height) = $\frac{1}{2}(0.8) \times 1 = 0.4$ ms^{-1}

Now add the areas together — Total area = 9.5 + 6.3 + 2.2 + 0.4 = 18.4 ms^{-1}

The estimated change in velocity of the car is 18.4 ms^{-1} = **20 ms^{-1} (to 1 s.f.)**

You can use the same method to find the area under any non-linear graph.

Practice Questions

Q1 How do you calculate acceleration from a velocity-time graph?

Q2 How do you calculate the displacement travelled from a velocity-time graph?

Q3 Sketch velocity-time graphs for constant velocity and constant acceleration.

Q4 Sketch velocity-time and acceleration-time graphs for a boy bouncing on a trampoline.

Q5 What does the area under an acceleration-time graph tell you?

Exam Question

Q1 A skier accelerates uniformly from rest at 2 ms^{-2} down a straight slope for 5 seconds. He then reaches the bottom of the slope and continues along the flat ground, decelerating at 1 ms^{-2} until he stops.

a) Sketch the velocity-time and acceleration-time graphs for his journey. [4 marks]

b) Use your *v-t* graph from part a) to find the distance travelled by the skier during the first 5 seconds. [2 marks]

Still awake — I'll give you five more minutes...

There's a really nice sunset outside my window. It's one of those ones that makes the whole landscape go pinky-yellowish. And that's about as much interest as I can muster on this topic. Normal service will be resumed on page 21.

Motion with Uniform Acceleration

Uniform Acceleration is Constant Acceleration

Acceleration could mean a change in speed or direction or both.

Uniform means **constant** here. It's nothing to do with what you wear.
There are **four main equations** that you use to solve problems involving **uniform acceleration**.
To help you learn them, it can be useful to know how they are derived:

1) **Acceleration is the rate of change of velocity.**
From this definition you get:

$a = \frac{(v - u)}{t}$ so $\boldsymbol{v = u + at}$

where:
$\boldsymbol{u}$ = initial velocity, $\boldsymbol{a}$ = acceleration
$\boldsymbol{v}$ = final velocity, $\boldsymbol{t}$ = time taken

2) **s = average velocity × time**
If acceleration is constant, the average velocity is just the average of the initial and final velocities, so:

$\boldsymbol{s = \frac{(u + v)}{2} \times t}$ s = displacement

3) Substitute the expression for $\boldsymbol{v}$ from equation **1** into equation **2** to give:

$s = \frac{(u + u + at) \times t}{2} = \frac{2ut + at^2}{2}$ $\boldsymbol{s = ut + ½at^2}$

4) You can **derive** the fourth equation from equations **1** and **2**:

These equations are also known as the constant acceleration equations.

Use equation **1** in the form: $a = \frac{(v - u)}{t}$

Multiply both sides by **s**, where: $s = \frac{(u + v)}{2} \times t$

This gives us: $as = \frac{(v - u)}{t} \times \frac{(u + v)t}{2}$

The *t*'s on the right cancel, so: $2as = (v - u)(v + u)$
$2as = v^2 - uv + uv - u^2$

so: $\boldsymbol{v^2 = u^2 + 2as}$

Example: A tile falls from a roof 25 m high. Calculate its speed when it hits the ground and how long it takes to fall. Acceleration due to gravity = 9.81 ms^{-2} (see p.22).

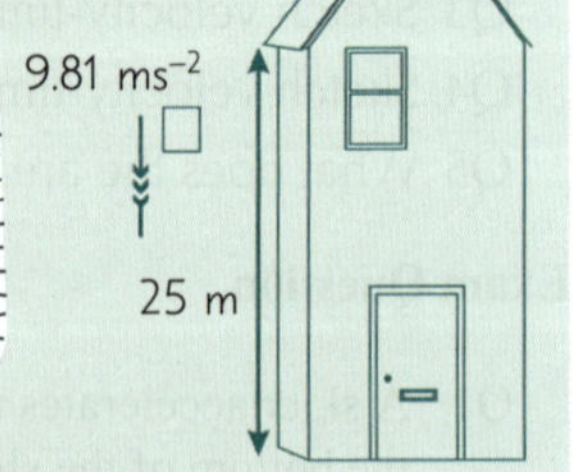

First of all, write out what you know:
s = 25 m
u = 0 ms^{-1} since the tile's stationary to start with
a = 9.81 ms^{-2}
v = ? t = ?

Usually you take upwards as the positive direction. In this question it's probably easier to take downwards as positive, so you get $a = +9.81\ ms^{-2}$ instead of $a = -9.81\ ms^{-2}$.

Then, choose an equation with only **one unknown quantity**.
So start with $v^2 = u^2 + 2as$
$v^2 = 0 + 2 \times 9.81 \times 25$
$v^2 = 490.5$
v = 22 ms^{-1} (to 2 s.f.)

Now, find t using:
$s = ut + ½at^2$
$25 = 0 + ½ \times 9.81 \times t^2$
$t^2 = \frac{25}{4.9}$
t = 2.3 (to 2 s.f.)

Final answers:
t = **2.3 s (to 2 s.f.)**
v = **22 ms^{-1} (to 2 s.f.)**

Motion with Uniform Acceleration

Example: A car accelerates steadily from rest at a rate of 4.2 ms^{-2} for 6.5 seconds.
a) Calculate the final speed.
b) Calculate the distance travelled in 6.5 seconds.

Remember — always start by writing down what you know.

a) $a = 4.2\ ms^{-2}$
$u = 0\ ms^{-1}$
$t = 6.5\ s$
$v = ?$

choose the right equation... $v = u + at$

$v = 0 + 4.2 \times 6.5$

Final answer: $v = 27.3\ ms^{-1}$
$= 27\ ms^{-1}$ (to 2 s.f.)

b) $s = ?$
$t = 6.5\ s$
$u = 0\ ms^{-1}$
$a = 4.2\ ms^{-2}$
$v = 27.3\ ms^{-1}$

you can use: $s = \frac{(u+v)t}{2}$ **or:** $s = ut + \frac{1}{2}at^2$

$s = \frac{(0 + 27.3) \times 6.5}{2}$ $s = 0 + \frac{1}{2} \times 4.2 \times (6.5)^2$

Final answer: $s =$ **89 m (to 2 s.f.)** $s =$ **89 m (to 2 s.f.)**

You Have to *Learn* How to *Use* the Uniform Acceleration *Equations*

Make sure you learn the equations. There are only four of them and these questions are always dead easy marks in the exam, so you'd be dafter than a hedgehog in a helicopter not to learn them...

Practice Questions

Q1 Write out the four uniform acceleration equations.

Q2 A small steel ball is dropped from a height of 1.5 m. Calculate its speed as it hits the ground.

Exam Questions

Q1 A skydiver jumps from a helicopter hovering at a height of 1500 m from the ground.
She accelerates due to gravity for 5.0 s.
a) Calculate her maximum vertical velocity. (Assume no air resistance.) [2 marks]
b) Calculate how far she falls in this time. [2 marks]

Q2 A motorcyclist slows down uniformly as he approaches a red light.
He takes 3.2 seconds to come to a halt and travels 40 m (to 2 s.f.) in this time.
a) Calculate how fast he was initially travelling. [2 marks]
b) Calculate his acceleration. (N.B. a negative value shows a deceleration.) [2 marks]

Q3 A stream provides a constant acceleration of 6 ms^{-2}. A toy boat is pushed directly against the current and then released from a point 1.2 m upstream from a small waterfall. Just before it reaches the waterfall, it is travelling at a speed of 5 ms^{-1}.
a) Calculate the initial velocity of the boat. [2 marks]
b) Calculate the maximum distance upstream from the waterfall the boat reaches. [2 marks]

Constant acceleration — it'll end in tears...

If a question talks about "uniform" or "constant" acceleration, it's a dead giveaway they want you to use one of these equations. The tricky bit is working out which one to use — start every question by writing out what you know and what you need to know. That makes it much easier to see which equation you need. To be sure. Arrr.

Acceleration Due To Gravity

Ahhh acceleration due to gravity. The reason falling apples whack you on the head.

Free Fall is when there's Only *Gravity* and Nothing Else

1) All objects on Earth experience a **force** due to the Earth's **gravitational field** that depends on their **mass**.
2) The **gravitational field strength** near the surface of the Earth is called g and is equal to **9.81 Nkg^{-1}**. g is given by the equation: $g = \frac{F}{m}$
3) The force on an object due to gravity is called its **weight** (see page 40).
4) When the only force acting on an object is its weight, it will undergo **free fall**, and will **accelerate towards the ground**.
5) Acceleration and force are related by $F = ma$.
   So, for an object in free fall, $a = \frac{F}{m} = g$.

Where m is the mass of the object and F is the force due to gravity.

Force and acceleration are both vector quantities (p.14). In free fall, they act vertically downwards.

> Objects undergoing free fall on Earth have an acceleration of $g = 9.81$ **ms^{-2}**.

6) Objects can have an initial velocity in any direction and still undergo **free fall** as long as the **force** providing the initial velocity is **no longer acting**.

It seems weird that g is both a gravitational field strength, in Nkg^{-1}, and an acceleration, in ms^{-2}. However, if you break down Nkg^{-1} into SI units, they are actually the same as ms^{-2}.

All Objects in Free Fall *Accelerate* at the *Same* Rate

Another gravity experiment.

1) For over 1000 years the generally accepted theory was that heavier objects would fall towards the ground quicker than lighter objects. It was challenged a few times, but it was finally overturned when **Galileo** came on the scene.
2) The difference with Galileo was that he set up **systematic** and **rigorous experiments** to **test** his theories — just like in modern science. These experiments could be repeated and the results described **mathematically** and compared.
3) Galileo believed that all objects fall at the same rate. The problem in trying to prove it was that free-falling objects **fell too quickly** for him to be able to take any accurate measurements (he only had a water clock), and **air resistance** affects the rate at which objects fall.
4) Galileo measured the time a ball took to roll down a **smooth** groove in an inclined plane. He killed two birds with one stone by rolling it down a plane, which **slows** the ball's fall as well as reducing the effect of **air resistance**.
5) By rolling the ball along different fractions of the total length of the slope, he found that the distance the ball travelled was proportional to the square of the time taken. The ball was **accelerating** at a **constant rate**.
6) In the end it took **Newton** to bring it all together to show and explain why **all** free falling objects have the same acceleration. He showed **mathematically** that all objects are attracted towards the Earth due to a force he called **gravity**. Ah, good ol' Newton...

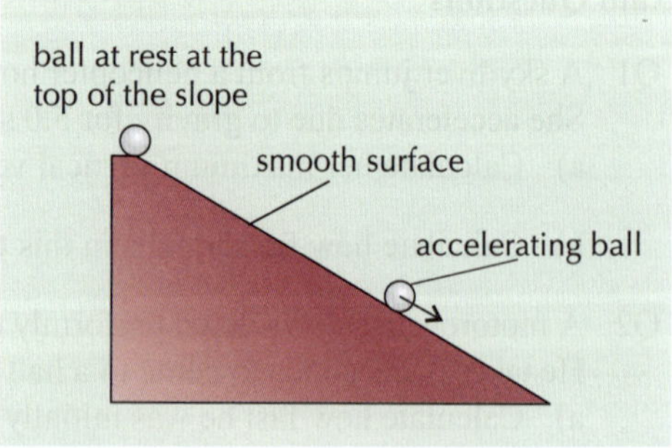

You can *Replace a* with *g* in the *Equations of Motion*

You need to be able to work out **speeds**, **distances** and **times** for objects in **free fall**. Since g is a **constant acceleration** you can use the **uniform acceleration equations**.

g acts **downwards**, so you need to be careful about directions. To make it clear, there's a sign convention: **upwards is positive**, **downwards is negative**.

$$v = u + gt \qquad s = ut + \tfrac{1}{2}gt^2$$
$$v^2 = u^2 + 2gs \qquad s = \frac{(u+v)t}{2}$$

> g is always **downwards** so it's **usually negative** — t **is always positive**
> u and v can be **positive or negative** — s can be **positive or negative**

If an object is dropped, not thrown, u is zero.

Acceleration Due To Gravity

You Can Calculate g By Performing an Experiment...

See pages 6-13 for more on doing experiments.

This is just one way of **measuring g** — there are loads of different experiments you could do.

1) Set up the equipment shown in the diagram on the right.
2) Measure the height ***h*** from the **bottom** of the ball bearing to the **trapdoor**.
3) Flick the switch to simultaneously **start the timer** and **disconnect the electromagnet**, releasing the ball bearing.
4) The ball bearing falls, knocking the trapdoor down and **breaking the circuit** — which **stops the timer**. Record the time ***t*** shown on the timer.
5) **Repeat** this experiment three times and **average** the time taken to fall from this height. Repeat this experiment but drop the ball from several **different heights**.
6) You can then use these results to find *g* using a **graph** (see below).

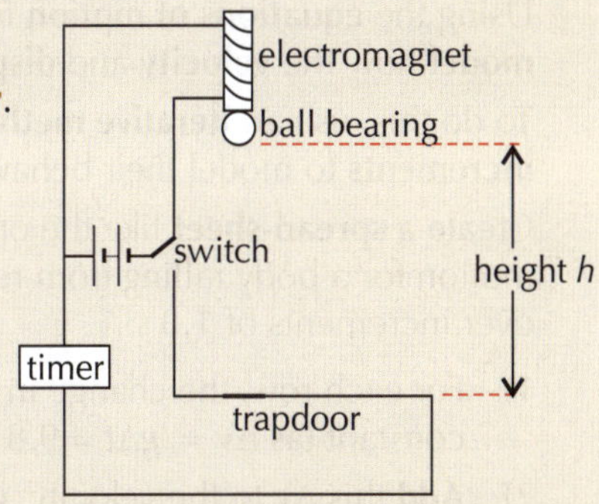

- Using a **small** and **heavy** ball bearing means you can assume air resistance is so small you can **ignore it**.
- Having a computer **automatically release** and **time** the ball-bearing's fall can measure times with a **smaller uncertainty** than if you tried to drop the ball and time the fall using a stopwatch.
- The most significant source of **error** in this experiment will be in the measurement of ***h***. Using a ruler, you'll have an uncertainty of about ±1 mm. This dwarfs any error from switch delay or air resistance. By making the values of ***h*** as **large** as possible, you can reduce the **percentage uncertainty** in your measurement of the height, (see p.10) as well as the percentage uncertainty caused by the **resolution** of the **timer** (as the bigger *h* is, the longer the ball-bearing will take to fall).

You could use video analysis techniques to investigate acceleration due to gravity — see p.28.

...and Plotting a Graph of Your Results

1) Use your data from the experiment above to plot a graph of **height** (*s*) against the **time** it takes the ball to fall, **squared** (t^2). Then draw a **line of best fit**.

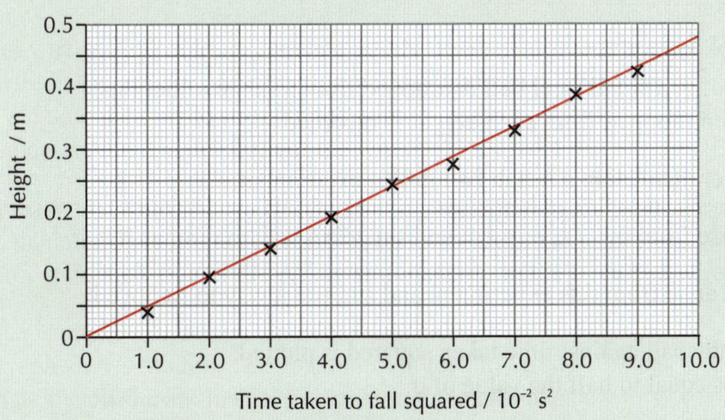

You could plot error bars on your data graph to find the error in your final value for *g*. See page 11 for more on error bars.

In the exam you might be asked to find *g* from a displacement-time graph (see p.16-17). *g* is an acceleration and the gradient of the graph will be velocity, so you can find *g* by finding the change in gradient between two points on the graph (as $a = \Delta v \div \Delta t$).

2) You know that with constant acceleration, $s = ut + \frac{1}{2}at^2$. If you drop the ball, initial speed $u = 0$, so $s = \frac{1}{2}at^2$.
3) Rearranging this gives $\frac{1}{2}a = \frac{s}{t^2}$, or $\frac{1}{2}g = \frac{s}{t^2}$ (remember the acceleration is all due to gravity).
4) So the gradient of the line of best fit, $\frac{\Delta s}{\Delta t^2}$, is equal to $\frac{1}{2}g$:

$$g = 2 \times \frac{\Delta s}{\Delta t^2} = 2 \times \frac{0.43}{0.090} = 9.555...$$
$$= \mathbf{9.6\ ms^{-2}\ (to\ 2\ s.f.)}$$

5) As you know *g* (9.8 ms^{-2} to 2 s.f.) you can calculate the **percentage difference** (see p.12) between your value of *g* and the true value, and use this to evaluate the **accuracy** of your results.

$$\%\ \text{difference} = \frac{9.6 - 9.8}{9.8} \times 100\,\% = -2.040...$$
$$= \mathbf{-2.0\%\ (to\ 2\ s.f.)}$$

Acceleration Due To Gravity

Motion can be **Modelled** Using **Iterative Methods**

You can **Model** Acceleration Due to Gravity Using a **Spreadsheet**

Using the **equations of motion** for **uniform acceleration** adapted for **acceleration due to gravity** (p.22) you can **model** how the **velocity** and **displacement** of a falling object vary over time.

To do this, use an **iterative method**, where **velocity** and **displacement** are calculated over **lots** of **small time increments** to model their behaviour over a **longer time period**.

Create a **spread-sheet** like the one shown (which models motion for a body falling from rest at 9.8 ms^{-2} (to 2 s.f.) over increments of 1.0 s).

t (s)	Δt (s)	Δv (ms^{-1})	v (ms^{-1})	Δs (m)	s (m)
0	-	-	0	-	0
1.0	1.0	9.8	9.8	4.9	4.9
2.0	1.0	9.8	19.6	14.7	19.6
3.0	1.0	9.8	29.4	24.5	44.1

1) For each row, the change in velocity, $\boldsymbol{\Delta v}$ will be constant (as $\Delta \boldsymbol{v} = \boldsymbol{g}\Delta \boldsymbol{t} = 9.8 \times 1.0 = 9.8$ ms^{-1}).
2) **Add** this Δv to the velocity, $\boldsymbol{v}$, from the **previous** row, giving you the **new velocity** for each increment.
3) To find the **change in displacement**, Δs, for each row use $\Delta \boldsymbol{s} = \boldsymbol{u}\Delta t + \frac{1}{2}\boldsymbol{g}(\Delta t)^2$ (where $\boldsymbol{u}$ is the value of $\boldsymbol{v}$ from the previous row). Add this to the previous **s** value to get the new **displacement** at the **end** of each time interval.
4) **Repeat** this to build up a full model of the body's motion over time. If you enter the correct formulas into the spreadsheet, then it can automatically complete as many rows (iterations) as you want.
5) You can then **plot a graph** of $\boldsymbol{v}$ against t (which should be a **straight line** through the origin with **gradient** equal to the body's **acceleration**) and **s** against t (which should be a **curve** similar to the one shown on page 16).

Practice Questions

Q1 What is meant by free fall? What is the value of the acceleration of a free-falling object on Earth?

Q2 How does the velocity of a free-falling object change with time?

Q3 What is the main reason Galileo's ideas became generally accepted in place of the old theory?

Q4 Describe an experiment that could be used to calculate the value of g.

Q5 Describe how you could use a spreadsheet to model displacement and velocity of an object in free fall.

Exam Questions

Q1 In an experiment to determine the value of g, a small steel ball is dropped from a range of heights. The time it takes to reach the ground when dropped from each height is recorded.

a) Explain why using a steel ball is better than using a beach ball in this experiment. [1 mark]

b) State one random error that could arise from this experiment and suggest a way to remove it. [2 marks]

c) State one systematic error that could arise from this experiment and suggest a way to remove it. [2 marks]

d) A graph of the distance travelled by the ball against time taken squared is plotted. Show that the gradient of the graph is equal to half the value of g. [3 marks]

Q2 Jan bounces on a trampoline, reaching a point 5.0 m above the trampoline's surface. Assume there is no air resistance.

a) Calculate the speed with which she leaves the trampoline surface. [2 marks]

b) Calculate how long it takes Jan to reach her highest point. [2 marks]

c) State her velocity as she lands back on the trampoline. [1 mark]

So it's this "Galileo" geezer who's to blame for my practicals...

Hmmm... I wonder what Galileo would be proudest of — insisting on the systematic, rigorous experimental method on which modern science hangs... or getting in a Queen song? Magnificoooooo...

Projectile Motion

Any object given an initial velocity and then left to move freely under gravity is a projectile, which handily you can predict the motion of by resolving its velocity. If you're doing Maths, you might have to cover this again there. Double the fun.

You have to think of **Horizontal** and **Vertical** Motion **Separately**

In projectiles, the **horizontal** and **vertical** components of the object's motion are **completely independent**. Projectiles follow a **curved path** because the horizontal velocity remains **constant**, while the vertical velocity is affected by the **acceleration due to gravity**, g.

Example: Sharon fires a scale model of a TV talent show presenter horizontally with a velocity of 100 ms^{-1} (to 2 s.f.) from 1.5 m above the ground. How long does it take to hit the ground, and how far does it travel horizontally? Assume the model acts as a particle, the ground is horizontal and there is no air resistance.

Think about vertical motion first:

1) It's **constant acceleration** under gravity...
2) You know $u_v = 0$ (no vertical velocity at first), $s_v = -1.5$ m and $a = g = -9.81$ ms^{-2}. You need to find t.
3) Use $s = \frac{1}{2}gt^2 \Rightarrow t = \sqrt{\frac{2s}{g}} = \sqrt{\frac{2 \times -1.5}{-9.81}} = 0.553...$ s
4) So the model hits the ground after **0.55 (to 2 s.f.)** seconds.

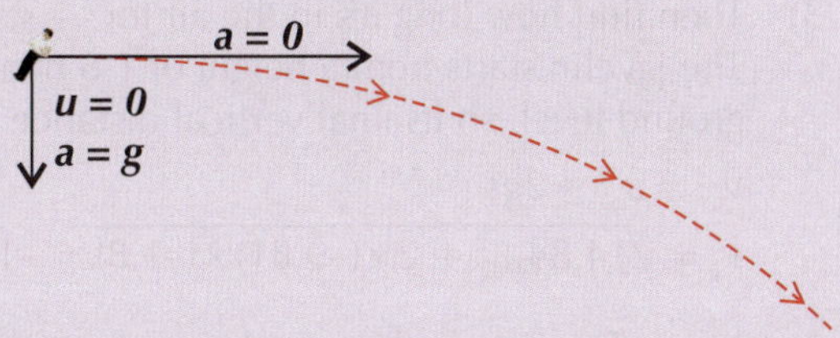

Then do the horizontal motion:

1) The horizontal motion isn't affected by gravity or any other force, so it moves at a **constant speed**.
2) That means you can just use good old **speed = distance / time**.
3) Now $v_h = 100$ ms^{-1}, $t = 0.553...$ s and $a = 0$. You need to find s_h.
4) $s_h = v_h t = 100 \times 0.553... =$ **55 m (to 2 s.f.)**

Where v_h is the horizontal velocity, and s_h is the horizontal distance travelled (rather than the height fallen).

It's **Slightly Trickier** if it **Starts Off** at an **Angle**

1) Velocities can be in **any direction**, so they're not always at right angles to each other. This is sometimes a bit **awkward** for **calculations**.
2) If something's projected at an **angle** (like, say, a javelin) you start off with both **horizontal** and **vertical velocity**:

Method:
1) **Resolve** the initial velocity into horizontal and vertical components (see below).
2) Use the vertical component to work out **how long** it's in the air and/or **how high** it goes.
3) Use the horizontal component to work out **how far** it goes while it's in the air.

Resolving a **Velocity** means **Splitting** it into **Components**

You'll also find examples of resolving on p.15, p.36-37 and p.45.

Example: A box is pushed at a velocity of 10.0 ms^{-1} at an angle of 0.52 radians above the horizontal. Find the vertical and horizontal components of the velocity.

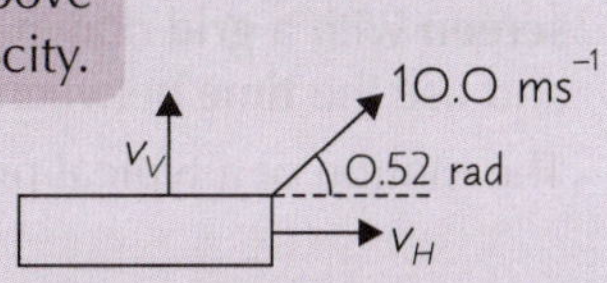

Use these formulas when resolving velocities:

$$\frac{v_H}{v} = \cos\theta \quad \text{or} \quad v_H = v\cos\theta$$

And

$$\frac{v_V}{v} = \sin\theta \quad \text{or} \quad v_V = v\sin\theta$$

Remember that θ is the angle from the **horizontal**.

$v_H = 10.0 \times \cos(0.52) = 8.67... =$ **8.7 ms^{-1} (to 2 s.f.)**

$v_V = 10.0 \times \sin(0.52) = 4.96... =$ **5.0 ms^{-1} (to 2 s.f.)**

To convert from degrees to radians (rad), multiply by $\pi/180°$. To convert from radians to degrees, multiply by $180°/\pi$.

Projectile Motion

Time for a Resolving Example

Example: An athlete throws a javelin from a height of 1.8 m with a velocity of 21 ms^{-1} at an upward angle of 45° to the ground. How far is the javelin thrown? Assume the javelin acts as a particle, the ground is horizontal and there is no air resistance.

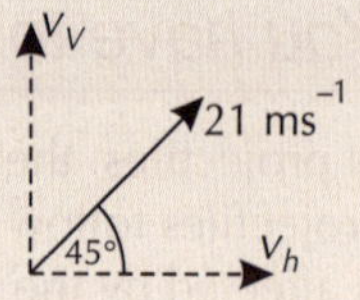

1) Draw a quick sketch of the information given in the question.

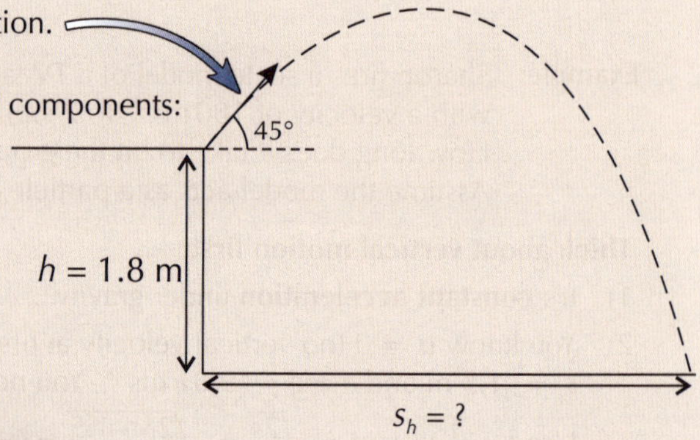

2) Start by resolving the velocity into horizontal and vertical components:
   $u_h = \cos 45° \times 21 = 14.84...\ ms^{-1}$
   $u_v = \sin 45° \times 21 = 14.84...\ ms^{-1}$

3) Then find how long it's in the air for — start by finding v_v. The javelin starts from a height of 1.8 m and finishes at ground level, so its final vertical distance $s_v = -1.8$ m:
   $v_v^2 = u_v^2 + 2gs$
   $v_v = \sqrt{14.84...^2 + 2 \times (-9.81) \times (-1.8)} = -15.99...\ ms^{-1}$

   You need the negative square root, as this is a velocity towards the ground.

   Now you can use this v_v value and $s = \left(\frac{u+v}{2}\right)t$ to find the time it stays in the air:

   $$s_v = \frac{(u_v + v_v)}{2} \times t \Rightarrow t = \frac{s_v}{(u_v + v_v)} \times 2 = \frac{-1.8}{14.84... - 15.99...} \times 2 = 3.144...\ s$$

4) Finally, as $a_h = 0$, you can use **speed = distance / time** to work out how far it travels horizontally in this time. The horizontal velocity is just u_h, so: $s_h = u_h t = 14.84... \times 3.144... =$ **47 m (to 2 s.f.)**

You can Investigate Projectile Motion Using a *Video Camera*...

If you **video** a projectile moving, you can use **video analysis software** to investigate its motion:

1) You can **plot the course** taken by an object by recording its **position** in **each frame**.
2) If you know the **frame rate**, and your video includes a metre ruler or grid lines that you can use as a scale, you can calculate the **velocity** of the projectile between **different points** in its motion, by looking at how far it travels **between frames**.

A video camera records a series of pictures, or frames (typically around 25 frames per second). Video analysis software lets you view videos frame by frame — see p.28.

...or *Strobe Photography*

In strobe photography, a camera is set to take a **long exposure**. While the camera is taking the photo, a **strobe light** flashes repeatedly and the projectile is released. The strobe light **lights up** the projectile at regular intervals. This means that the projectile appears **multiple times** in the same photograph, in a **different position** each time.

Again, if you've got a **reference object** in the photo (for example, you might throw an object in front of a **screen** with a **grid** drawn on it), you can calculate **how far** the object travels **between flashes** of the strobe, and use the **time** between flashes to calculate the **velocity** of the projectile between the flashes.

The motion of a typical projectile captured with strobe photography is shown below.

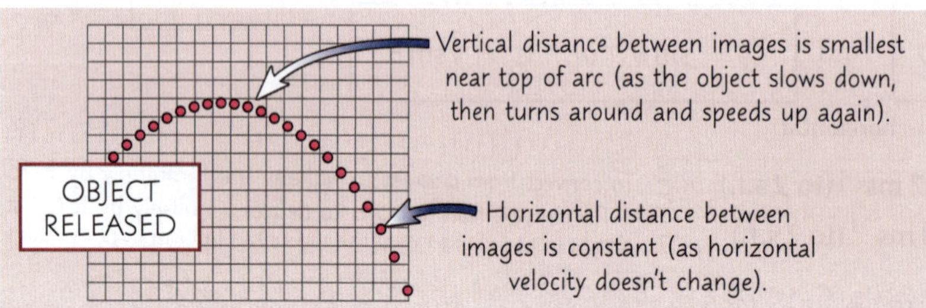

Strobe photography and video cameras give you more information than using light-gates to study an object's projectile motion. They can be used whatever the size of the object, unlike a light-gate.

Projectile Motion

You can **Model** Projectile Motion by **Drawing Vectors**

1) Consider an object fired horizontally, so that it begins to move with **projectile motion** under the influence of gravity.
2) Draw a grid (or use graph paper). Split the projectile's motion up into **even time intervals**, Δt. For each time interval Δt, draw starting and finishing vector arrows to **scale**. The **difference** between these vectors is the **increase in vertical velocity**, $\mathbf{\Delta v}$.

Drawing vectors is useful for visualising motion in 2 dimensions.

3) Initially, the object has horizontal speed v_0 and vertical speed 0. After the first time interval, the object still has horizontal speed v_0, but its **vertical speed** has **increased by** $\mathbf{\Delta v}$. This gives a resultant velocity of $\mathbf{v_1}$, as shown.
4) After the second time interval, the horizontal speed has remained constant and the vertical speed has again **increased by** $\mathbf{\Delta v}$.

The new resultant velocity is $\mathbf{v_1} + \mathbf{\Delta v} = \mathbf{v_2}$.

5) This pattern **continues** for each time interval. The resultant velocity **increases** during each interval (the blue arrow gets longer each time), as the object is **accelerating**.

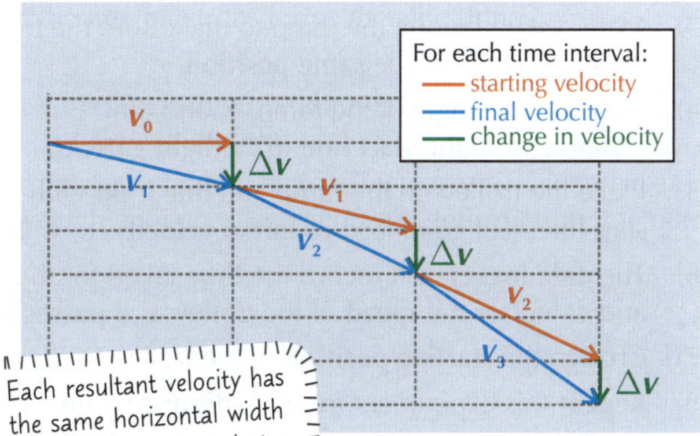

Each resultant velocity has the same horizontal width since Δt is constant, but a greater vertical component than the previous one.

Practice Questions

Q1 What is the initial vertical velocity for an object projected horizontally with a velocity of 5 ms^{-1}?

Q2 How does the horizontal velocity of a projectile change with time?

Q3 What is the horizontal component of the velocity of a stone, hurled at 30 ms^{-1} at 35° to the horizontal?

Q4 Explain how you might use video analysis software to analyse the motion of a projectile.

Q5 Explain how strobe photography can be used investigate projectile motion.

Q6 Describe how to model projectile motion by considering the changes in velocity over set time intervals and drawing vectors.

Exam Questions

Q1 Jason stands on a vertical cliff edge throwing stones into the sea below. He throws a stone horizontally with a velocity of 20.0 ms^{-1}, 560 m above sea level.

a) Calculate how long it takes for the stone to hit the water from leaving Jason's hand. Use $g = 9.81$ ms^{-2} and ignore air resistance. [2 marks]

b) Calculate the distance of the stone from the base of the cliff when it hits the water. [2 marks]

Q2 Robin fires an arrow into the air with a vertical velocity of 30.0 ms^{-1}, and a horizontal velocity of 20.0 ms^{-1}, from 1.0 m above horizontal ground.

Choose the correct option which shows the maximum height from the ground (to 2 significant figures) reached by his arrow. Use $g = 9.81$ ms^{-2} and ignore air resistance. [1 mark]

A	45 m	C	47 m
B	46 m	D	48 m

All this physics makes me want to create projectile motions...

...by throwing my revision books out of the window. The maths can be tricky, but take it step by step and all will be fine. On the plus side, the next page is all about chucking trolleys down slides. It's like tobogganing for dolls...

Motion Experiments and Stopping Distances

It's all getting a bit hi-tech now — using light gates and video cameras to look at how an object's velocity changes as it rolls down a ramp or crashes into something. Who doesn't love a good motion experiment...

You Can **Investigate** What **Affects** the **Motion** of a Trolley on a Slope

1) To investigate how the **distance** a trolley has rolled affects its speed, set up the experiment shown in the diagram.
2) Measure the **length** of the trolley and enter this into your data logger's software.
3) Mark a **start line** on the ramp to make sure the trolley always starts from the **same position**.
4) Measure the **angle** of the ramp, θ, and the **distance** from the **start line** to the **light gate**, d.
5) Place the trolley on the **ramp** and **line it up** with the start line. Let go of it so its **initial velocity**, u, is **0**.
6) The **data logger** will record the **time** taken for the trolley to pass through the light gate and calculate the **speed** of the trolley as it passes through the gate.
7) Change the **starting position** of the trolley, so d is varied.
8) **Repeat** this experiment for each distance 3 times and average the recorded velocities to reduce the **error** in your final result.

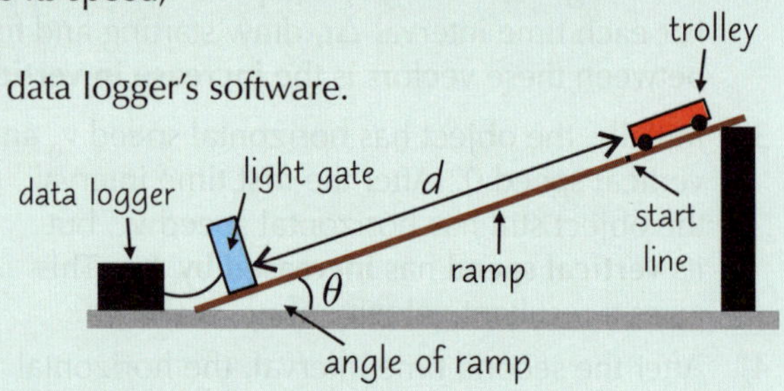

Using a light gate gives a much lower uncertainty in the measurement than using a stopwatch and calculating the velocity manually.

You can use the same set-up as above to investigate other factors. For example, you could change the **angle** of the ramp or the **material** it's made from, or change the **mass**, **size** or **shape** of the trolley.

You Can Use A **Ticker Timer** Instead of a **Data Logger**

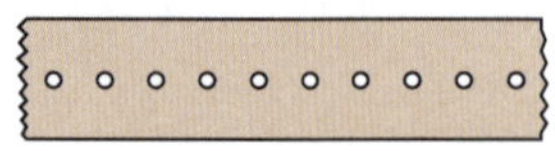

10 dots = 1/5th of a second.

Ticker timers create **holes** in (or make **dots** on) a long piece of **paper** (**ticker tape**) at **regular time intervals**. This is usually about **every 50th** of a second. You can **calculate** how long it takes a trolley to travel down a ramp by **attaching ticker tape** to the back of the trolley and threading it through the timer. Switch the ticker timer **on** when you **release** the trolley, and **off** when it reaches the **end** of the **ramp**. The **time taken** can then be calculated from the **number of holes** punched into the ticker tape (50 holes = 1 second).

Ticker timers are able to measure time very **accurately** but rely on **manually** turning on and off the machine, which will add **uncertainty** to the measurements e.g. human error/reaction time. Uncertainty can also be introduced when having to **count** the **total number** of dots.

Data loggers do not have this **human error** and can calculate the **velocity** and display it in **real time** — saving **time** and allowing **comparisons** between experiments to be easily made.

You Can Also **Investigate** How **Collisions** Affect the **Motion** of an Air Glider

Air gliders have **minimal friction**, as they're not in contact with a surface as they move. You can investigate what happens when **two** gliders collide using the experiment below.

- Set up the experiment shown in the diagram below, with a **video camera** positioned side-on to the motion of the gliders.
- Measure the **length** and **mass** of both gliders.
- Turn on the video camera and **start recording**.
- **Push** one glider so it hits the second glider.
- When both gliders have come to a stop, stop recording.

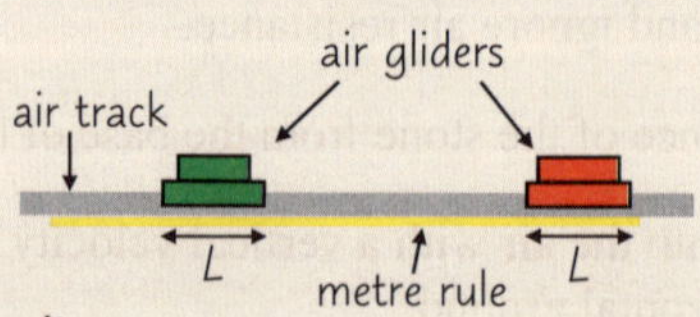

If you use the velocities you find to work out the total momentum before and after the collision, you should find it is the same (see p.44).

1) You can use this experiment to investigate how the **mass** and **velocity** of a glider just **before** a collision **affect** the **velocities** of both gliders after the **collision**.
2) Using **video analysis** software, you can view your videos **frame by frame**. Pick a **point of reference** on the metre rule and count how many **frames** it takes a glider to pass that point.
3) By knowing how many **frames per second** the video is shot at (the frame rate of the video), you can calculate the **time taken** (t) for a whole glider to pass that point. You recorded the **length** (L) of each glider, and so you can calculate their **velocities**.

$$\text{Time taken } (t) \text{ for a glider to pass a point} = \text{Number of frames for glider to pass a point} \times \frac{1 \text{ second}}{\text{Frame rate of camera}}$$

$$\text{velocity} = \frac{L}{t}$$

Motion Experiments and Stopping Distances

Many Factors Affect How Quickly a Vehicle Stops

From crashing trolleys into each other to trying to avoid crashing a car — there are lots of different factors that affect how quickly a car can stop (and how far it travels before it does). The biggest factor is its initial velocity.

The braking distance and thinking distance together make the **total distance you need to stop** after you see a problem:

Thinking distance + Braking distance = Stopping distance

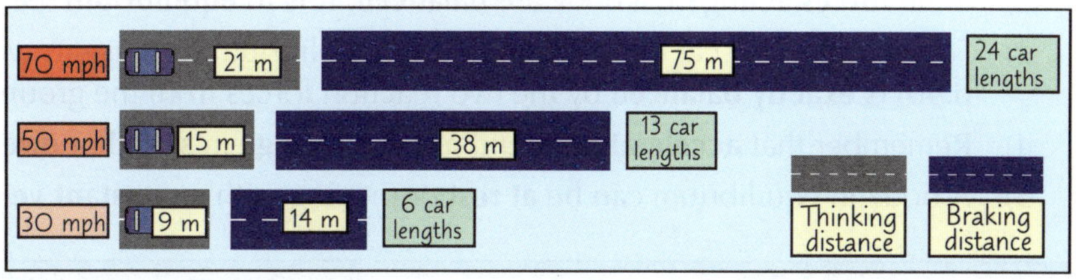

1) The **reaction time** is the time your body takes to **react** and **hit** the brakes after seeing a hazard.
2) The **thinking distance** is the distance the vehicle travels during the driver's **reaction time**.
3) The **braking distance** is the distance the vehicle travels after the **brakes are applied** until it comes to a complete **stop**.

There are many factors which can affect both the **thinking** and **braking** distances:

Thinking distance = speed × reaction time

Reaction time is increased by:

1) **Tiredness**.
2) **Alcohol** or other **drug** use.
3) **Illness**.
4) **Distractions** such as noisy children and loud music.

Braking distance depends on the **braking force**, **friction** between the tyres and the road, the **mass** and the **speed**.

1) **Braking force** is reduced by **reduced friction** between the brakes and the wheels (**worn** or **badly adjusted brakes**).
2) **Friction** between the tyres and the road is reduced by **wet** or **icy** roads, **leaves or dirt** on the road, **worn-out tyre treads**, etc.
3) **Mass** is affected by the size of the car and what you put in it.

Practice Questions

Q1 Describe an experiment you could do to investigate the motion of a trolley on a slope.

Q2 Explain how a ticker timer works. State one advantage of using a data logger instead of a ticker timer.

Q3 What is the formula for calculating stopping distance?

Q4 Name a factor which affects thinking distance.

Q5 Name a factor which affects braking distance.

Exam Questions

Q1 Sarah sees a cow step into the road 30.0 m ahead of her. Sarah's reaction time is 0.50 s. She is travelling at 20.0 ms^{-1}. Her maximum braking force is 1.0×10^4 N and her car (with her in it) has a mass of 850 kg.

a) How far does she travel before applying her brakes? [2 marks]

b) Calculate Sarah's braking distance. Assume she applies the maximum braking force until she stops. [3 marks]

c) Does Sarah hit the cow? Justify your answer with a suitable calculation. [1 mark]

Q2 A video recording is made of two 15 cm long trollies. Trolley 1 is pushed and then allowed to collide with trolley 2. The frame rate of the camera is 26 frames per second.

a) The video is analysed and after the collision, trolley 2 takes a little less than one frame to fully pass a reference point. It is assumed the trolley takes exactly one frame to pass the reference point. Estimate the velocity of trolley 2. [3 marks]

b) Suggest and explain one change which could be made to the experiment that could give more precise velocity measurements. [2 marks]

...Huh?... Slow reaction time... Don't know what you mean...

Being safe in a car is mainly common sense — don't drive if you're ill, drunk or just tired, and don't drive a car with dodgy brakes. But you still have to do exam questions, so don't go on till you're sure you know this all by heart.

Newton's Laws of Motion

You probably did most of this at GCSE, but that doesn't mean you can skip over it now. Don't miss out on easy marks...

A Force is Needed to Start, Stop, Accelerate or Decelerate

1) A **net** (or **resultant**) **force** is needed for an object to **accelerate** (start or stop moving, or change velocity). There is a net force when the forces acting on an object are **not balanced**.

This is known as Newton's First Law — see next page.

2) If the **forces** acting on a body are **balanced**, it is in **equilibrium** (i.e. not accelerating).
3) For example, the cat in the diagram is in equilibrium because its weight (see p.40) is **exactly balanced** by the two reaction forces from the ground (*R*).
4) Remember that **acceleration** could mean a **change** in **speed** or **direction**, or **both**.
5) A body in equilibrium can be **at rest** or moving with a **constant velocity**.

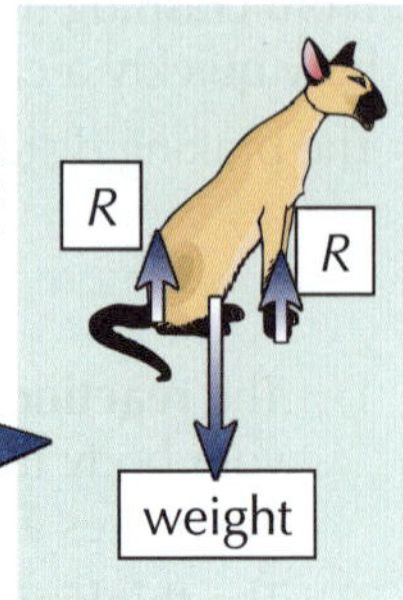

Free-Body Force Diagrams show All Forces on a Single Body

1) The best way to work out if there is a **net force** is to draw a **free-body force diagram**.
2) **Free-body force** diagrams show a **single body** on its own.
3) The diagram should include all the **forces** that **act on** the body, but **not** the **forces it exerts** on the rest of the world.
4) Remember **forces** are **vector quantities** (p.14) so the **size** and **direction** of the forces should be shown.

Drawing free-body force diagrams isn't too hard — you just need practice. Here are a few **examples**:

Gravity pulls apple downwards (weight = $m \times g$).

Gravity pulls man down. Air resistance pushes man up.

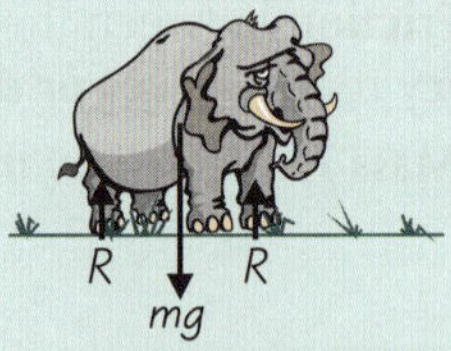

Gravity pulls elephant down. Earth pushes elephant up (reaction).

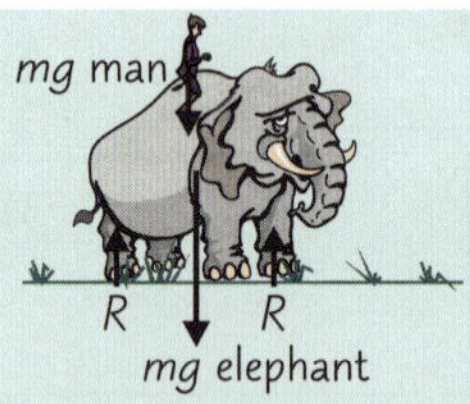

Gravity pulls elephant down. Man pushes elephant down. Earth pushes elephant up.

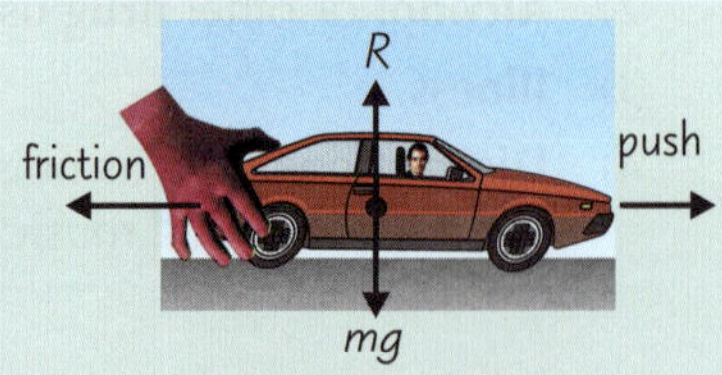

Hand pushes toy car forward. Friction pulls car backwards. Gravity pulls car down. Earth pushes car up.

You Need to Think About All the Forces and What Causes Them

When working out **force** problems, you have to make sure you've caught **every** force that could be acting on the body. Here's a few examples of **common forces** acting on objects:

1) **Weight** — every object has weight on Earth due to gravity, acting directly downwards (see p.40).
2) **Normal contact force** — if an object exerts a force on a surface, the surface exerts an equal but opposite force on the object. The force acts normal (perpendicular) to the surface. It is a reaction force (see page 32.)
3) **Tension** — if a string is pulled tight, tension is the force pulling equally on the objects at either end of the string.
4) **Friction** — if an object is moving, it usually has a friction force acting on it in the opposite direction to motion.

Example — Tension:

Each side pulls on the rope with force 410 N. Tension in the rope balances this force.

Examples — Normal Contact Force:

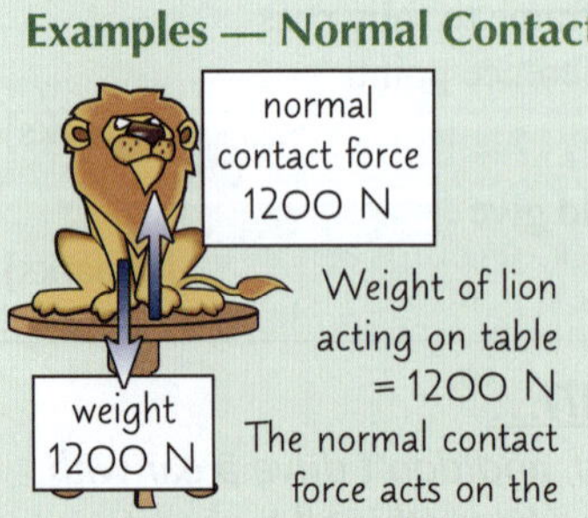

Weight of lion acting on table = 1200 N

The normal contact force acts on the lion and is also 1200 N.

The man pushes on the wall with a force of 300 N. The wall 'pushes back' with a normal contact force of 300 N.

Example — Friction:

Car is moving at a constant speed. Driving force is balanced by friction.

Note: These aren't all free-body force diagrams — the forces don't all act on one object.

Newton's Laws of Motion

Newton's **1st Law** Says That a **Force** is Needed to Change Velocity

1) **Newton's 1st law of motion** states that the **velocity** of an object will **not change** unless a **resultant force** acts on it.
2) In plain English this means a body will stay still or move in a **straight line** at a **constant speed**, unless there's a **resultant force** acting on it.
3) If the forces **aren't balanced**, the **overall resultant force** will make the body **accelerate**. This could be a change in **direction**, or **speed**, or both. (See Newton's 2nd law, below.)

An apple sitting on a table won't go anywhere because the **forces** on it are **balanced**.

reaction (R) = **weight** (mg)
(force of table pushing apple up) (force of gravity pulling apple down)

Newton's **2nd Law** Says That **Acceleration** is **Proportional** to the Force

...which can be written as the well-known equation:

resultant force (N) = mass (kg) × acceleration (ms^{-2})

$$F = m \times a$$

You might see this given as $\Sigma F = ma$. Σ is the symbol for 'sum of' — the resultant force is the vector sum of all the forces on an object.

Learn this — it crops up all over the place in Physics. And learn what it means too:

1) It says that the **more force** you have acting on a certain mass, the **more acceleration** you get.
2) It says that for a given force the **more mass** you have, the **less acceleration** you get.
3) There's more on this most excellent law on p.45.

REMEMBER:
1) The **resultant force** is the **vector sum** of all the forces.
2) The force is **always** measured in **newtons**.
3) The **mass** is always measured in **kilograms**.
4) The **acceleration** is always in the **same direction** as the **resultant force** and is measured in **ms^{-2}**.

Galileo said: **All Objects Fall** at the **Same Rate** (if You **Ignore Air Resistance**)

You need to understand **why** this is true. Newton's 2nd law explains it neatly — consider two balls dropped at the same time — ball **1** being heavy, and ball **2** being light. Then use Newton's 2nd law to find their acceleration.

mass = m_1
resultant force = F_1
acceleration = a_1
By Newton's Second Law:

$$F_1 = m_1a_1$$

Ignoring air resistance, the only force acting on the ball is weight, given by $W_1 = m_1g$ (where g = gravitational field strength = 9.81 Nkg^{-1}).

So: $F_1 = m_1a_1 = W_1 = m_1g$

So: $m_1a_1 = m_1g$, then m_1 cancels out to give: $a_1 = g$

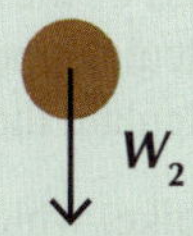

mass = m_2
resultant force = F_2
acceleration = a_2
By Newton's Second Law:

$$F_2 = m_2a_2$$

Ignoring air resistance, the only force acting on the ball is weight, given by $W_2 = m_2g$ (where g = gravitational field strength = 9.81 Nkg^{-1}).

So: $F_2 = m_2a_2 = W_2 = m_2g$

So: $m_2a_2 = m_2g$, then m_2 cancels out to give: $a_2 = g$

...in other words, the **acceleration** is **independent of the mass**. It makes **no difference** whether the ball is **heavy or light**. And I've kindly **hammered home the point** by showing you two almost identical examples.

Newton's Laws of Motion

Newton's **3rd Law** Says Each Force has an **Equal, Opposite Reaction Force**

There are a few different ways of stating Newton's 3rd law, but the clearest way is:

If an object A EXERTS a FORCE on object B, then object B exerts AN EQUAL BUT OPPOSITE FORCE on object A.

You'll also hear the law as "every action has an equal and opposite reaction". But this confuses people who wrongly think the forces are both applied to the same object. (If that were the case, you'd get a resultant force of zero and nothing would ever move anywhere...)

The two forces actually represent the **same interaction**, just seen from two **different perspectives**:

1) If you **push against a wall**, the wall will **push back** against you, **just as hard**. As soon as you stop pushing, so does the wall. Amazing...
2) If you **pull a cart**, whatever force **you exert** on the rope, the rope exerts the **exact opposite** pull on you (unless the rope's stretching).
3) When you go **swimming**, you push **back** against the water with your arms and legs, and the water pushes you **forwards** with an equal-sized force.

Newton's 3rd law applies in **all situations** and to all **types of force**. But the pairs of forces are always the **same type**, e.g. both gravitational or both electrical, and they act along the same line.

This looks like Newton's 3rd law...

But it's <u>NOT</u>.

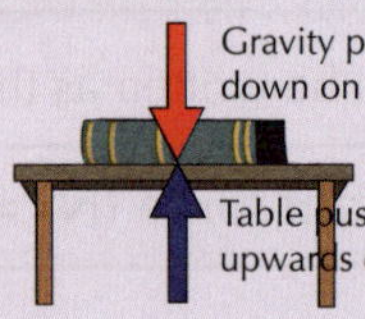

...because both forces are acting on the book, and they're not of the same type. They are two separate interactions. The forces are equal and opposite, resulting in zero acceleration, so this is showing Newton's 1st law.

Practice Questions

Q1 Sketch a free-body force diagram for an ice hockey puck moving across the ice (assuming no friction).

Q2 Give an example of: a) a tension force, b) a normal contact force.

Q3 State Newton's 1st, 2nd and 3rd laws of motion, and explain what they mean.

Q4 What are the two equal and opposite forces acting between an orbiting satellite and the Earth?

Exam Questions

Q1 A boat is moving across a river. The engines provide a force of 500 N at right angles to the flow of the river and the boat experiences a drag of 100 N in the opposite direction. The force on the boat due to the flow of the river is 300 N. The mass of the boat is 250 kg.

a) Calculate the magnitude of the resultant force acting on the boat. [2 marks]

b) Calculate the magnitude of the acceleration of the boat. [2 marks]

Q2 John's bike, which has a mass of m, breaks and he has to push it home. The bike has a constant acceleration a and a frictional force F opposes the motion. What force is John using to push his bike?

A	ma
B	$ma + F$
C	$m(a - F)$
D	$ma - F$

[1 mark]

Q3 Olivia and Tom are both keen on diving. They notice that they seem to take the same time to drop from the diving board to the water. Use Newton's second law to explain why this is the case. (Assume no air resistance.) [3 marks]

Newton's three incredibly important laws of motion...

These laws may not really fill you with a huge amount of excitement (and I could hardly blame you if they don't)... but it was pretty fantastic at the time — suddenly people actually understood how forces work, and how they affect motion. I mean arguably it was one of the most important scientific discoveries ever...

Drag, Lift and Terminal Speed

If you jump out of a plane at 1500 m, you want to know that you're not going to be accelerating all the way.

Friction is a *Force* that *Opposes Motion*

There are two main types of friction — **dry friction** (also called **contact friction**) between **solid surfaces**, and **fluid friction** (known as **drag**, fluid resistance or air resistance).

Fluid Friction or Drag:

1) 'Fluid' is a word that means either a **liquid or a gas** — something that can **flow**.
2) The force depends on the thickness (or **viscosity**, see p.38) of the fluid.
3) It **increases** as the **speed increases** (for simple situations it's directly proportional, but you don't need to worry about the mathematical relationship).
4) It also depends on the **shape** of the object moving through it — the larger the **area** pushing against the fluid, the greater the resistance force.
5) A **projectile** (see p.25) is **slowed down** by air resistance. If you calculate how far a projectile will travel without thinking about air resistance, your answer will be **too large**.

Things you need to remember about frictional forces:

1) They **always** act in the **opposite direction** to the **motion** of the object.
2) They can **never** speed things up or start something moving.
3) They convert **kinetic energy** into **heat** and **sound**.

Lift is *Perpendicular* to *Fluid Flow*

1) '**Lift**' is an **upwards force** on an object moving through a fluid.
2) It happens when the shape of an object causes the fluid flowing over it to **change direction**.
3) The force acts **perpendicular** to the direction the fluid flows in.

Example: Cross-section of a plane wing moving through air

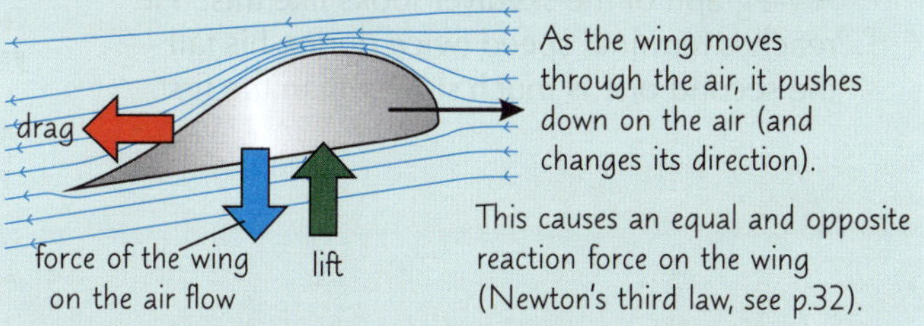

Terminal Speed — When the *Friction* Force *Equals* the *Driving Force*

You will reach a **terminal (maximum) speed** at some point, if you have:

1) a **driving force** that stays the **same** all the time,
2) a **frictional** or **drag force** (or collection of forces) that increases with speed.

Terminal speed can also be called terminal velocity provided you state the direction.

There are **three main stages** to reaching terminal speed:

The car **accelerates** from **rest** using a constant driving force.

As the **speed increases**, the **frictional forces increase** (you don't need to worry about why they do). This **reduces the resultant force** on the car and hence **reduces its acceleration**.

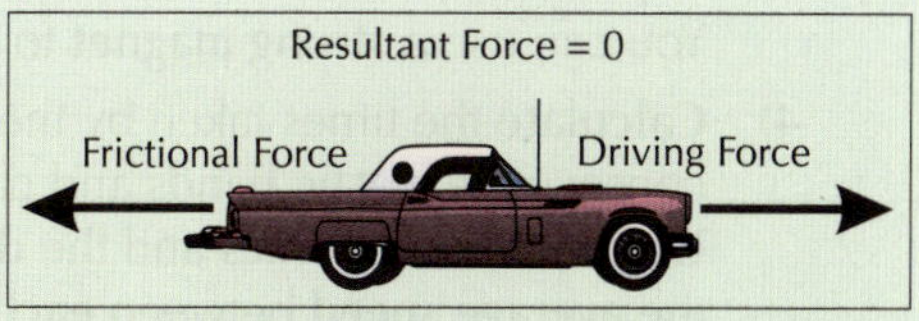

Eventually the car reaches a speed at which the **frictional forces are equal to the driving force**. There is now **no resultant force** and **no acceleration**, so the car carries on at **constant speed**.

Different factors affect a vehicle's maximum speed

As you just saw, a vehicle reaches maximum speed when the driving force is equal to the frictional force. So there are two main ways of increasing a vehicle's maximum speed:

1) **Increasing the driving force**, e.g. by increasing the engine size.
2) **Reducing the frictional force**, e.g. making the body more streamlined.

Drag, Lift and Terminal Speed

*Things **Falling** through **Air** or **Water** Reach a **Terminal Speed** too*

When something's falling through air, the weight of the object is a constant force accelerating the object downwards. Air resistance is a frictional force opposing this motion, which increases with speed.
So before a parachutist opens their parachute, exactly the same thing happens as with the car example:

1) A skydiver leaves a plane and will **accelerate** until the **air resistance** equals his **weight**.

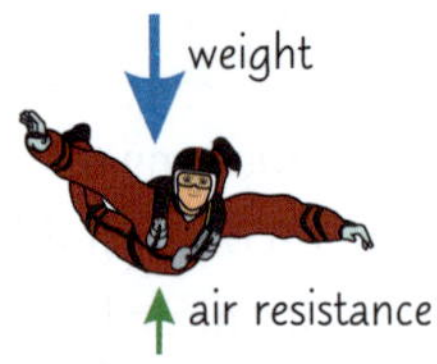

2) He will then be travelling at a **terminal speed**.

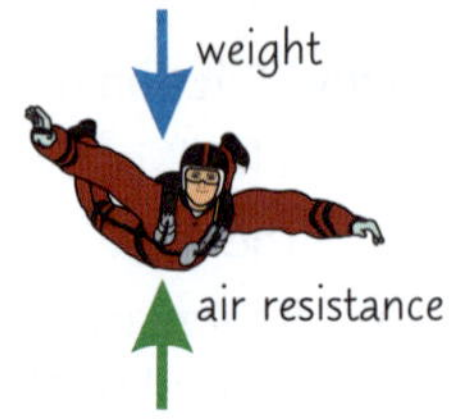

But... the terminal speed of a person in free fall is too great to land without dying a horrible death.
The **parachute increases** the **air resistance massively**, which slows him down to a lower terminal speed:

3) Before reaching the ground he will **open his parachute**, which immediately **increases the air resistance** so it is now **bigger** than his **weight**.

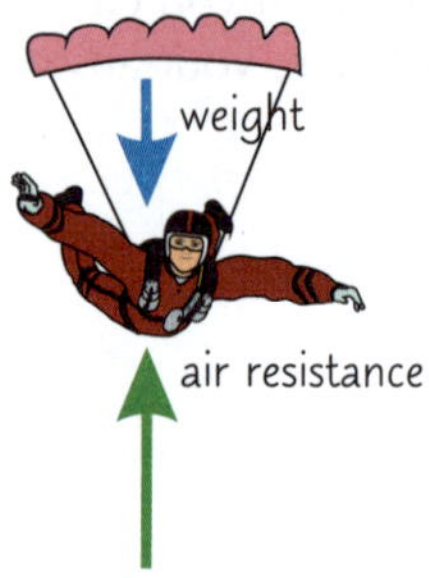

4) This **slows him down** until his speed has dropped enough for the **air resistance** to be **equal to his weight** again. This new terminal speed is small enough for him to land safely.

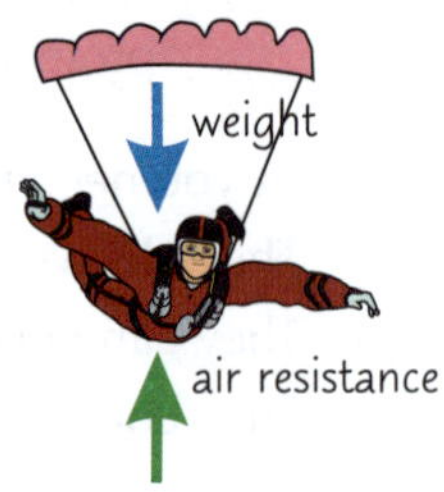

A *v-t* graph of the skydiver looks like this. He reaches terminal speed twice during his fall — the second one is much slower than the first.

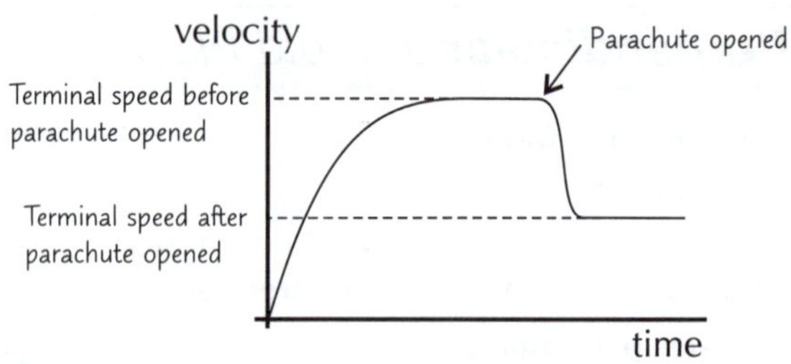

Measure** the **Terminal Speed** of a **Ball Bearing

You don't have to use elastic bands — you could also use insulation tape or another marker for your intervals.

You can calculate the terminal speed of a **ball bearing** (a little steel ball) in a **viscous** (thick) liquid by setting up an experiment like this:

1) Put **elastic bands** around the tube of viscous liquid at **fixed distances** using a **ruler**.
2) **Drop** a ball bearing into the tube, and use a **stopwatch** to record the time at which it reaches **each band**. Record your results in a **table** (see below).
3) **Repeat** this a few times to reduce the effect of **random errors** on your results. You can use a **strong magnet** to remove the ball bearing from the tube.
4) **Calculate** the times taken by the ball bearing to travel between consecutive elastic bands and calculate an **average** for each reading. Use the **average times** and the **distance between bands** to calculate the **average speed** between **each pair** of elastic bands.
5) You should find that the average speed **increases** at first, then **stays constant** — this is the ball bearing's **terminal speed** in the viscous liquid used.

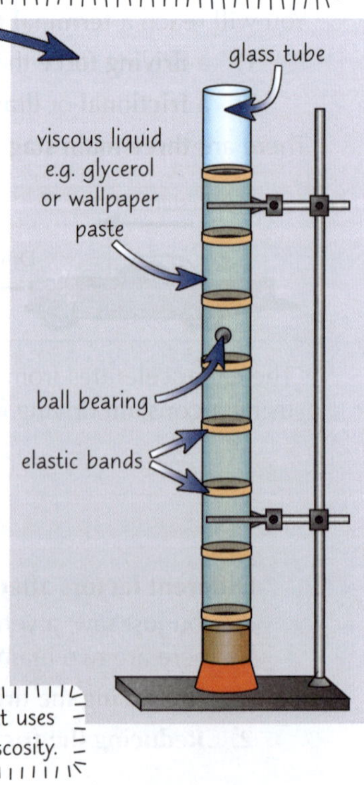

Elastic band	Time / s	Time from last band / s	Average time / s	Average speed / ms^{-1}
2	1			
	2			
	3			
3	1			
	2			
	3			

If your stopwatch has a lap timer, you might be able to measure these times directly.

See p.39 for an experiment that uses this apparatus to investigate viscosity.

Drag, Lift and Terminal Speed

You can Work Out what **Affects** Terminal Speed

Prof. Fraise dedicated his life to investigating terminal speed in fluids.

1) Use your average speed data to plot a graph of **speed** against **time**. Draw a smooth curve and use it to estimate the terminal speed.
2) You might be asked to draw **force diagrams** as the ball bearing falls. Remember that the forces are balanced when the ball reaches terminal speed.
3) You can change parts of your experiment to see what effect they have on terminal speed and the time taken to reach terminal speed. For example you could:
   - Change the **liquid** — the terminal speed will be **lower** in more viscous (thicker) liquids because the drag is **greater**. Try mixing water into wallpaper paste and see by how much the terminal speed increases when the drag is lower.
   - Change the **size** of the ball. What happens if the ball is larger? Or smaller?
   - Change the **shape** of the thing you are dropping. The drag force will be greater on **less streamlined** shapes.
   - Change the **mass** of the thing you are dropping, while keeping the **size** the **same** (this might be a bit tricky). You should find that **heavier objects** reach a **faster** terminal speed because a **greater drag force** is needed to balance the extra weight. (Remember, objects with different masses only fall at the same rate if drag is ignored — p.31.)

Turn back to p.18 for more on velocity-time graphs.

Practice Questions

Q1 What forces limit the speed of a skier going down a slope?

Q2 What causes a lift force on a plane wing as it moves through air?

Q3 Suggest two ways in which the maximum speed of a car can be increased.

Q4 What conditions cause a terminal speed to be reached?

Q5 Sketch a graph to show how the acceleration changes with time for an object falling through air.

Exam Questions

Q1 A space probe free-falls towards the surface of a planet.
The graph on the right shows the velocity data recorded by the probe as it falls.

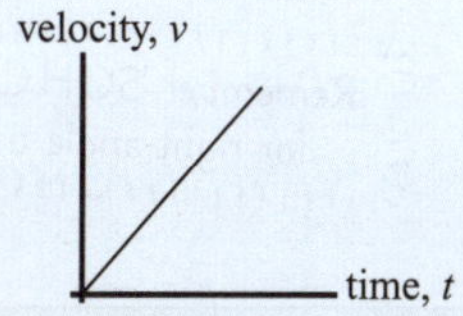

a) The planet does not have an atmosphere. Explain how you can tell this from the graph. [2 marks]

b) On the graph, sketch the line you would expect to see if the planet did have an atmosphere. [2 marks]

c) Explain the shape of the graph you have drawn. [3 marks]

Q2 A student is investigating how the terminal speed of paper cones varies with cone size.
She drops weighted cones of base diameter 5 cm, 10 cm and 15 cm point-down from a height of 2 m and uses a video camera and video analysis software to obtain data on the displacement of the cone at certain times. She then plots a displacement-time graph to calculate the terminal speed. You may assume that the weights of the cones are negligible compared to that of the weights used to stabilise them.

a) State which size of cone you expect to have the lowest terminal speed. Explain your answer. [2 marks]

b) Sketch a graph of speed against time for the three cones. Put all three curves on the same axes. [3 marks]

c) Suggest one factor the student must keep the same in her experiment and explain why. [1 mark]

d) Describe and explain how the velocity-time graph would change if the largest cone was crushed into a rough ball shape. [3 marks]

You'll never understand this without going parachuting...*

When you're doing questions about terminal speed, remember the frictional forces reduce acceleration, not speed. They usually don't slow an object down, apart from in the parachute example, where the skydiver is travelling faster when the parachute opens than the terminal speed for the parachute-skydiver combination.

* No. 37 in a series of the 100 least convincing excuses for an interesting holiday.

Resolving Forces

Remember the vector stuff from pages 14-15...? Good, you're going to need it...

Resolving a *Force* means *Splitting* it into *Components*

1) **Forces** are **vector quantities** and so when you draw the forces on an object, the **arrow labels** should show the **size** and **direction** of the forces.
2) Forces can be in **any direction**, so they're not always at right angles to each other. This is sometimes a bit **awkward** for **calculations**.
3) To make an 'awkward' force easier to deal with, you can think of it as **two separate forces**, acting at **right angles** to **each other**.

The force ***F*** has exactly the same effect as the horizontal and vertical forces, F_H and F_V. Replacing ***F*** with F_H and F_V is called **resolving the force *F*.**

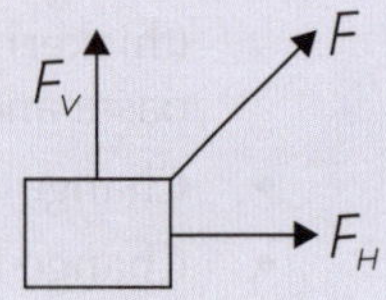

4) To find the size of a component of a force in a particular direction, you need to use **trigonometry**. Forces are vectors, so you treat them in the same way as velocities — put them end to end.

So this...

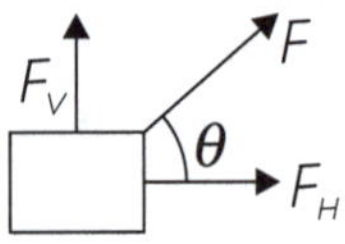

...could be drawn like this:

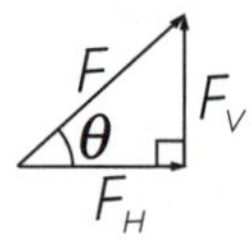

Using trig. you get:

$$\frac{F_H}{F} = \cos\theta \quad \text{or} \quad F_H = F\cos\theta$$

And:

$$\frac{F_V}{F} = \sin\theta \quad \text{or} \quad F_V = F\sin\theta$$

Remember 'SOH CAH TOA' for right-angle triangles.

Example:

A tree trunk is pulled along the ground by an elephant exerting a force of 1200 N at an angle of 25° to the horizontal. Calculate the components of this force in the horizontal and vertical directions.

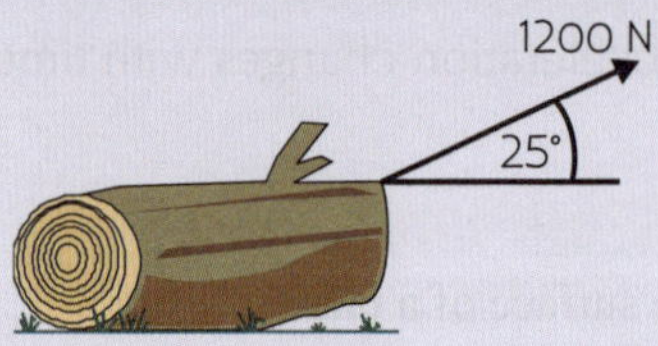

Horizontal force =
1200 × cos 25° = **1100 N (to 2 s.f.)**

Vertical force =
1200 × sin 25° = **510 N (to 2 s.f.)**

Three Forces Acting on a *Point* in *Equilibrium* form a *Triangle*

1) When **three** coplanar (all in the same plane) forces all act on an object in **equilibrium**, you know there is **no net force** on the object — the sum of the forces is zero.

$$F_1 + F_2 + F_3 = 0$$

2) You can draw the forces as a triangle, forming a **closed loop** like these:

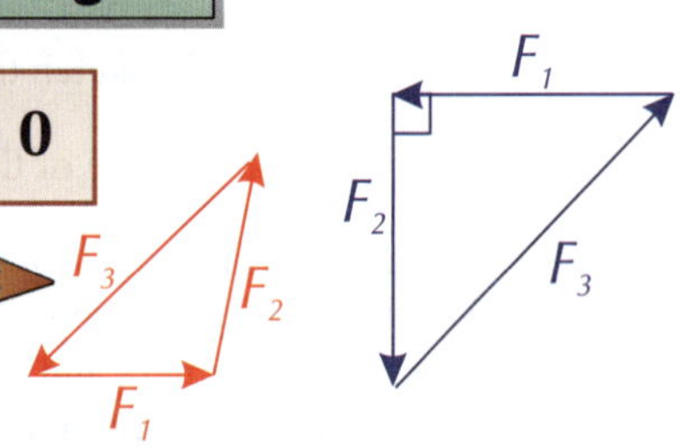

3) Be careful when you draw the triangles not to go into autopilot and draw F_3 as the sum of F_1 and F_2 — it has to be in the **opposite** direction to balance the other two forces.
4) You can then use the triangles to while away Saturday nights in with your friends working out the magnitude or direction of a missing force, either by using Pythagoras, or by measuring on a scale diagram (p.14).

The sum of all the angles in a triangle is 180° — you might need to use this to answer exam questions.

Example:

Tim hangs a picture of his brother up using a piece of string. The picture weighs 0.3 N and is in equilibrium, as shown. Find the magnitude of *F*.

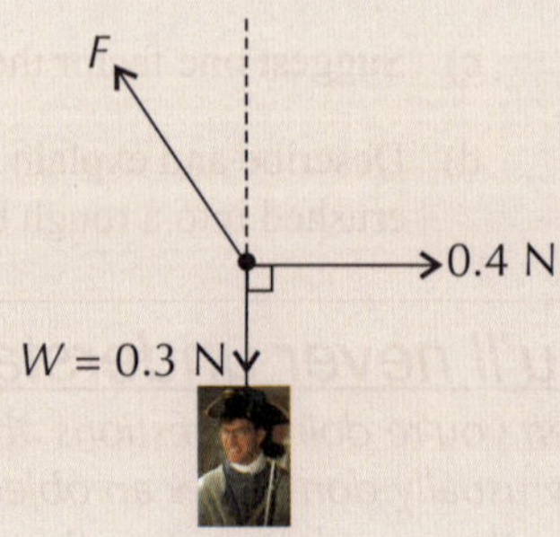

Draw a vector triangle.

Then you need to use **Pythagoras** to find the magnitude of *F*.

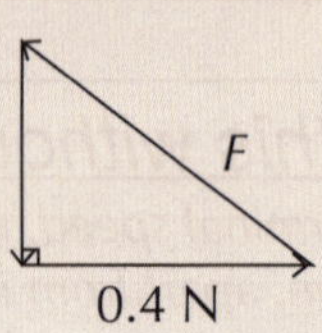

$$F = \sqrt{0.3^2 + 0.4^2} = \sqrt{0.25^2} = \mathbf{0.5\,N}$$

Resolving Forces

You **Add Components Back Together** *to get the* **Resultant Force**

1) If **two forces** act on an object, you find the **resultant** (net) **force** by adding the **vectors** together and creating a **closed triangle**, with the resultant force represented by the **third side**.
2) Forces are vectors (as you know), so you use **vector addition** — draw the forces as vector arrows put 'tip-to-tail'.
3) Then it's yet more trigonometry to find the **angle** and the **length** of the third side.

Example: Two dung beetles roll a dung ball along the ground at constant velocity. Beetle A applies a force of 0.50 N northwards, beetle B exerts a force of 0.20 N eastwards. Find the resultant force on the dung ball.

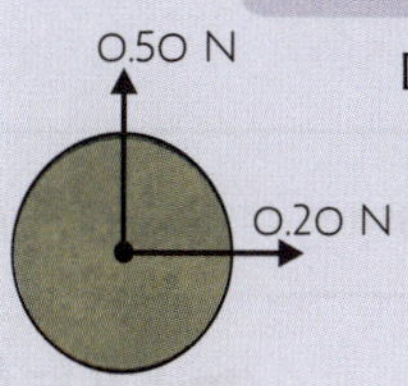

Draw a triangle:

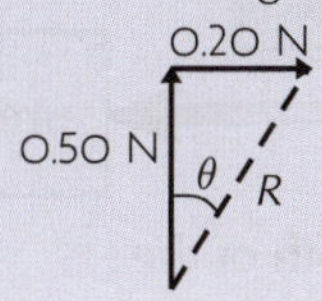

By Pythagoras, $R^2 = 0.50^2 + 0.20^2$, so $R = \sqrt{0.29} = 0.54$ N (to 2 s.f.)

$\tan\theta = \frac{0.20}{0.50}$ so $\theta = \tan^{-1}0.4 = 22°$ (to 2 s.f.)

So the resultant force is **0.54 N** at an angle of **22°** from north.

Choose *sensible* **Axes** *for* **Resolving**

Use directions that **make sense** for the situation you're dealing with. If you've got an object on a slope, choose your directions **along the slope** and **at right angles to it**. You can turn the paper to an angle if that helps.

Always choose sensible axes

Examiners like to call a slope an "inclined plane".

Example:

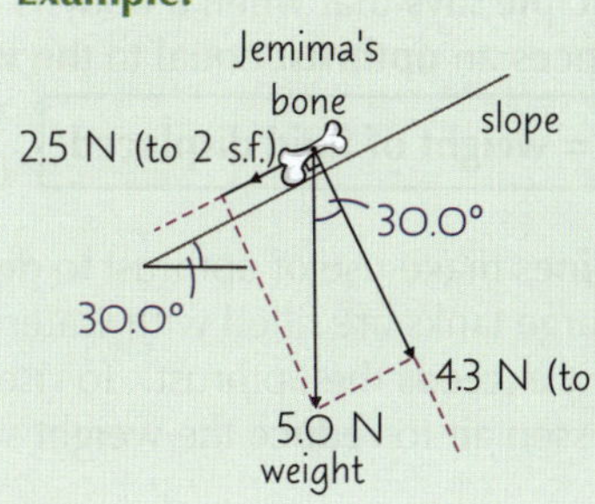

The component of the bone's weight down the slope is 2.5 N so you'd need 2.5 N of friction to stop it sliding away.

Practice Questions

Q1 What are the horizontal and vertical components of the force *F*?

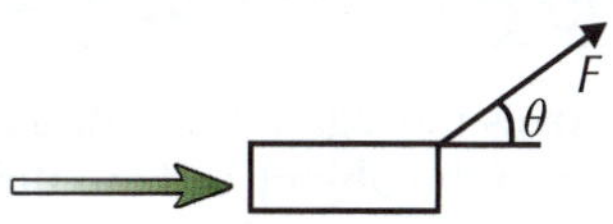

Q2 How could you use a scale drawing to show that three coplanar forces are producing no net force?

Q3 Three forces act on an object in equilibrium. What is the sum of the forces?

Exam Questions

Q1 A picture is suspended from a hook as shown in the diagram.
Calculate the tension force, *T*, in the string.

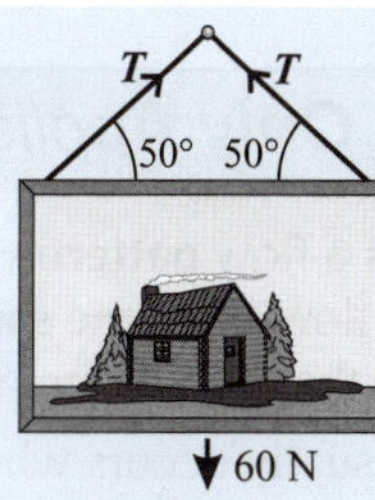

[2 marks]

Q2 Two elephants pull a tree trunk as shown in the diagram.
Calculate the resultant force on the tree trunk.

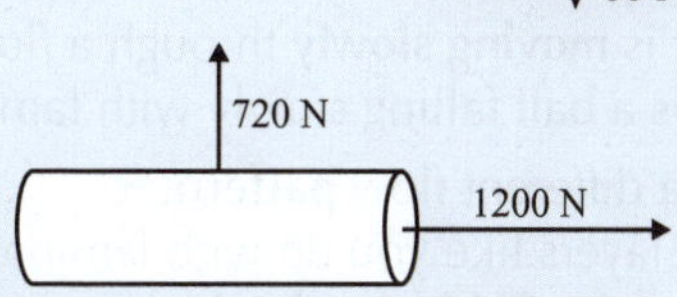

[2 marks]

I'm in e-quill-ibrium — if I lose my pen, I always order a new one right away...

Remember those $F\cos\theta$ and $F\sin\theta$ bits. Write them on bits of paper and stick them to your wall. Scrawl them on your pillow. Tattoo them on your brain. Whatever it takes — you just have to learn them.

Density, Upthrust and Viscosity

Yikes, three whole things in one go. Grab a coffee, take a deep breath and put on your snazziest learning hat...

Density *is* ***Mass*** *per* ***Unit Volume***

Density is a measure of the '**compactness**' (for want of a better word) of a substance. It relates the **mass** of a substance to how much **space** it takes up.

1) The density of an object depends on what it's made of. Density **doesn't vary** with **size or shape**.
2) The **average density** of an object determines whether it **floats** or **sinks**. A solid object will **float** on a fluid if it has a **lower density** than the **fluid**.

The symbol for density is a Greek letter rho (ρ) — it looks like a p but it isn't.

$$\textbf{density} = \frac{\textbf{mass}}{\textbf{volume}} \qquad \rho = \frac{m}{V}$$

The units of density are g cm^{-3} or kg m^{-3}
1 g cm^{-3} = 1000 kg m^{-3}

Bodies *in* ***Fluids*** *Experience* ***Pressure*** *and* ***Upthrust***

1) Pressure is the **amount of force** applied per **unit area**. It is measured in **pascals (Pa)**, which are equivalent to newtons per square metre (Nm^{-2}).

$$\textbf{pressure} = \frac{\textbf{force}}{\textbf{area}} \qquad p = \frac{F}{A}$$

2) The extra pressure acting on an object due to a fluid depends on the **depth** of the object in the fluid (h), the **density** of the fluid (ρ) and the **gravitational field strength** (g).

$$p = h\rho g$$

3) Upthrust is an **upward force** that fluids exert on objects that are **completely** or **partially submerged** in the fluid. It's caused because the top and bottom of a submerged object are at different depths. Since $p = h\rho g$, there is a difference in pressure which causes an overall upwards force known as upthrust.
4) **Archimedes' principle** says that when a body is completely or partially immersed in a fluid, it experiences an **upthrust** equal to the **weight** of the fluid it has **displaced**.

upthrust = weight of fluid displaced

This is just because $(h_2 - h_1)\rho gA = V\rho g = mg = W$

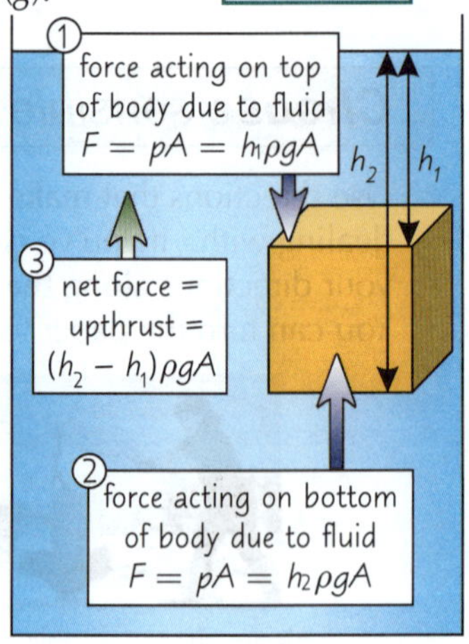

Example: Submarines make use of upthrust to dive underwater and return to the surface. To sink, large tanks are filled with water to **increase** the weight of the submarine so that it **exceeds** the upthrust. To rise to the surface, the tanks are filled with compressed air to **reduce** the weight so that it's **less** than the upthrust.

Viscous Drag *Acts on* ***Objects Moving Through Fluids***

When an object moves through a fluid, or a fluid moves past an object, you get **friction** between the surface of the **object** and the **fluid**. This is **viscous drag.** You can calculate the **force due to viscous drag** on a **spherical** object moving through a fluid using **Stokes's law**.

$$F = 6\pi\eta rv$$

Viscous drag depends on the **viscosity** (or "thickness") of a fluid, η. Viscosity is **temperature-dependent** — **liquids** get **less viscous** as temperature **increases**, but **gases** (like **dry air**) get **more viscous** as temperature **increases.**

F is the viscous drag (N), η (eta) is the viscosity of the fluid (Nsm^{-2} or Pa s), r is the radius of the object (m) and v is the speed the object is moving at (ms^{-1}).

Stokes's Law ***Only*** *Applies to* ***Small*** *Objects* ***Moving Slowly*** *with* ***Laminar Flow***

1) **Laminar flow** is a flow **pattern** where all the parts of the fluid are flowing in the **same direction** — the layers in the fluid **do not mix**.
2) Laminar flow usually occurs when a fluid is **flowing slowly**, or when an object is **moving slowly** through a fluid. The diagram shows a ball falling slowly with laminar flow.
3) **Turbulent flow** is a different flow **pattern**. You don't get nice layers like you do with laminar flow, all the parts of the **fluid** get **mixed up**.
4) Turbulent flow usually occurs when a **fluid** is flowing **quickly**, or an object is **moving quickly** through a fluid.
5) **Stokes's law only** applies to **small**, **spherical** objects moving **slowly** with **laminar flow**. It **doesn't** apply to **turbulent flow**.

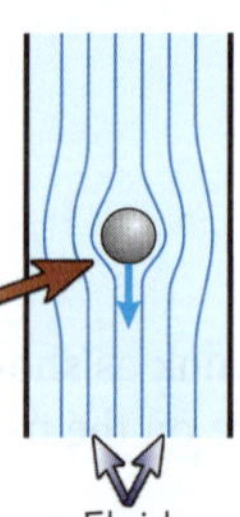

Fluid 'layers'

What do you get if you cross a baby sheep with a river? A lamb in a flow.

Density, Upthrust and Viscosity

Measure the Viscosity of a Liquid

You saw this apparatus back on page 34 — it's just as useful for measuring viscosity as it is for finding terminal speed (or terminal velocity).

You can calculate the **viscosity** of a liquid by timing the **falls** of **different sizes** of ball bearings of **known density**, ρ, using an experiment like this:

1) Fill a wide, clear tube with the liquid you want to investigate (e.g. washing up liquid). Make sure you know the **density** of the liquid, σ.
2) Put one rubber band about halfway down the tube at a position such that the ball bearings will have achieved **terminal velocity** (see page 33-35) when they reach it.
3) Place two more elastic bands **below** the first so that the distance between each band is **equal** and the lowest band is near the **bottom** of the tube. Record the distance between them. These are the points at which you will record t_1 and t_2.
4) Measure the diameter of your ball bearing and halve it to get the **radius**.
5) **Drop** the ball bearing into the tube. Start a **stopwatch** when the ball reaches the first band, and record the time at which it reaches each of the other bands. Record your results in a **table** (see below).
6) **Repeat** this at least three times for your ball bearing, then repeat the whole thing for different sizes of ball bearing.
7) If the ball falls **close** to the **wall**, **re-do the reading** — the flow will no longer be laminar and Stokes's law will not apply.
8) **Calculate** the **average time taken** for each size of ball bearing to fall between the elastic bands. Use the **average time** and the **distance between bands** to calculate the **average (terminal) velocity** of the ball bearing between the elastic bands.
9) You can then calculate the **viscosity**, η, of the liquid. The ball bearings are falling at terminal velocity, so the **sum** of the forces acting on the ball is **zero**:

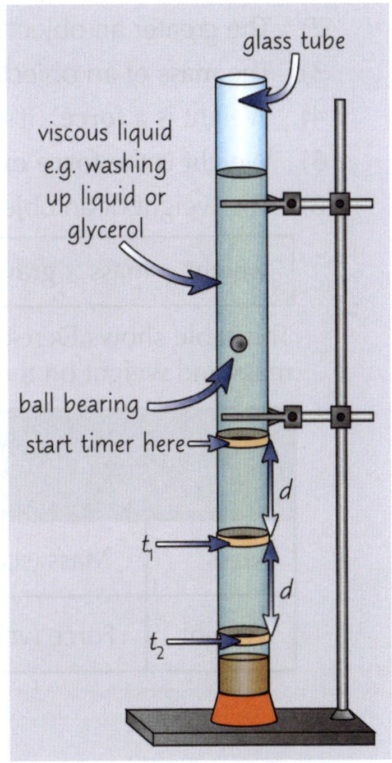

weight – drag – upthrust = 0

Substituting the equation for each force and rearranging gives $\eta = \dfrac{2r^2g(\rho - \sigma)}{9v}$, where r is the radius of the ball bearing, g is the gravitational field strength due to gravity, ρ is the density of the ball bearing, σ is the density of the liquid and v is the average terminal velocity between the elastic bands.

10) Your table could look like this:

radius of ball bearing / mm	density of ball bearing / kg m^{-3}	time to t_1 / s	velocity at t_1 / ms^{-1}	time between t_1 and t_2 / s	velocity at t_2 / ms^{-1}	average velocity / ms^{-1}	viscosity / Nsm^{-2}	average viscosity / Nsm^{-2}
2.0		1		1				
		2		2				
		3		3				
3.0		1						
		2						

If the velocities at t_2 are consistently larger than those at t_1, the ball bearing might not have reached terminal velocity when you started timing — move the elastic bands down the tube a bit and try again.

Practice Questions

Q1 Show that the mass of 3.8 m^3 of a liquid with density 1.5 kg m^{-3} is 5.7 kg.

Q2 State Archimedes's principle and explain how it is used by submarines.

Q3 What is meant by laminar flow?

Q4 Describe an experiment you could carry out to determine the viscosity of a liquid using ball bearings.

Exam Question

Q1 a) Stokes's law is given by $F = 6\pi\eta rv$. Explain what each term represents. [1 mark]

b) State three conditions that must be true for Stokes's law to be apply. [3 marks]

c) A ball bearing of radius 3.0 mm is falling at a constant velocity of 0.040 ms^{-1} through a liquid of viscosity 3.0 Pa s. The density of the liquid is 1400 kg m^{-3}. Calculate the weight of the ball bearing. ($V_{sphere} = \frac{4}{3}\pi r^3$) [5 marks]

My discosity increases with temperature — warm me up, I'll dance for hours...

Sorry about the 'lamb-in-a-flow' joke — it really is awful. My boss Joe thought it up, I'm just not that funny. If you're struggling to remember this page, try explaining it to someone else (little brothers are useful for this) — it really helps.

Mass, Weight and Centre of Mass

I'm sure you know all this 'mass' and 'weight' stuff from GCSE. But let's just make sure...

The Mass of a Body makes it Resist Changes in Motion

1) The **mass** of an object is the **amount of 'stuff'** (or **matter**) in it. It's measured in **kg**.
2) The greater an object's mass, the greater its **resistance** to a **change in velocity** (called its **inertia**).
3) The **mass** of an object **doesn't change** if the strength of the **gravitational field** changes.
4) Weight is a **force**. It's measured in **newtons** (N), like all forces.
5) Weight is the **force experienced by a mass** due to a **gravitational field**.
6) The weight of an object **does vary** according to the size of the **gravitational field** acting on it.

weight = mass × gravitational field strength ($W = mg$) where g = 9.81 Nkg^{-1} on Earth.

This table shows Derek (the lion)'s mass and weight on the Earth and the Moon.

Name	Quantity	Earth (g = 9.81 Nkg^{-1})	Moon (g = 1.6 Nkg^{-1})
Mass	Mass (scalar)	150 kg	150 kg
Weight	Force (vector)	1500 N (to 2 s.f.)	240 N (to 2 s.f.)

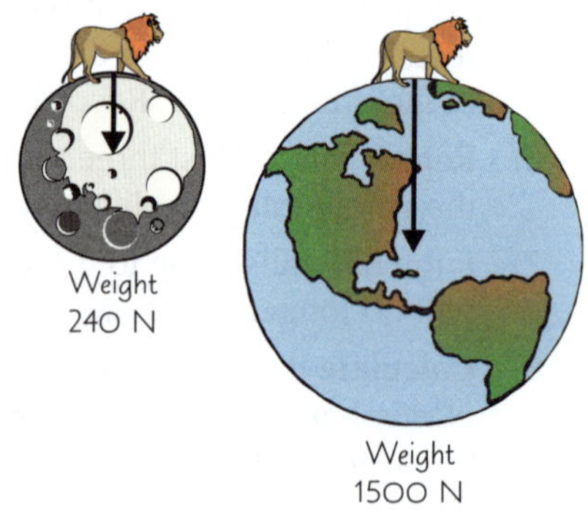

Centre of Mass — Assume All the Mass is in One Place

1) The **centre of mass** of an object is the **single point** that you can consider its **whole weight** to **act through** (whatever its orientation).
2) The object will always **balance** around this **point**, although in some cases the **centre of mass** will **fall outside** the object.
3) The centre of mass of a **uniform, regular solid** (e.g. a sphere, a cube) is at its **centre**.

The 'centre of mass' is sometimes called the 'centre of gravity' instead — it means the same thing.

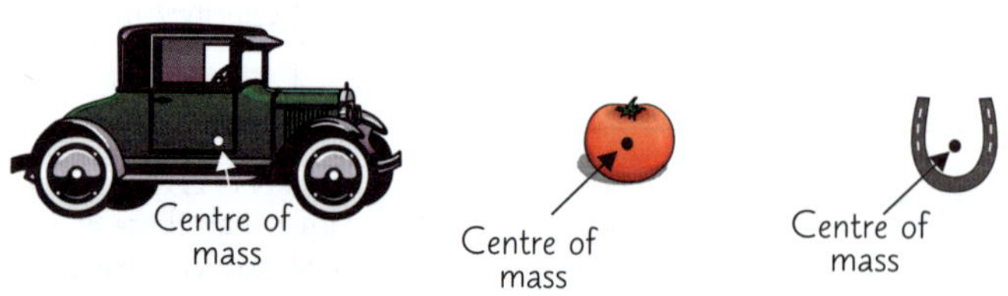

Find the Centre of Mass Either by Symmetry...

1) To find the centre of mass for a **regular** object you can just use **symmetry**.
2) The centre of mass of any regular shape is at its **centre** — where the lines of symmetry will cross.
3) The centre of mass is **halfway** through the **thickness** of the object at the point the lines meet.

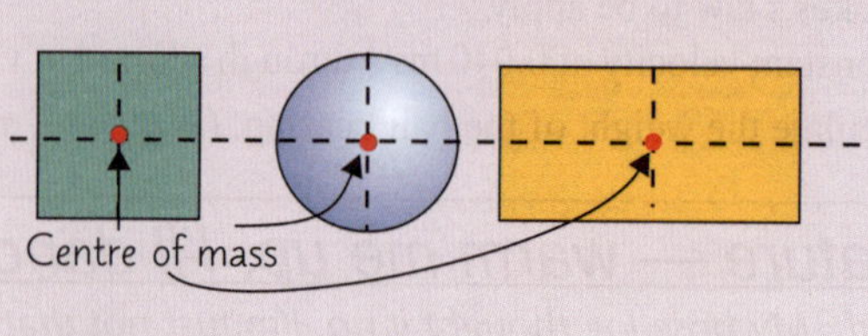

The symmetry in this picture shows the centre of cuteness.

Mass, Weight and Centre of Mass

... Or By *Experiment*

Experiment to find the Centre of Mass of an Irregular Object

1) **Hang** the object freely from a point (e.g. one corner).
2) Draw a **vertical line** downwards from the point of suspension — use a plumb bob to get your line exactly vertical.
3) Hang the object from a different point.
4) Draw another vertical line down.
5) The centre of mass is where the two lines **cross**.

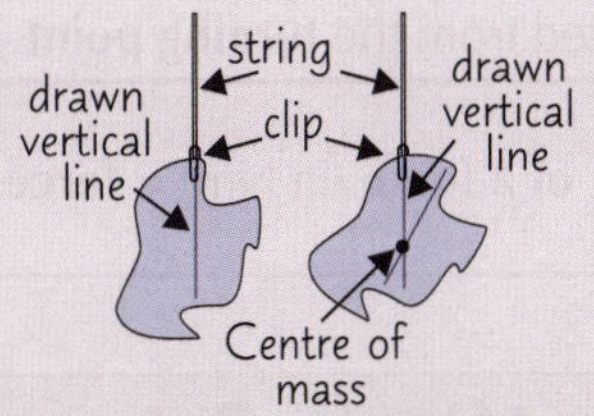

A plumb bob is just a weight on a string — when suspended, the string will be exactly vertical.

How *Stable* an Object is Depends on its *Centre of Mass* and Base Area

1) An object will topple over if a **vertical line** drawn **downwards** from its **centre of mass** (i.e. the line of action of its weight) falls **outside** its **base area**.
2) This is because a **resultant moment** (page 42) occurs, which provides a **turning force**.
3) An object will be nice and **stable** if it has a **low centre of mass** and a **wide base area**. This idea is used a lot in design, e.g. Formula 1® racing cars.
4) The higher the centre of mass and the smaller the base area, the less stable the object is. Think of unicyclists...

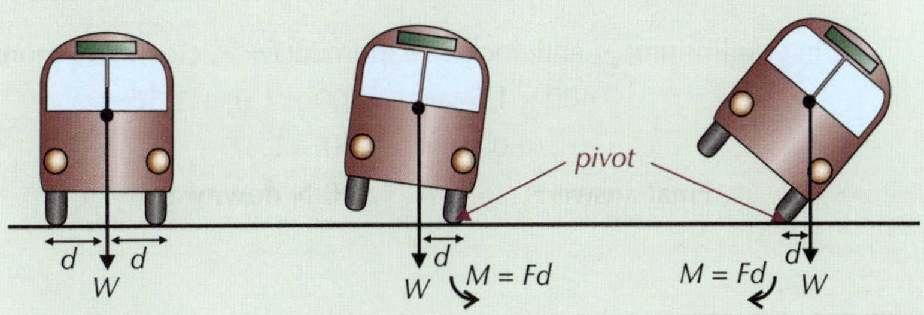

No resultant moment. Bus is stable and stationary.

Net anticlockwise moment about pivot. Bus will rotate to the left and right itself.

Net clockwise moment about pivot. Bus will rotate to the right and fall over.

Practice Questions

Q1 What are the differences between mass and weight?

Q2 A lioness has a mass of 200 kg. What would be her mass and weight on the Earth (where $g = 9.81\ \text{Nkg}^{-1}$) and on the Moon (where $g = 1.6\ \text{Nkg}^{-1}$)?

Q3 What is meant by the centre of mass of an object?

Q4 Why will an object topple if its centre of mass is not over the object's base?

Exam Question

Q1 a) Describe an experiment to find the centre of mass of an object of uniform density with a constant thickness and irregular cross-section. Identify one major source of uncertainty and suggest a way to reduce its effect on your result. [5 marks]

b) Explain why you would not need to conduct this experiment for a regular, uniform solid. [1 mark]

The centre of mass of this book should be round about page 74...

This is a really useful area of physics. To would-be nuclear physicists it might seem a little dull, but if you want to be an engineer — something a bit more useful (no offence Einstein) — then things like centre of mass are dead important things to understand. You know, for designing things like cars and submarines... yep, pretty useful I'd say.

Moments

*This is not a time for jokes. There is not a moment to lose. Oh ho ho ho ho *bang*. (Ow.)*

A Moment is the Turning Effect of a Force

The line of action of a force is a line along which it acts.

The **moment** of a **force** depends on the **size** of the force and **how far** the force is applied from the **turning point** (also called the **axis of rotation**):

moment of a force (in Nm) = **force** (in N) × **perpendicular distance from the line of action of the force to the axis of rotation** (in m)

$$M = Fd$$

Moments Must be Balanced or the Object will Turn

The **principle of moments** states that for a body to be in **equilibrium**, the **sum of the clockwise moments** about any point **equals** the **sum of the anticlockwise moments** about the same point.

Remember Σ means "the sum of".

Example:

Two children sit on a seesaw as shown in the diagram. An adult balances the seesaw at one end. Find the size and direction of the force that the adult needs to apply.

In equilibrium, Σ anticlockwise moments = Σ clockwise moments

$400 \times 1.5 = (300 \times 1.0) + 1.5F$

$600 = 300 + 1.5F$

Final answer: F = **200 N downwards**

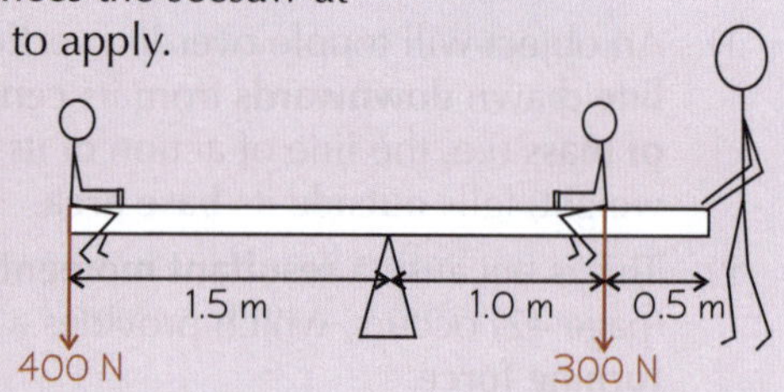

Muscles, Bones and Joints Act as Levers

1) In a lever, an **effort force** (in this case from a muscle) acts against a **load force** (e.g. the weight of your arm) by means of a **rigid object** (the bone) rotating around a **pivot** (the joint).
2) You can use the **principle of moments** to answer lever questions:

Example:

Find the force, E, exerted by the biceps in holding a bag of gold still. The bag of gold weighs 100 N and the forearm weighs 20 N.

Take moments about **A**.

In equilibrium:

Σ anticlockwise moments = Σ clockwise moments

$(100 \times 0.4) + (20 \times 0.2) = 0.04E$

$40 + 4 = 0.04E$

Final answer: E = 1100 = **1000 N (to 1 s.f.)**

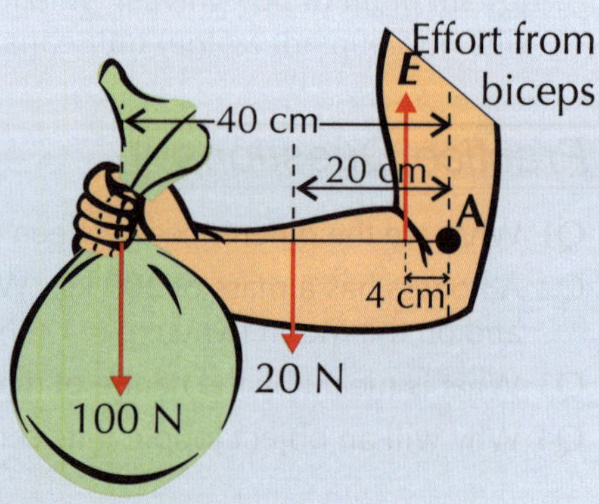

The Principle of Moments can Explain why Things Fall Over

1) As you saw on page 40, you can assume that **all** the weight of an object **acts through** its **centre of mass**. This is important when dealing with moments.
2) If the moments acting on an object are **not balanced** (as shown here) the object will **fall over**. There is a net **clockwise** moment on the broomstick so it will **rotate** clockwise and fall off the pivot.

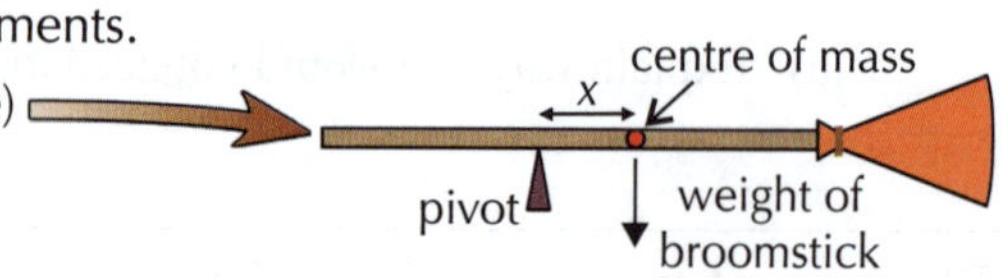

3) Balanced objects have **no resultant force** acting on them — their **moments** are **balanced**, and they **won't fall over** until an external force acts on them.

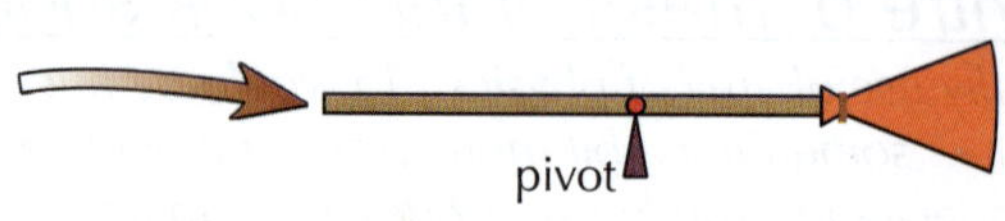

Moments

A Couple is a Pair of Coplanar Forces

1) A couple is a **pair** of **forces** of **equal size** which act **parallel** to each other, but in **opposite directions**. The forces are **coplanar** (they're all in the same plane).
2) A couple doesn't cause any resultant linear force, but **does** produce a **turning effect** (i.e. a moment).

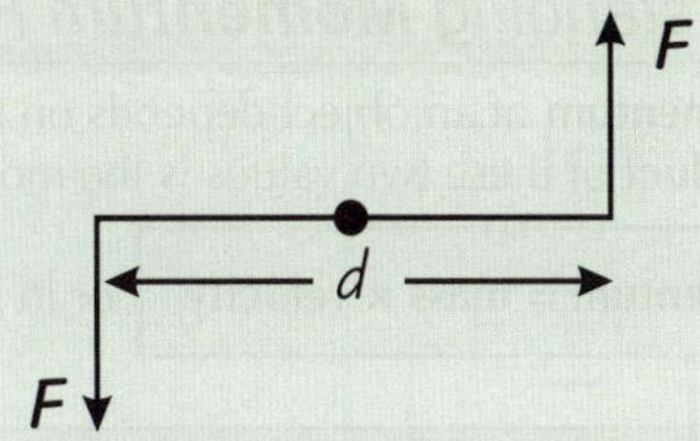

The **size** of this **moment** depends on the **size** of the **forces** and the **distance** between them.

moment of a couple (in Nm) **= size of one of the forces** (in N) × **perpendicular distance between the lines of action of the forces** (in m)

Again, in symbols, that's: $M = Fd$

Another word for turning force is 'torque'. The turning force caused by a couple is often called the torque of a couple, T.

Example:

A cyclist turns a sharp right corner by applying equal but opposite forces of 20 N to the ends of the handlebars.

The length of the handlebars is 0.60 m. Calculate the moment applied to the handlebars.

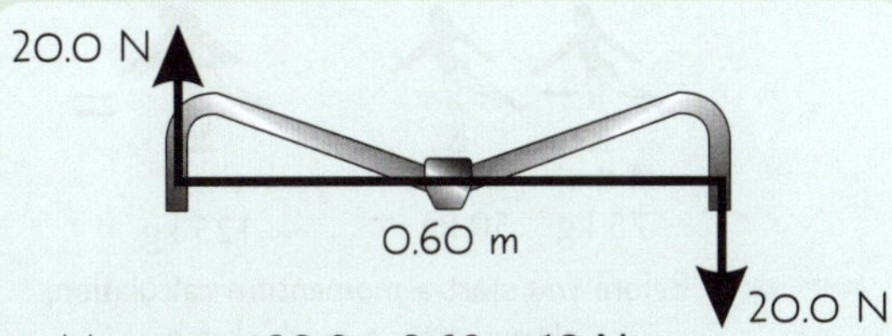

Moment = 20.0 × 0.60 = **12 Nm**

Practice Questions

Q1 A force of 54 N acts at a perpendicular distance of 84 cm from a pivot. Calculate the moment of the force.

Q2 A girl of mass 40 kg sits 1.5 m from the middle of a seesaw. Show that her brother, mass 50 kg, must sit 1.2 m from the middle if the seesaw is to balance.

Q3 What is meant by the word 'couple'?

Q4 A racing car driver uses both hands to apply equal and opposite forces of 65 N to the edge of a steering wheel with radius 20 cm. Calculate the moment of the forces.

Exam Questions

Q1 A driver is changing his flat tyre. The moment required to undo the nut is 60.0 Nm. He uses a 0.40 m long double-ended wheel wrench. Calculate the force that he must apply at each end of the wrench. [2 marks]

Q2 A diver of mass 60 kg stands on the end of a diving board 2.0 m from the pivot point. Calculate the downward force exerted on the board by the retaining spring 30 cm from the pivot. ($g = 9.81$ Nkg^{-1})

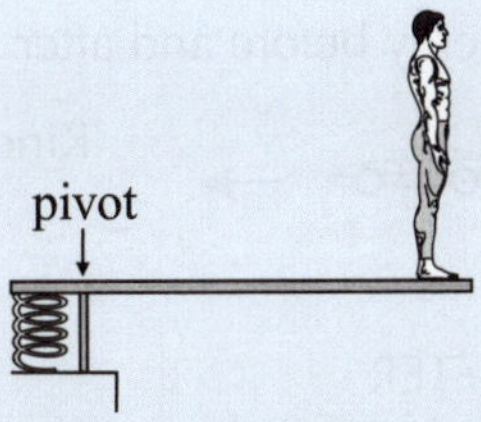

[2 marks]

It's all about balancing — just ask a tightrope walker...

They're always boring questions aren't they — seesaws or bicycles. It'd be nice if just once, they'd have a question on... I don't know, rotating knives or something. Just something unexpected. It'd make physics a lot more fun, I'm sure.

Momentum and Impulse

These pages are about linear momentum — that's momentum in a straight line (not a circle).

Understanding **Momentum** Helps You Do **Calculations** on **Collisions**

The **momentum** of an object depends on two things — its **mass** and **velocity**.
The **product** of these two values is the momentum of the object.

momentum = mass × velocity or in symbols: p **(in kg ms^{-1}) =** m **(in kg) ×** v **(in ms^{-1})**

Remember, momentum is a vector quantity, so it has size **and** direction.

Momentum is Always **Conserved**

You might see momentum referred to as 'linear momentum'. The other kind is 'angular momentum', but you don't need to know about that here.

1) Assuming **no external forces** act, momentum is always **conserved**.
2) This means the **total momentum** of two objects **before** they collide **equals** the total momentum **after** the collision.
3) This is really handy for working out the **velocity** of objects after a collision (as you do...):

Example: A skater of mass 75 kg and velocity 4.0 ms^{-1} collides with a stationary skater of mass 50 kg (to 2 s.f.). The two skaters join together and move off in the same direction. Calculate their velocity after impact.

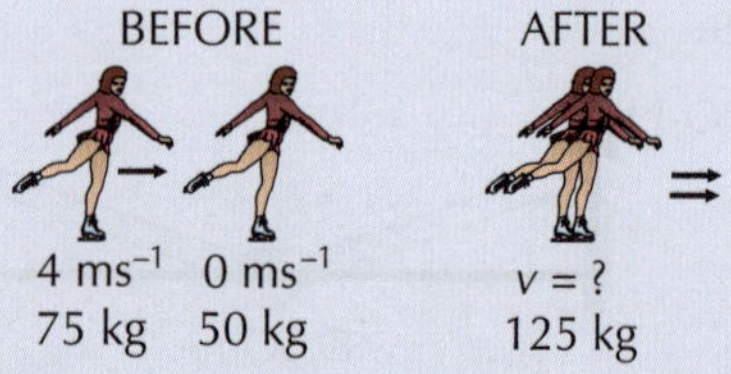

Before you start a momentum calculation, always draw a quick sketch.

Momentum of skaters before = Momentum of skaters after
$(75 \times 4.0) + (50 \times 0) = 125v$
$300 = 125v$
So v **= 2.4 ms^{-1}**

4) The same principle can be applied in **explosions**. E.g. if you fire an **air rifle**, the **forward momentum** gained by the pellet **equals** the **backward momentum** of the rifle, and you feel the rifle recoiling into your shoulder.

Example: A bullet of mass 0.0050 kg is shot from a rifle at a speed of 200 ms^{-1} (to 2 s.f.). The rifle has a mass of 4.0 kg. Calculate the velocity at which the rifle recoils.

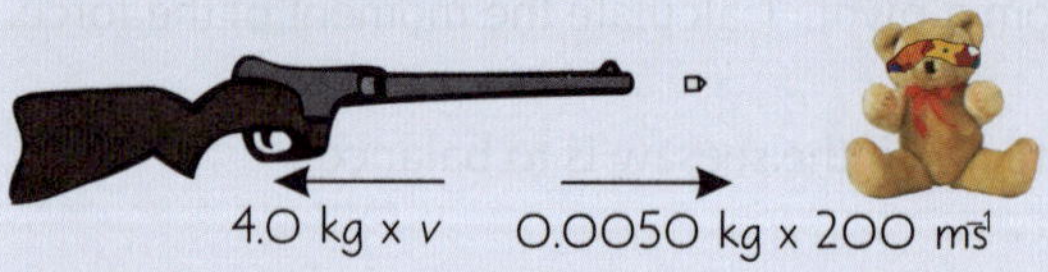

Momentum before explosion = Momentum after explosion
$0 = (0.0050 \times 200) + (4.0 \times v)$
$0 = 1 + 4v$
v **= –0.25 ms^{-1}**

Collisions Can be **Elastic** or **Inelastic**

An **elastic collision** is one where **momentum** is **conserved** and **kinetic energy** (see p.50) is **conserved** — i.e. no energy is dissipated as heat, sound, etc. If a collision is **inelastic** it means that some of the kinetic energy is converted into other forms during the collision. But **momentum is always conserved**.

Example: A toy lorry (mass 2.0 kg) travelling at 3.0 ms^{-1} crashes into a smaller toy car (mass 800 g (to 2 s.f.)), travelling in the same direction at 2.0 ms^{-1}. The velocity of the lorry after the collision is 2.6 ms^{-1} in the same direction. Calculate the new velocity of the car and the total kinetic energy before and after the collision.

Kinetic energy $= \frac{1}{2}mv^2$ — see p.50.

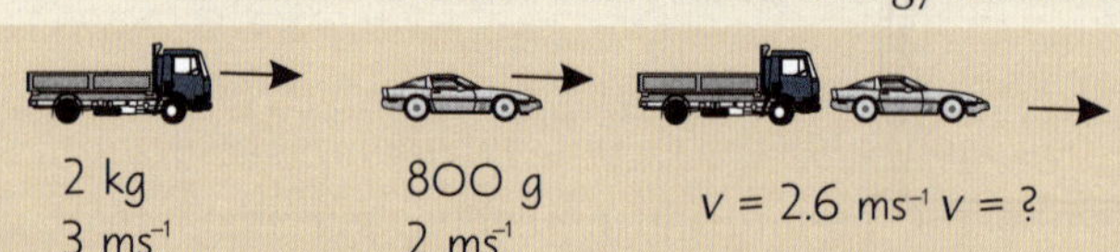

Momentum before collision = Momentum after collision
$(2 \times 3) + (0.8 \times 2) = (2 \times 2.6) + (0.8v)$
$7.6 = 5.2 + 0.8v$
$2.4 = 0.8v$
v **= 3 ms^{-1}**

Kinetic energy before = KE of lorry + KE of car
$= \frac{1}{2}mv^2$ (lorry) $+ \frac{1}{2}mv^2$ (car)
$= \frac{1}{2}(2 \times 3^2) + \frac{1}{2}(0.8 \times 2^2)$
$= 9 + 1.6$
= 11 J (to 2 s.f.)

Kinetic energy after $= \frac{1}{2}(2 \times 2.6^2) + \frac{1}{2}(0.8 \times 3^2)$
$= 6.76 + 3.6$
= 10 J (to 2 s.f.)

The difference in the two values is the amount of kinetic energy dissipated as heat or sound, or in damaging the vehicles — so this is an **inelastic** collision.

Momentum and Impulse

Collisions and Explosions Usually Happen in More Than One Dimension

In **two-dimensional collisions**, momentum is conserved in **both dimensions**.
You can solve two-dimensional collision problems by **resolving vectors** (see page 15).

Example: Ball A collides with stationary ball B, as shown in diagram 1. After the collision, the two balls move off as shown in diagram 2. Ball A has a mass of 40 g. Calculate the mass, m, of ball B.

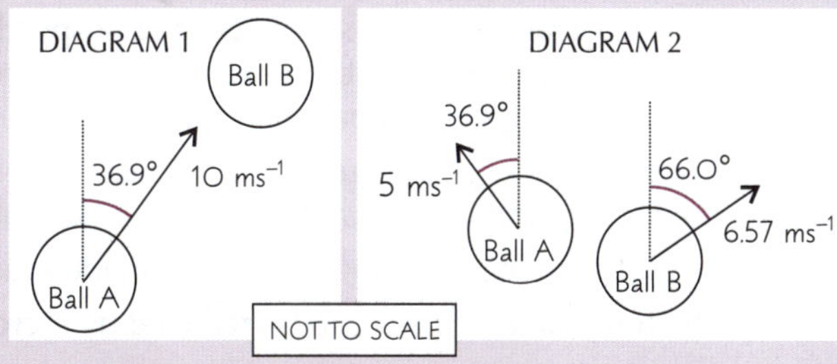

You can work this using conservation of momentum in just the horizontal direction (or just the vertical).
Start by picking the positive direction — let's say right is positive.

horizontal momentum before the collision = horizontal momentum after the collision

$(0.04 \times 10 \sin 36.9°) + (m \times 0) = -(0.04 \times 5 \sin 36.9°) + (m \times 6.57 \sin 66.0°)$

so, $m = \dfrac{0.04 \times (10 \sin 36.9° + 5 \sin 36.9°)}{6.57 \sin 66.0°}$ = 0.0600... kg = **60 g (to 2 s.f.)**

The horizontal component of the velocity of ball A is negative after the collision — it's to the left.

Newton's 2nd Law Says That Force is the Rate of Change in Momentum

The **rate of change of momentum** of an object is **directly proportional** to the **resultant force** which acts on the object.

So: $F = \dfrac{\Delta(mv)}{\Delta t}$ or $F\Delta t = \Delta(mv)$ (where F is constant)

Remember that acceleration is equal to the rate of change of velocity (page 16), so if mass is constant then this formula gives you that mechanics favourite, $F = m \times a$. This is the equation of the law that you met on p.31.

Impulse = Change in Momentum

1) **Rearranging** Newton's 2nd law gives: →
Where **impulse** is defined as **force × time**, Ft.
The units of impulse are **newton seconds**, Ns.

$Ft = mv - mu$
so impulse = change in momentum

Where v is the final velocity and u is the initial velocity.

2) **Impulse** is the **area under** a **force-time graph** — this is really handy for solving problems where the force changes.

Example: The graph shows the resultant force acting on a toy car. If the car is initially at rest, what is its momentum after 3 seconds?

Impulse = **change in momentum** = $mv - mu$. The **initial momentum** (mu) is **zero** because the toy car is stationary to begin with. So, **impulse** = mv.

Impulse is the **area under the graph**, so to find the **momentum** of the car after 3 seconds, you need to find the **area under the graph** between 0 and 3 seconds.

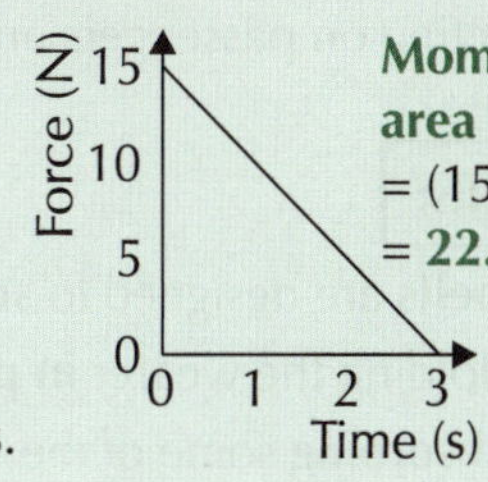

Momentum = area under graph
= (15 × 3) ÷ 2
= **22.5 Ns**

Practice Questions

Q1 Give two examples of conservation of momentum in practice.

Q2 Describe how to find the impulse of a collision from a force-time graph.

Exam Question

Q1 A ball of mass 0.60 kg moving at 5.0 ms^{-1} collides with a larger stationary ball of mass 2.0 kg. The smaller ball rebounds in the opposite direction at 2.4 ms^{-1}.
a) Calculate the velocity of the larger ball immediately after the collision. [3 marks]
b) State and explain whether this is an elastic or inelastic collision. Support your answer with calculations. [3 marks]

Momentum'll never be an endangered species — it's always conserved...

*Impulse is the change of momentum of **one** object, but conservation of momentum is talking about the **whole** system.*

Car Safety

Newton's laws aren't all about apples on tables — they're also particularly important for designing cars...

If a Car Crashes, the Forces are Very Large

Impulse isn't just important in car safety — e.g. when you kick a football, you want to have the largest impulse possible (i.e. be in contact with the ball for as little time as possible assuming change in momentum is the same).

1) Newton's second law of motion (p.45) says: $F = \frac{\Delta(mv)}{\Delta t} = \frac{\Delta p}{\Delta t}$
2) When a car **crashes**, there is a big **change in speed** — possibly from 70 mph (about 110 km h^{-1}) to zero. This means there is a big change in **momentum**.
3) If the car comes to a stop **quickly** (Δt is small), then the **force acting** on the car, anyone inside it and whatever it has collided with is very **large**.
4) If Δt can be **increased**, this force can be made **much smaller** for a given change of momentum.
5) So, the **force** of an impact can be **reduced** by **increasing the time** the impact takes place over.

Example: A toy car with a mass of 1 kg, travelling at 5 ms^{-1}, hits a wall and stops in a time of 0.5 seconds.

The average force on the car is: $F = \frac{\Delta p}{\Delta t} = \frac{(1 \times 5) - (1 \times 0)}{0.5} = 10\,\text{N}$

But if the time of impact is doubled to **1 second**, the force on the car is halved.

Most Car Safety Features are Designed to Slow You Down Gradually

Modern cars have lots of built-in **safety features.** A lot of them reduce the forces acting on passengers in a crash by **increasing the time** over which the **change in momentum** takes place.

Crumple Zones:

Crumple zones are areas at the front and rear of a car that are designed to **crumple on impact**. They have two effects:

1) They absorb some of the car's **kinetic energy** when they deform, which would otherwise be **transferred to the passengers** and whatever the car had collided with (e.g. other cars).
2) They **increase time** taken for the car to slow down, (Δt in the equation above) which **reduces the forces** acting on passengers and whatever the car has hit.

Air Bags:

Air bags are 'cushions' in the dashboard and elsewhere that **inflate** very quickly on collision.
They protect the passengers by:

1) Making passengers **slow down more gradually** (increasing Δt).
2) Stopping passengers from hitting the dashboard, steering wheel, etc. during a crash.

Seat Belts:

Seat belts are designed to **stretch** slightly. They protect wearers in a crash by:

1) Holding the wearer **in place** in the car, stopping them from being thrown from their seats.
2) **Absorbing** some of the wearer's **kinetic energy** by stretching (p.53).
3) **Increasing the time** that the wearer comes to a stop over, Δt (again, by stretching).

Example: Giles's car bumps into the back of a stationary bus. The car was travelling at 2.0 ms^{-1} and comes to a stop in 0.20 s. Giles was wearing his seatbelt and takes 0.80 s to stop. Giles's mass is 75 kg.

a) Calculate the average force acting on Giles during the accident.

b) Calculate the average force that would have acted on Giles if he had stopped as quickly as the car.

a) $F = \frac{\Delta p}{\Delta t}$ So, for Giles: $F = \frac{75 \times 2.0 - 75 \times 0}{0.80} = 187.5 =$ **190 N (to 2 s.f.)**

b) Again, $F = \frac{\Delta p}{\Delta t}$ but $\Delta t = 0.20$ s. So: $F = \frac{75 \times 2.0 - 75 \times 0}{0.20} =$ **750 N (to 2 s.f.)**

Car Safety

Crash Tests *are Used to Analyse the* ***Forces*** *in a Crash*

1) Early cars didn't have many safety features in the event of a crash. But as cars became **faster** and the **number** of cars on the roads increased, car manufacturers started to think about how to make cars **safer**.
2) This led to the development of the **three-point seat belt**, **crumple zones** and **air bags** back in the 1950's.
3) At the same time, engineers were developing better ways of **understanding** the **forces** acting on passengers in a collision. They developed increasingly sophisticated **crash test dummies** that could **measure** all the forces that different passengers experience in different **kinds** of collisions.
4) **Crash tests** also allowed car manufacturers to test the **effectiveness** of certain **safety features**, so they could see how much they reduced the likelihood of injury.
5) Wearing a seat belt is now **mandatory** (everyone has to do it), and features such as crumple zones and air bags have become **standard features** of modern cars (along with many more sophisticated safety features).
6) Understanding the forces involved in road collisions has **saved many lives** — it's estimated that the three-point seat belt alone has saved **over a million lives worldwide** in the last 50 years.

Car Safety Features *Have Some* ***Risks***

1) Seat belts can cause **bruising** during a crash.
2) They can also be dangerous for **small children** — if the top part of the belt lies across the child's **neck** this can cause injury in a crash and if the child is **too small** for a seat belt it may not **secure them properly**. In the UK, children must use **booster seats** or **cushions** to reduce these risks.

June didn't care how many safety features the car had — if Fido was driving, she wasn't getting in.

3) **Air bags** are designed for use with seat belts and can be **dangerous** if you're not wearing one. Air bags inflate very **rapidly**, with a lot of **force**. If a passenger seat belt isn't **secured properly** then the passenger can keep moving forwards quickly as the car slows down and hit the air bag as it is inflating with a **force** big enough to cause **injury**.
4) Air bags are also dangerous when using **rear-facing child seats** — the air bag inflates **behind** the child and can throw the child seat towards the car seat with some force. It is now **illegal** to use a rear-facing child seat in the UK in a seat fitted with an air bag.

Most of these risks are caused by not using a car's **safety features properly**.
If you use a seat belt and air bags as you are meant to, you are **far safer** in a car than you would be without them.

Practice Questions

Q1 Give three safety features of a car that are designed to protect passengers in the event of a crash.

Q2 Give one feature of a car's design that might protect passengers in another vehicle in the event of a collision.

Exam Question

Q1* Explain the ways that air bags protect passengers in the event of a crash, and describe some of the risks associated with their use and how these risks can be reduced. [6 marks]

*The quality of your extended response will be assessed in this question.

Never mind the physics — the best car safety feature is a careful driver...

Some real, applied physics here to get your head round, but it's important stuff. Understanding forces has had a huge effect on how cars are designed, and in doing so has saved a lot of people's lives. Not bad for a few equations, hey?

Work and Power

As everyone knows, work in Physics isn't like normal work. It's harder. Work also has a specific meaning that's to do with movement and forces. You'll have seen this at GCSE — it just comes up in more detail for A level.

Work** is Done Whenever **Energy** is **Transferred

This table gives you some examples of **work being done** and the **energy changes** that happen.

1) Usually you need a force to move something because you're having to **overcome another force**.
2) The thing being moved has **kinetic energy** while it's **moving**.
3) The kinetic energy is transferred to **another form of energy** when the movement stops.

ACTIVITY	WORK DONE AGAINST	FINAL ENERGY FORM
Lifting up a box.	gravity	gravitational potential energy
Pushing a chair across a level floor.	friction	heat
Pushing two magnetic north poles together.	magnetic force	magnetic energy
Stretching a spring.	stiffness of spring	elastic strain energy

The word '**work**' in Physics means the **amount of energy transferred** from one form to another when a force causes a movement of some sort. It is measured in **joules**, J.

Work = Force × Distance

Distance is sometimes written as *x* instead of *s*.

When a car tows a caravan, it applies a force to the caravan to move it to where it's wanted. To **find out** how much **work** has been **done**, you need to use the **equation**:

work done (W) **= force causing motion** (F) **× distance moved** (s), **or** $W = Fs$
...where W is measured in joules (J), F is measured in newtons (N) and s is measured in metres (m).

You might also see this written as $\Delta W = F\Delta s$, where delta stands for 'change in'.

Points to remember:

1) **Work** is the **energy** that's been **changed** from one form to another — it's not necessarily the **total** energy. E.g. moving a book from a low shelf to a higher one will increase its gravitational potential energy, but it had some potential energy to start with. Here, the **work done** would be the **increase** in potential energy, **not the total** potential energy.
2) Remember the distance needs to be measured in metres — if you have **distance in centimetres or kilometres**, you need to **convert** it to metres first.
3) The force ***F*** will be a **fixed** value in any calculations, either because it's **constant** or because it's the **average** force.
4) The equation assumes that the **direction of the force** is the **same** as the **direction of movement**.
5) The equation gives you the **definition** of the joule (symbol J):
'One joule is the work done when a force of 1 newton moves an object through a distance of 1 metre'.

*The **Force** isn't Always in the **Same Direction** as the **Movement***

Sometimes the **direction of movement** is **different** from the **direction of the force**.

Example:

1) To **calculate the work done** in a situation like the one on the right, you need to consider the **horizontal** and **vertical components** of the **force**.
2) The only **movement** is in the **horizontal** direction. This means the **vertical force** is not causing any motion (and hence not doing any work) — it's just **balancing** out some of the **weight**, meaning there's a **smaller reaction force**.
3) The horizontal force is causing the motion — so to **calculate** the **work done**, this is the **only force** you need to consider. Which means we get:

$$W = Fs\cos\theta$$

Where θ is the **angle** between the **direction of the force** and the **direction of motion**. See page 36 for more on resolving forces.

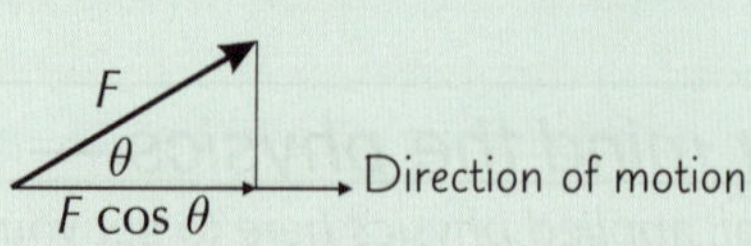

Work and Power

The Area Under a Force-Displacement Graph Tells You the Work Done

1) For a **variable force**, you can't just use the formula $W = Fs$ — nightmare.
2) Luckily, plotting a **graph** of **force against distance moved** lets you calculate the work done by just finding the **area under the graph**.
3) You might need to **split it up** into sections that make shapes you can work out the area for, e.g. trapeziums.

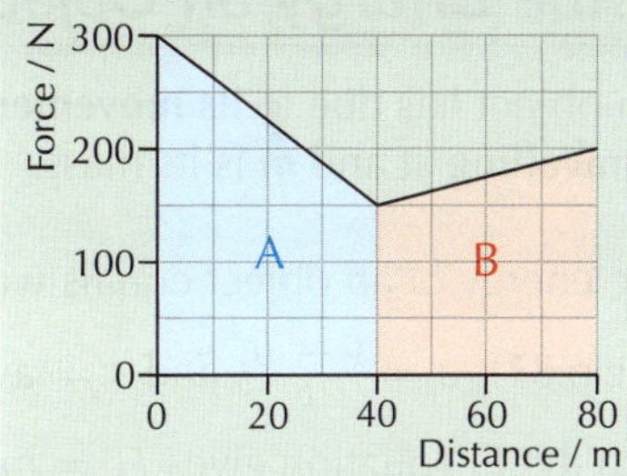

The graph shows the force exerted by Tibalt the circus monkey as he cycled up a hill.

Work done in section A:

$40 \times \frac{300 + 150}{2} = 9000$ J

Work done in section B:

$40 \times \frac{200 + 150}{2} = 7000$ J

Total work done = **16 000 J**

Power = Work Done per Second

Remember, power = energy ÷ time.

Power means many things in everyday speech, but in Physics (of course!) it has a special meaning. Power is the **rate of doing work** — in other words it is the **amount of energy transferred** from one form to another **per second**. You **calculate power** from this equation:

Power (P) = change in energy (or work done) (ΔW) / change in time (Δt), or $P = \frac{\Delta W}{\Delta t}$

...where P is measured in watts (W), ΔW is measured in joules (J) and Δt is measured in seconds (s).

The **watt** (symbol W) is defined as a **rate of energy transfer** equal to **1 joule per second** (Js^{-1}).

Power is also Force × Velocity

Sometimes, it's **easier** to use **this version** of the power equation:

1) You know $P = \Delta W / \Delta t$.
2) You also know $\Delta W = F\Delta s$, which gives $P = F\Delta s / \Delta t$.
3) But $v = \Delta s / \Delta t$, which you can substitute into the above equation to give: $P = Fv$

It's easier to use this if you're given speed in the question.

4) If the force and motion are in different directions, you can replace F with $F\cos\theta$ to get: $P = Fv\cos\theta$

Practice Questions

Q1 Write down the equation used to calculate work done if the force and motion are in the same direction.

Q2 Write down the equation for work done if the force is at an angle to the direction of motion.

Q3 What does the area under a force-distance graph represent?

Q4 Write down the equations relating a) power and work and b) power and speed.

Exam Questions

Q1 A traditional narrowboat is drawn by a horse walking along the towpath. The horse pulls the boat at a constant speed between two locks which are 1500 m apart. The tension in the rope is exactly 100 N (to 2 s.f.) at exactly 40° to the direction of motion.

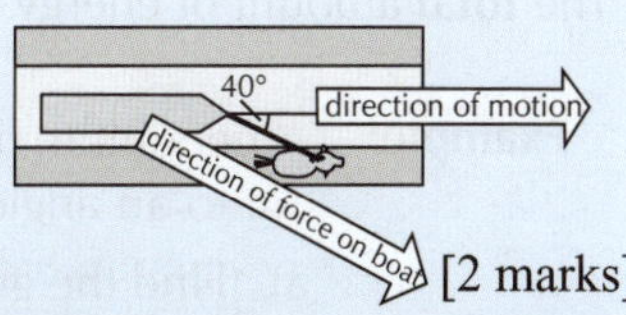

a) Calculate the work done on the boat. [2 marks]

b) The boat moves at 0.80 ms^{-1}. Calculate the power supplied to the boat in the direction of motion. [2 marks]

Q2 A motor is used to lift a 20 kg (to 2 s.f.) load a height of 3.0 m. (Take $g = 9.81$ Nkg^{-1}.)

a) Calculate the work done in lifting the load. [2 marks]

b) The speed of the load during the lift is 0.25 ms^{-1}. Calculate the power delivered by the motor. [2 marks]

Work — there's just no getting away from it...

Loads of equations to learn. Well, that's what you came here for, after all. Can't beat a good bit of equation-learning, as I've heard you say quietly to yourself when you think no one's listening. Aha, can't fool me. Ahahahahahahahahaha.

Conservation of Energy and Efficiency

***Energy** can never be **lost**. I repeat — **energy** can **never** be lost.*

Kinetic Energy — the Energy an Object has Because it is Moving

Kinetic energy is energy an object has due to its **movement**. You can calculate it using the equation $\boldsymbol{E_k = \frac{1}{2}mv^2}$, where $\boldsymbol{v}$ is the velocity it's travelling at and $\boldsymbol{m}$ is its mass. Here's how to **derive** this equation...

- The change in kinetic energy of an object equals **work done** (Fs, see p.48), so $E_k = Fs$.
- You know $F = ma$ (see p.31), $a = \frac{v-u}{t}$, and s = average $v \times t = \frac{(u+v)}{2}t$ (see p.20).
- Substituting these into the E_k equation gives $E_k = mas = m \times \frac{(v-u)}{t} \times \frac{(u+v)}{2} \times t$.
- **Cancelling** the 't's and **expanding** the brackets gives you a slightly more familiar looking $E_k = \frac{1}{2}mv^2 - \frac{1}{2}mu^2$.
- The kinetic energy of an object is the kinetic energy compared to when the object is at rest, i.e. when $\boldsymbol{u = 0}$ and the object has **no kinetic energy**.
- Substituting $u = 0$ into the equation above gives you: → $\boldsymbol{E_k = \frac{1}{2}mv^2}$

An Object's Gravitational Potential Energy Depends on its Position in a Field

1) **Gravitational potential energy** ($\boldsymbol{E_p}$) is the energy an object has due to its **position** in a gravitational field. The **greater** the **height** of the object, the **greater** its gravitational potential energy.
2) You normally want to find the change in an object's E_p, which you can work out using the equation $\boldsymbol{\Delta E_p = mg\Delta h}$, where $\boldsymbol{m}$ is the mass of the object, $\boldsymbol{\Delta h}$ is the change in height and $\boldsymbol{g}$ is the gravitational field strength.
3) The derivation for this equation is a bit shorter: →

- The gravitational energy gained is equal to the work done in moving the object a distance $\boldsymbol{h}$ upwards, so $\Delta E_p = W = F \times \Delta h$.
- The force that work is done against is the force of **gravity**, which is equal to $\boldsymbol{mg}$. So $\boldsymbol{\Delta E_p = mg\Delta h}$.

Learn the Principle of Conservation of Energy

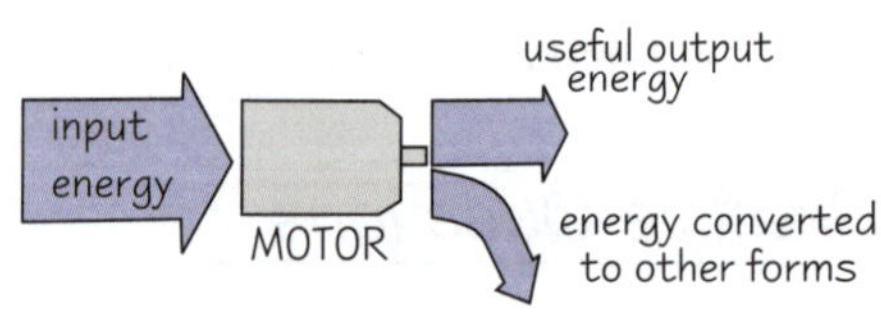

The **principle of conservation of energy** says that:

> Energy **cannot be created** or **destroyed**. Energy **can be transferred** from one form to another but the total amount of energy in a closed system **will not change**.

Total energy in = Total energy out

The key point to remember when answering questions is that the **total energy** will always be the same. E.g. after a **ball** is thrown **upwards**, its **kinetic energy** is converted into **gravitational potential energy**. When it **comes down** again, that **gravitational potential** energy is all **converted back** into **kinetic** energy (assuming no air resistance). The **total** amount of energy stays the **same**, but how much of it is kinetic or gravitational potential energy changes.

Example: A pendulum has a mass of 700 g and a length of 50 cm. It is pulled out to an angle of 30° from the vertical. (All values are given to 2 s.f.)
a) Find the gravitational potential energy stored in the pendulum bob.

As well as kinetic and gravitational potential energy, questions might also involve elastic strain energy — see p.53.

Start by drawing a diagram.
You can work out the increase in height, h, of the end of the pendulum using trig.

$\Delta E_p = mg\Delta h = 0.7 \times 9.81 \times (0.5 - 0.5\cos 30°) = 0.460... = \mathbf{0.46\ J\ (to\ 2\ s.f.)}$

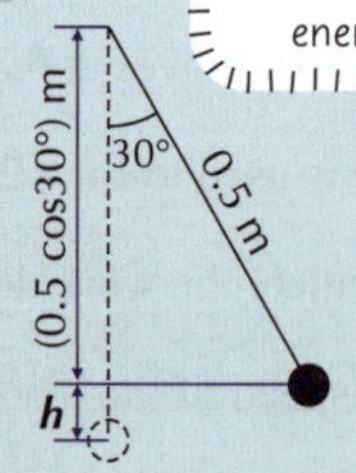

b) The pendulum is released. Find the maximum speed of the pendulum bob as it passes the vertical position. Assume there is no air resistance.

When travelling at its maximum speed, $mg\Delta h = \frac{1}{2}mv^2$.

So $\frac{1}{2}mv^2 = 0.460...$, so $v = \sqrt{\frac{2 \times 0.460...}{0.7}} = \mathbf{1.1\ ms^{-1}\ (to\ 2\ s.f.)}$

Or, you could cancel the 'm's and rearrange:
$v = \sqrt{2g\Delta h} = \sqrt{2 \times 9.81 \times (0.5 - 0.5\cos 30°)}$
$= \mathbf{1.1\ ms^{-1}\ (to\ 2\ s.f.)}$

Conservation of Energy and Efficiency

All Energy Transfers Involve Losses

You saw on the last page that **energy can never be created or destroyed**. But whenever **energy** is **converted** from one form to another, some is always **'lost'**. It's still there (i.e. it's **not destroyed**) — it's just not in a form you can **use**.

Most often, **energy** is lost as **heat** — e.g. **computers** and **TVs** are always **warm** when they've been on for a while. In fact, **no device** (except possibly a heater) is ever **100% efficient** (see below) because some energy is **always** lost as **heat**. (You want heaters to give out heat, but in other devices the heat loss isn't useful.) Energy can be **lost** in other forms too (e.g. **sound**) — the important thing is the lost energy **isn't** in a **useful** form and you **can't** get it back.

Often the heat that is lost is caused by **friction**. Luckily you can usually assume that **friction** is **zero** in exams.

Efficiency is the Ratio of Useful Energy Output to Total Energy Input

Efficiency is one of those words we use all the time, but it has a **specific meaning** in Physics. It's a measure of how well a **device** converts the **energy** you put **in** into the energy you **want** it to give **out**. So, a device that **wastes** loads of **energy** as heat and sound has a really **low efficiency**.

$$\text{Efficiency} = \frac{\text{useful energy output}}{\text{total energy input}}$$

$$\text{Efficiency} = \frac{\text{useful power output}}{\text{total power input}}$$

You can also give efficiency as a percentage if you multiply the ratio by 100.

Energy, as always, is measured in **joules** (J). Efficiency has **no units** because it's a **ratio**.

Some questions will be kind and **give you** the **useful output energy** — others will tell you how much is **wasted**. You just have to **subtract** the **wasted energy** from the **total input energy** to find the **useful output energy**, so it's not too tricky if you keep your wits about you.

Practice Questions

Q1 What are the equations for calculating kinetic energy and gravitational potential energy?

Q2 State the principle of conservation of energy.

Q3 Show that, if there's no air resistance and the mass of the string is negligible, the speed of a pendulum is independent of the mass of the bob.

Q4 Why can a device never be 100% efficient?

Q5 What is the equation for efficiency in terms of power input and output?

Exam Questions

gravitational field strength, $g = 9.81\ \text{N kg}^{-1}$

Q1 A skateboarder is skating on a half-pipe. He lets the board run down one side of the ramp and up the other. The height of the ramp is 2.0 m.

a) Calculate his speed at the lowest point of the ramp. Assume friction is negligible. [3 marks]

b) State how high he will rise up the other side of the half-pipe. [1 mark]

c) Real ramps are not frictionless. Describe what the skater must do to reach the top on the other side. [1 mark]

Q2 A 20.0 g rubber ball is released from a height of 8.0 m. Assume that the effect of air resistance is negligible.

a) Calculate the kinetic energy of the ball just before it hits the ground. [2 marks]

b) The ball strikes the ground and rebounds to a height of 6.5 m. Calculate the amount of energy that is converted to heat and sound in the impact with the ground. [2 marks]

Q3 Calculate the efficiency of a device that wastes 65 J for every 140 J of input energy. [1 mark]

Energy is never lost — it just sometimes prefers the scenic route...

Remember to check your answers — I can't count the number of times I've forgotten to square the velocities or to multiply by the ½... I reckon it's definitely worth the extra minute to check. You never know what you might find.

Hooke's Law

Hooke's law applies to all materials, but only up to a point. For some materials that point is so tiny you wouldn't notice...

Hooke's Law Says that Extension is Proportional to Force

If a **metal wire** is supported at the top and then a weight is attached to the bottom, the wire **stretches**. The weight pulls down with force ΔF, producing an equal and opposite force at the support.

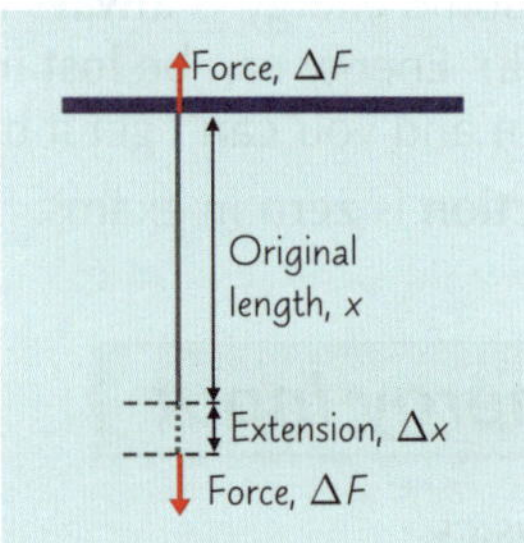

1) **Robert Hooke** discovered in the 17th century that the extension of a stretched wire, Δx, is proportional to the change in load or force, ΔF. This relationship is now called **Hooke's law**.
2) Hooke's law can be written:

$$\Delta F = k\Delta x$$

Where k is the stiffness of the object that is being stretched. k is called the **force constant** (or **stiffness constant**) and has units Nm^{-1}.

The material will only deform (stretch, bend, twist etc.) if there's a pair of opposite forces acting on it.

You might see this formula without any deltas, or with a ΔL instead of the Δx.

Hooke's Law Also Applies to Springs

A metal spring also changes length when you apply a **pair of opposite forces**.

1) The **extension** or **compression** of a spring is **proportional** to the **force** applied — so Hooke's law applies.
2) For springs, k in the formula $\Delta F = k\Delta x$ can also be called the **spring stiffness** or **spring constant**.

Hooke's law works just as well for **compressive** forces as **tensile** forces. For a spring, k has the **same value** whether the forces are tensile or compressive (that's not true for all materials).

3) **Hooke's Law** doesn't just apply to metal **springs** and **wires** — most **other materials** obey it up to a point.

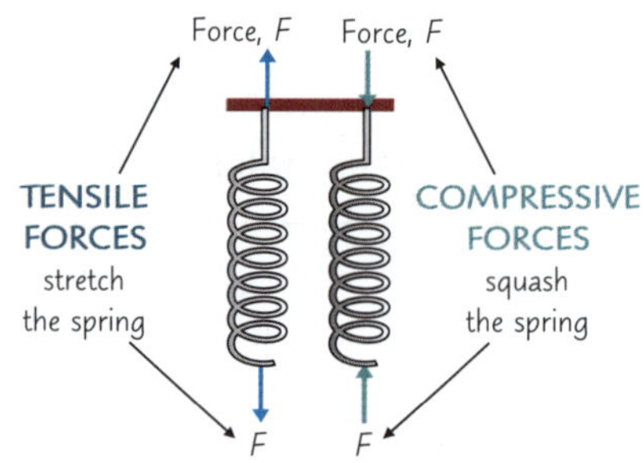

Tensile forces create tension in a stretched spring. Compressive forces create compression in a squashed spring. Tensile or compressive forces in the spring act in the opposite direction to the tensile or compressive forces stretching or squashing it.

Hooke's Law Stops Working when the Load is Great Enough

There's a **limit** to the force you can apply for Hooke's law to stay true.

1) The graph shows force against extension for a **typical metal wire** or **spring**.
2) The first part of the graph (up to point P) shows Hooke's law being obeyed — there's a **straight-line relationship** between **force** and **extension**.
3) When the force becomes great enough, the graph starts to **curve**. **Metals** generally obey Hooke's law up to the **limit of proportionality**, **P**.
4) The point marked **E** on the graph is called the **elastic limit**. If you exceed the elastic limit, the material will be **permanently stretched**. When all the force is removed, the material will be **longer** than at the start.

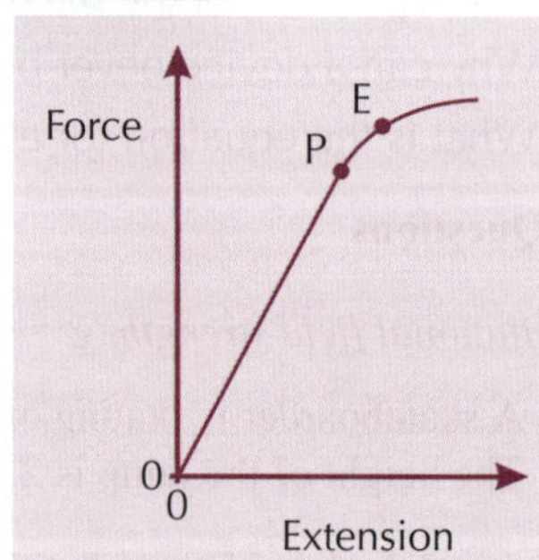

5) Be careful — there are some materials, like **rubber**, that only obey Hooke's law for **really small** extensions.

A Deformation can be Elastic or Plastic

A material will show elastic deformation **up to** its **elastic limit**, and plastic deformation **beyond** it. If a **deformation** is **elastic**, the material returns to its **original shape** once the forces are removed.

1) When the material is put under **tension**, the **atoms** of the material are **pulled apart** from one another.
2) Atoms can **move** slightly relative to their **equilibrium positions**, without changing position in the material.
3) Once the **load** is **removed**, the atoms **return** to their **equilibrium** distance apart.

If a deformation is **plastic**, the material is **permanently stretched**.

1) Some atoms in the material move position relative to one another.
2) When the load is removed, the **atoms don't return** to their original positions.

Extension and compression are sometimes called tensile deformation and compressive deformation.

Hooke's Law

Energy is Always Conserved When Stretching

There's more about elastic strain energy on the next two pages (plus some formulas — hurrah!).

When a material is **stretched**, **work** has to be done in stretching the material.

1) If a deformation is **elastic**, all the work done is **stored** as **elastic strain energy** in the material.
2) When the stretching force is removed, this **stored energy** is **transferred** to **other forms** — e.g. when you remove a mass from a spring, the elastic strain energy is **converted** to **kinetic** and **gravitational potential energy** as the spring **contracts**.
3) If a deformation is **plastic**, work is done to **separate atoms**, and energy is **not** stored as strain energy (it's mostly dissipated as heat).
4) This fact is used in **transport design** — **crumple zones** are designed to deform **plastically** in a **crash**. Some energy goes into **changing the shape** of the vehicle's **metal body** (and so less energy is transferred to the people inside).

Investigating Extension

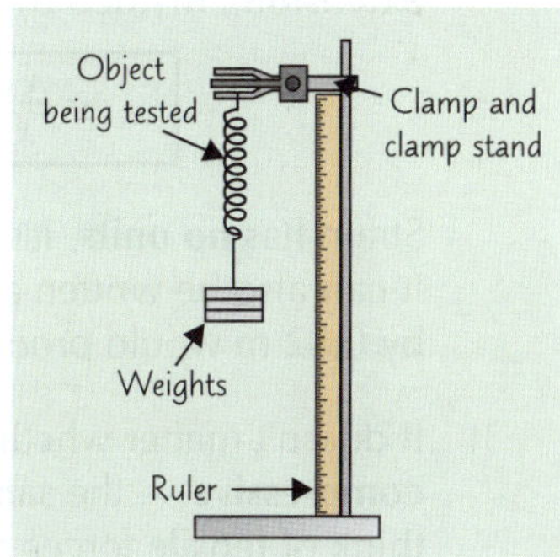

1) Set up the experiment shown in the **diagram**. **Support** the object being tested at the top (e.g. with a clamp) and measure its **original length** with a **ruler**.
2) Add **weights** one at a time to the bottom of the object.
3) After each weight is added, measure the new length of the object, then **calculate the extension**:

   extension = new length – original length

4) Plot a graph of **force** (weight) against **extension** for your results. Where the line of best fit is **straight**, then the object obeys Hooke's law and the gradient = $\boldsymbol{k}$ (as $\Delta F = k\Delta x$). If you've loaded the object beyond its limit of proportionality (see previous page), the graph will start to curve.
5) Make sure you carry out the experiment **safely**. You should be **standing up** so you can get out of the way quickly if the weights fall, and wearing **safety goggles** to protect your eyes in case the object snaps.

Practice Questions

Q1 State Hooke's law and explain what is meant by the elastic limit of a material.

Q2 Define tensile forces and compressive forces.

Q3 From studying the force-extension graph for a material as it is loaded and unloaded, how can you tell if Hooke's law is being obeyed?

Q4 What is meant by plastic deformation of a material?

Q5 Describe how you could determine the stiffness constant of an object being stretched, k, experimentally.

Exam Questions

Q1 A metal guitar string stretches 4.0 mm when a 10.0 N force is applied.

a) If the string obeys Hooke's law, calculate how far the string will stretch when a 15 N force is applied. [1 mark]

b) Calculate the force constant for this string in Nm^{-1}. [1 mark]

c) The string is then stretched beyond its elastic limit. Describe the effect this will have on the string. [1 mark]

Q2 A rubber band is 6.0 cm long. When it is loaded with 2.5 N, its length becomes 10.4 cm. Further loading increases the length to 16.2 cm when the force is 5.0 N.

State whether the rubber band obeys Hooke's law when the force on it is 5.0 N. Explain your answer. [2 marks]

Sod's Law — if you don't learn it, it'll be in the exam...

Three things you didn't know about Robert Hooke — he was the first person to use the word 'cell' (as in biology, not prisons), he helped Christopher Wren with his designs for St. Paul's Cathedral and no-one's sure what he looked like. I'd like to think that if I did all that stuff, then someone would at least remember what I looked like — poor old Hooke.

Stress and Strain

How much a material stretches for a particular applied force depends on its dimensions. If you want to compare it to another material, you need to use stress and strain instead. A stress-strain graph is the same for any sample of a particular material — the size of the sample doesn't matter.

A Stress Causes a Strain

A material subjected to a pair of **opposite forces** might **deform** (i.e. **change shape**).
If the forces **stretch** the material, they're **tensile**. If the forces **squash** the material, they're **compressive**.

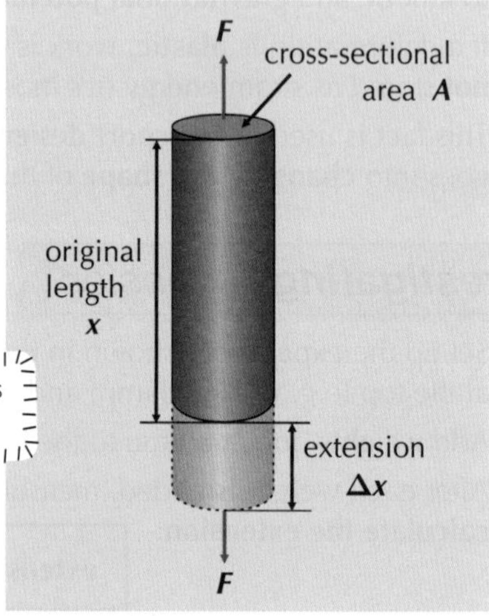

1) **Tensile stress**, σ, is defined as the **force applied**, F, divided by the **cross-sectional area**, A:

$$\sigma = \frac{F}{A}$$

The **units** of stress are **Nm^{-2}** or pascals, **Pa**.

2) **Tensile strain**, ε, is defined as the **change in length** (i.e. the **extension**), divided by the **original length** of the material:

$$\varepsilon = \frac{\Delta x}{x}$$

Again, you might see this with L instead of x.

Strain has **no units**, it's just a **ratio** and is usually written as a **number**. It can also be written as a **percentage**, e.g. extending a 0.5 m wire by 0.02 m would produce a strain of (0.02 ÷ 0.5) × 100 = 4%.

3) It doesn't matter whether the forces producing the **stress** and **strain** are **tensile** or **compressive** — the **same equations** apply. The only difference is that you tend to think of **tensile** forces as **positive**, and **compressive** forces as **negative**.

A Stress Big Enough to Break the Material is Called the Breaking Stress

As a greater and greater tensile **force** is applied to a material, the **stress** on it **increases**.

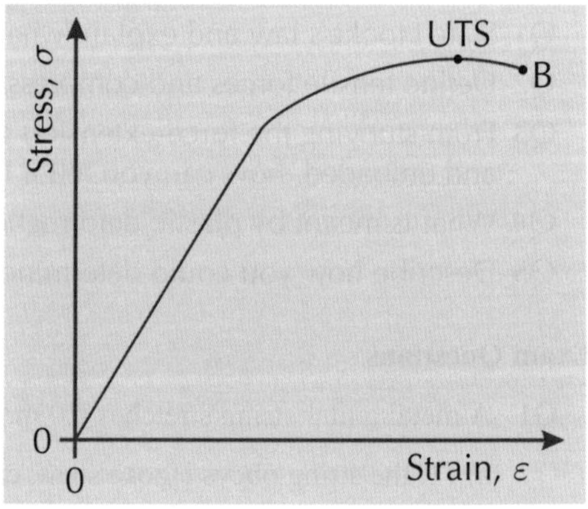

1) The effect of the **stress** is to start to **pull** the **atoms apart** from one another.
2) Eventually the stress becomes **so great** that atoms **separate completely**, and the **material breaks**. This is shown by point **B** on the graph. The stress at which this occurs is called the **breaking stress**.
3) The point marked **UTS** on the graph is called the **ultimate tensile stress** (or the **ultimate tensile strength**). This is the **maximum stress** that the material can withstand.
4) Both **UTS** and **B** depend on conditions like the **temperature** of the material.
5) **Engineers** have to consider the **UTS** and **breaking stress** of materials when designing a **structure** — e.g. they need to make sure the stress on a material won't reach the **UTS** when the **conditions change**.

Elastic Strain Energy is the Area under a Force-Extension Graph

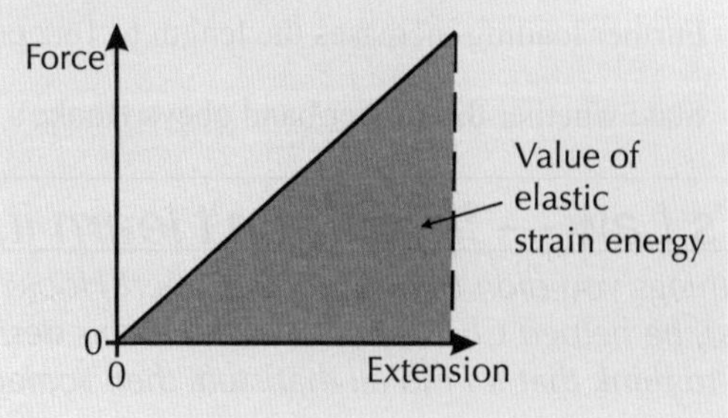

1) You saw on the previous page that the **work done** on stretching a material before it reaches its **elastic limit** is stored as **elastic strain energy** in the material.
2) On a **graph** of **force against extension**, the elastic strain energy is given by the **area under the graph**.
3) If the force-extension graph is **non-linear**, you'll need to **estimate** the area by **counting squares** or dividing the curve into **trapeziums** (see page 19).

Stress and Strain

You can Calculate the Energy Stored in a Stretched Material

Provided a material obeys Hooke's law, the **potential energy** stored inside it can be **calculated** quite easily.

1) The work done on the material in stretching it is equal to the energy stored.
2) **Work done** equals **force × displacement** (see p.48).
3) However, the **force** on the material **isn't constant**. It rises from zero up to force F. To calculate the **work done**, use the **average force** between zero and F, i.e. $\frac{1}{2}F$: → **work done = $\frac{1}{2}F\Delta x$**
4) So the **elastic strain energy** stored, $\Delta E_{elastic}$, is: $\Delta E_{elastic} = \frac{1}{2}F\Delta x$

This is the triangular area under the force-extension graph — see previous page.

5) Because Hooke's law is being obeyed, $\Delta F = k\Delta x$, which means F can be replaced in the equation to give: $\Delta E_{elastic} = \frac{1}{2}k\Delta x^2$

Example: A metal wire is 55.0 cm long. A force of 550 N is applied to the wire, and the wire stretches. The length of the stretched wire is 56.5 cm. Calculate the elastic strain energy stored in the wire.

The extension of the wire is Δx = 56.5 cm – 55.0 cm = 1.5 cm = 0.015 m

So the elastic strain energy $\Delta E_{elastic} = \frac{1}{2} \times F \times \Delta x$
$= \frac{1}{2} \times 550 \times 0.015 = 4.125$ J = **4.1 J (to 2 s.f.)**

Practice Questions

Q1 Write a definition for tensile stress.

Q2 Explain what is meant by the tensile strain on a material.

Q3 What is meant by the breaking stress of a material?

Q4 How can the elastic strain energy in a material under load be found from its force-extension graph?

Q5 The work done is usually calculated as force multiplied by displacement. Explain why the work done in stretching a wire is $\frac{1}{2}F\Delta x$.

Exam Questions

Q1 A steel wire is 2.00 m long. When a 300 N (to 3 s.f.) force is applied to the wire, it stretches by 4.0 mm. The wire has a circular cross-section with a diameter of 1.0 mm.
a) Calculate the tensile stress in the wire. [2 marks]
b) Calculate the tensile strain of the wire. [1 mark]

Q2 A copper wire (which obeys Hooke's law) is stretched by 3.0 mm when a force of 50.0 N is applied.
a) Calculate the stiffness constant for this wire in Nm^{-1}. [2 marks]
b) Calculate the value of the elastic strain energy in the stretched wire. [1 mark]

Q3 A force is applied to stretch an unknown material. The graph shows the force-extension graph for this stretch. Estimate the elastic strain energy stored in the material as a result of this stretch.

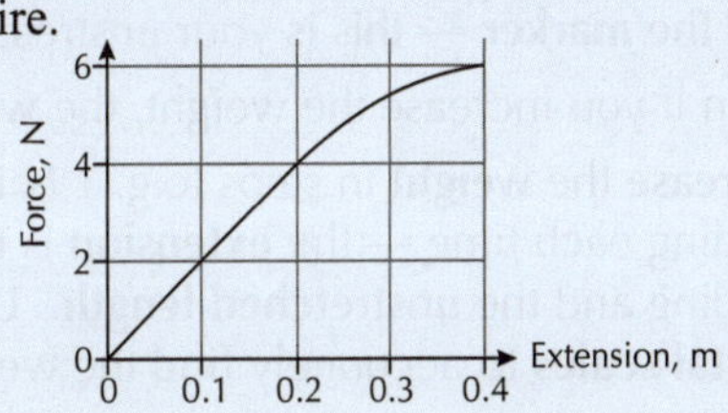

[2 marks]

Q4 A pinball machine contains a spring which is used to fire a small, light metal ball to start the game. The spring has a stiffness constant of 40.8 Nm^{-1}. It is compressed by 5.00 cm and then released to fire the ball. Calculate the maximum possible kinetic energy of the ball. [3 marks]

UTS a laugh a minute, this stuff...

...or it would be, if there were more jokes. But you know what they say — a stress causes a strain, and nobody likes strained jokes. So you'd better just get on with learning this stuff proper good like. And before you ask, no — grabbing a slingshot and shooting the neighbours' begonias doesn't count as 'revising the energy stored in a stretched material'.

The Young Modulus

Busy chap, Thomas Young. He did this work on tensile stress as something of a sideline. Light was his main thing. He proved that light behaved like a wave, explained how we see in colour and worked out what causes astigmatism.

The **Young Modulus** is Stress ÷ Strain

When you apply a **load** to stretch a material, it experiences a **stress** and a **strain**.

1) Up to a point called the **limit of proportionality** (see p.52), the stress and strain of a material are proportional to each other.
2) So below this limit, for a particular material, stress divided by strain is a constant. This constant is called the **Young modulus, *E*.**

$$\textbf{Young modulus} = E = \frac{\textbf{stress}}{\textbf{strain}}$$

The Young Modulus is a measurement of stiffness (p.58).

3) The **units** of the Young modulus are the same as stress (**Nm^{-2}** or pascals), since strain has no units.
4) The Young modulus is used by **engineers** to make sure their materials can withstand sufficient forces.

To **Find** the Young Modulus, You Need a **Very Long Wire**

This is the experiment you're most likely to do in class:

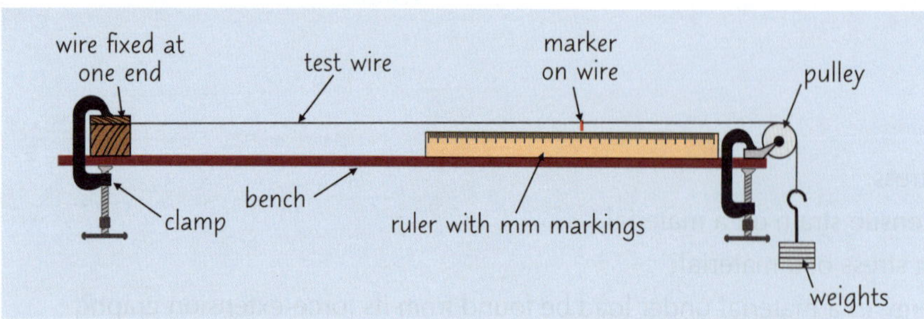

"Okay, found one. Now what?"

1) The test wire should be thin, and as long as possible. The **longer and thinner** the wire, the more it **extends** for the same force. This reduces the **uncertainty** (p.10) in your measurements.
2) First you need to find the **cross-sectional area** of the wire. Use a **micrometer** to measure the **diameter** of the wire **three times** in different places along the wire, before taking an **average** of your results. By assuming that the cross-section is **circular**, you can use the formula for the area of a circle: → **area of a circle = πr^2**
3) **Clamp** the wire to the bench (as shown in the diagram above) so you can hang **weights** off one end of it. Start with the **smallest weight** necessary to **straighten** the wire. (**Don't** include this weight in your final calculations.)
4) Measure the **distance** between the **fixed end of the wire** and the **marker** — this is your unstretched length.
5) Then if you increase the weight, the **wire stretches** and the **marker moves**.
6) **Increase** the **weight** in steps (e.g. 1 N intervals), recording the marker reading each time — the **extension** is the **difference** between this reading and the **unstretched length**. Use a **mass meter** or a set of **digital scales** to accurately find the weight you add at each step.
7) You can use your results from this experiment to calculate the **stress** and **strain** on the wire and plot a stress-strain graph (see next page).

If you're doing this experiment, make sure you're standing up so you can get out of the way quickly if the weights fall. And wear safety goggles — if the wire snaps, it could get very messy...

To reduce random errors you should use a thin marker on the wire, and always look from directly above the marker and ruler when measuring the extension.

As you unload the wire, re-measure the extension for each weight to make sure you haven't gone past the wire's elastic limit.

(The other standard way of measuring the Young modulus in the lab is using **Searle's apparatus**. This is a bit more accurate, but it's harder to do and the equipment's more complicated.)

You can also use this apparatus to find the **breaking stress** of a material. Do a preliminary experiment where you add weights to the wire to **roughly** find the force required to **break** the wire. **Repeat** the experiment with an **identical wire**, but this time add weights in small increments as the force applied approaches the force that previously broke the wire. This will help you find the breaking force (and breaking stress) more **accurately**. Then all you need to do is calculate the breaking stress from the breaking force.

The Young Modulus

Use a Stress-Strain Graph to Find E

You can plot a **graph** of **stress against strain** from your results.

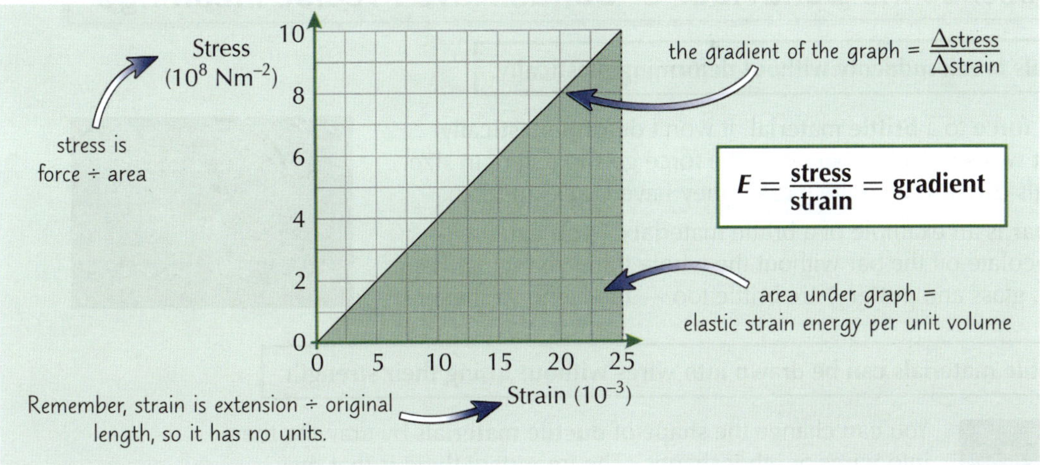

1) The **gradient** of the graph gives the Young modulus, ***E***.
2) The **area under the graph** gives the **strain energy** (or energy stored) per unit volume, i.e. the energy stored per 1 m³ of wire.
3) The stress-strain graph is a **straight line** provided that Hooke's law is obeyed, so you can also calculate the energy per unit volume (Jm^{-3}) as:

energy per unit vol = ½ × stress × strain

Example: The stress-strain graph above is for a thin metal wire. Find the Young modulus of the wire from the graph.

E = stress ÷ strain = gradient

The gradient of the graph = $\frac{\Delta\text{stress}}{\Delta\text{strain}} = \frac{10 \times 10^8}{25 \times 10^{-3}}$

$= \mathbf{4 \times 10^{10}\ Nm^{-2}}$

Practice Questions

Q1 Define the Young modulus for a material. What are the units for the Young modulus?

Q2 Describe an experiment to find the Young modulus of a test wire. Explain why a thin test wire should be used.

Q3 What is given by the area contained under a stress-strain graph?

Exam Questions

Q1 A steel wire is stretched elastically. For a load of 80.0 N, the wire extends by 3.6 mm. The original length of the wire was 2.50 m and its average diameter is 0.60 mm. Calculate the value of the Young modulus for steel. [4 marks]

Q2 Two wires, A and B, are stretched elastically under a load of 50.0 N. The original length and the extension of both wires under this load are the same. The Young modulus of wire A is found to be 7.0×10^{10} Nm^{-2}. The cross-sectional area of wire B is half that of wire A. Calculate the Young modulus of wire B. [2 marks]

Q3 The Young modulus for copper is 1.3×10^{11} Nm^{-2}.

a) The stress on a copper wire is 2.6×10^8 Nm^{-2}. Calculate the strain on the wire. [2 marks]

b) The load applied to the copper wire is 100 N (to 3 s.f.). Calculate the cross-sectional area of the wire. [2 marks]

c) Calculate the strain energy per unit volume for this loaded wire. [2 marks]

Learn that experiment — it's important...

Getting back to the good Dr Young... As if ground-breaking work in light, the physics of vision and materials science wasn't enough, he was also a well-respected physician, a linguist and an Egyptologist. He was one of the first to try to decipher the Rosetta stone (he didn't get it right, but nobody's perfect). Makes you feel kind of inferior, doesn't it...

Structures and Properties of Matter

You wouldn't try doing surgery with scalpels made out of marshmallows, just as you wouldn't make a glass crash mat. It's important to know the mechanical properties of materials so you can select the best ones to suit your needs.

Terms to Describe the **Behaviour of Solids** Have **Precise Meanings**

Brittle materials break suddenly without deforming plastically.

If you apply a **force** to a **brittle material**, it won't **deform plastically** (see p.52), but will suddenly **snap** when the force gets to a certain size. Brittle materials can also be quite **weak** if they have **cracks** in them.

A **chocolate bar** is an example of a brittle material — you can break chunks of chocolate off the bar without the whole thing changing shape. **Ceramics** (e.g. **glass** and **pottery**) are brittle too — they tend to shatter.

Ductile materials can be drawn into wires without losing their strength.

You can change the **shape** of **ductile materials** by drawing them into **wires** or other shapes. The important thing is that they **keep their strength** when they're deformed like this.

Copper is ductile, and with its high electrical conductivity this means that it's ideal for **electric wires**. A **ductile material** has been used for the cables supporting the **ski lift** in the photo — it's been drawn into long wires, but kept its strength.

Strong materials can withstand high stresses without deforming or breaking.

Strength is a measure of how much a material can **resist** being **deformed** (bent, stretched, fractured etc.) by a force without breaking. This can be resisting a **pulling** force (tensile strength) or a **squeezing** force (compressive strength).

Steel beams used to create structures like **bridges** are very strong — they withstand the force caused by lots of vehicles going over them without bending or breaking.

Hard materials are very resistant to cutting, indentation and abrasion.

If you try to cut, dent or scratch a hard material, you'll probably have very little effect. Their structure means **hard materials** are **resistant** to **cutting, indentation** (becoming dented) and **abrasion** (scratching).

Cutting tools (e.g. chisels) need to be harder than the stuff they're cutting — they're often made from **hardened steel**. **Diamond** is just about the hardest material there is — it's often used to reinforce the tips of drill bits.

Stiff materials have a high resistance to bending and stretching.

Changing the shape of **stiff materials** is really difficult as they are **resistant** to both **bending** and **stretching**. Stiffness is measured by the **Young modulus** (see p.56) — the higher the value, the stiffer the material.

The outer protective casing of **safety helmets** and **safety boots** need to be very stiff so that they keep their shape and don't **crush** onto your body when something impacts on them.

Tough materials are really difficult to break.

Toughness is a measure of the **energy** a material can **absorb** before it breaks. Really **tough materials** can absorb a lot of energy so are very **difficult** to **break**.

Some **polymers**, including certain types of **polythene**, are very tough. The hull of this **kayak** is made of a tough material so it won't break on rocks.

Structures and Properties of Matter

Metals — Positive Ions in a 'Sea' of Free Electrons

1) The atoms in a metal usually form a **crystalline** lattice — where the metal atoms are arranged in a **regular repeating pattern**. (They can also be **polycrystalline** — see below).
2) The outer electrons of the metal atoms don't need much energy to be able to desert their atoms in this crystalline structure. They form a 'sea' of **free electrons**, leaving behind a lattice of ions. It's these free electrons that make metals such **good conductors** of heat and electricity.
3) The electrostatic attraction between the ion lattice and the free electrons forms metallic bonds. It's these **strong** bonds that make metals **stiff** materials.
4) The strongly bonded lattice structure of a metal makes it **tough**. The ions within the lattice can **move** when you apply a force to the metal — making it **ductile**.
5) When a **force** is applied to the metal, the **interatomic spacing** between the ions increases. This increase is **uniform** during **elastic deformation**. Once the **stress** is high enough to cause **plastic deformation**, the **planes** ('sheets' of metal ions) within the metal **slip** over each other.
   If there is a **dislocation** (imperfection) in the metal, the **stress** needed to cause slipping is **lower** than the stress needed to cause slipping in a **perfect metal**.
6) **Atoms** of a **second metal can be placed inside dislocations** to **pin them down**. This **increases** the **stress** needed to cause **slipping**. This process is called **alloying** and makes the metal **harder** and less **ductile**.

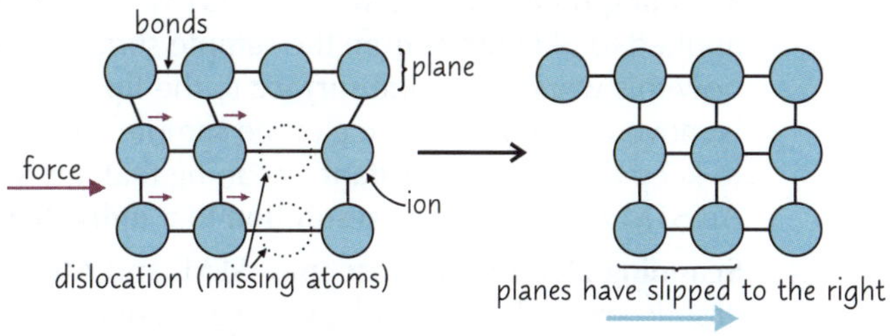

Ceramics — Giant Rigid Structures

1) Ceramics like **pottery**, **brick** and **glass** are made by melting certain materials, and then letting them cool.
2) The arrangement of atoms in a ceramic can be **crystalline** or **polycrystalline** — where there are many regions (or **grains**) of crystalline structure. The atoms in each grain line up in the same direction.
3) Some ceramics like **glass** are **amorphous** — there's no overall pattern; the atoms are arranged at **random**. The **quicker** a molten ceramic material is cooled, the more likely it is to be amorphous.

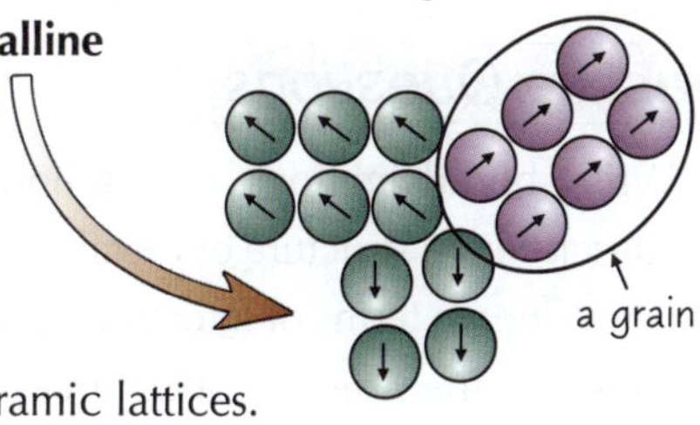

4) This **random** atomic bonding means that there are **no slip planes** in ceramic lattices. They also don't have **mobile dislocations** (dislocations which can **move**) — meaning that ceramic materials rarely deform **plastically** before they **fracture** (p.61).
5) However they're arranged, the atoms in a ceramic are either **ionically** or **covalently** bonded in a **giant rigid structure**. The **strong bonds** between the atoms make ceramics **stiff**, while the **rigid** structure means that ceramics are very **brittle** materials.
6) Ceramics being **brittle** means that **cracks spread** through them when they **fracture**. This is because the applied **force** acts on a **small area** (the **tip** of the crack) so the **stress** is high.

Polymers — Lots of Monomers Joined Together

1) A **polymer** is a molecular **chain**, made up of a **single repeating unit** called a **monomer**.
2) You get **natural** polymers like rubber, as well as a whole host of **man-made** ones like polythene.
3) The monomers in a polymer chain are **covalently** bonded together, and so are very hard to separate. This means even the thin polymer material used to make carrier bags is still pretty **strong**.
4) The **polymer chains** are often **entangled** (twisted and scrunched together), and can be **unravelled** by **rotating** about their bonds when you pull them. This is what makes polymer materials **flexible**. The more easily the monomers can **rotate**, the more the **chains** will **untangle** and the more **flexible** the polymer will be.
5) The strength and number of bonds **between** the chains also affect a polymer's flexibility. The stronger the cross-linking bonds, and the more cross-linking bonds you've got, the more **rigid** the material.

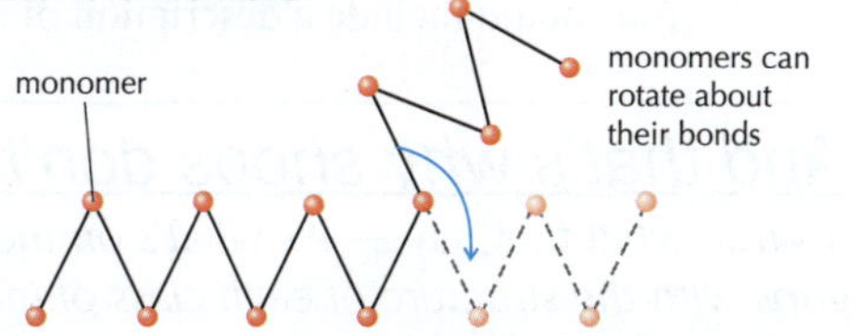

Structures and Properties of Matter

Rayleigh Estimated Atom Sizes with Oil Drops...

1) One of the first experiments used to calculate atomic size was **Rayleigh's oil drop experiment** in 1890.
2) Olive oil was released, one drop at a time, into a **tub of water** until it **just** covered the **entire surface**. Rayleigh assumed the oil would **spread** as much as it could, so this **thickness** would be the **size** of one **molecule of oil**.
3) By knowing the **surface area** of the tub and the **volume** of the oil dropped, the **thickness** of the oil film could be found. The **size of an atom** is roughly equal to the size of an **oil molecule** divided by the **number of atoms** in an oil molecule, so Rayleigh could use his measurements to find an **upper limit** for the size of the **individual atoms**.

...Now We Use Other Methods

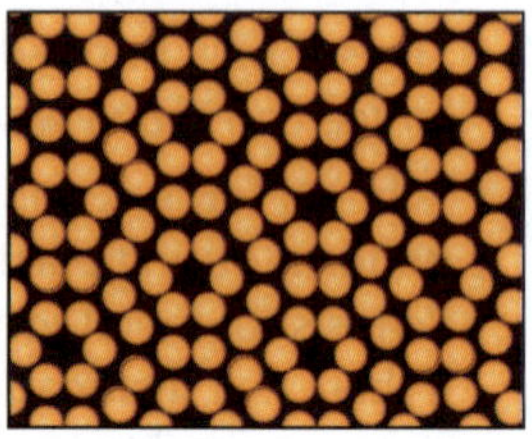

Scanning Tunnelling Microscopes can produce images of atoms.

1) There are loads of ways of measuring both the size of atoms and their spacing.
2) **X-ray crystallography** involves firing X-rays at a sample and using their **diffraction patterns** (see page 95) to investigate atomic spacing and structure.
3) **Scanning Tunnelling Microscopes** (**STM**) have a very fine tip which a voltage is applied to. **Electrons** from the sample surface **tunnel** (you don't need to know how this works) from the **surface** to the **tip** and cause a **current** to flow. The tip is moved across the surface of the sample, and the **height** of the tip is adjusted to keep the current **constant**, meaning that any small **bumps** or **dips** in the surface can be detected. An STM has such a **fine resolution** that **individual atoms** can be resolved and their **size** and **spacing** measured.
4) **Scanning Electron Microscopes** (**SEM**) and **Atomic Force Microscopes** (**AFM**) can also be used to measure atomic sizes. They don't let you 'see' a material's surface directly, but can be used to build up an **atom-by-atom** image of the surface on a **computer screen**. By knowing the **magnification** of the **image** on the computer screen and the size of the 'blobs' representing each atom, the sizes of the **atoms** can be calculated.
5) Modern techniques give the diameter of an atom as **0.1–0.5 nm**, depending on the atom being measured. This is much more **accurate** than the values Rayleigh **estimated** from his experiment.

Practice Questions

Q1 Write short definitions of the following terms: ductile, stiff, tough, strong.

Q2 Describe the structure of metals.

Q3 Describe, in terms of interatomic spacing, what happens to the ions in a metal as it undergoes elastic deformation.

Q4 Describe the structure of ceramics and the directions of atoms in a polycrystalline ceramic.

Q5 Explain why there are no slip planes in ceramic materials.

Q6 Why do cracks spread through ceramics when a force is applied to them?

Exam Questions

Q1 Stress is applied to a sample of iron which contains dislocations.
The sample begins to plastically deform at a lower stress than expected.

a) Explain what is happening to the ions in the iron as it plastically deforms and comment on why plastic deformation occurred at a lower stress than expected. [2 marks]

b) Describe alloying and what effect it would have on the sample. [4 marks]

Q2 Select the statement that does **not** correctly describe polymers:

A Monomers are covalently bonded together to create a polymer.
B Polymer chains are tangled together.
C The more cross-linking bonds there are, the more flexible a polymer is.
D Polymers are unravelled when you pull on them, causing the monomers to rotate. [1 mark]

Q3 Describe how our knowledge of atomic size and spacing have changed over the past 150 years.
You should include a description of at least two methods of measuring atomic size in your answer. [6 marks]

And that's why shops don't make their bags out of clay...

It's true what they say — it's what's on the inside that counts... and it's no different for bricks. Make sure you get to grips with the structure of each class of materials so you can explain why materials have certain properties.

Stress-Strain and Force-Extension Graphs

I hope the stresses and strains of this section aren't getting to you too much.
Don't worry, though — there's just these three pages to go before you're on to the electrifying world of electricity.

There are **Three Important Points** on a **Stress-Strain Graph**

In the exam you could be given a **stress-strain graph** and asked to **interpret** it. Luckily, most stress-strain graphs share **three** important points — as shown in the **diagram**.

Point **Y** is the **yield point** — here the material suddenly starts to **stretch** without any extra load. The **yield point** (or yield stress) is the **stress** at which a large amount of **plastic deformation** takes place with a **constant** or **reduced load**.

Point **E** is the **elastic limit** — at this point the material starts to behave **plastically**. From point E onwards, the material would **no longer** return to its **original shape** once the stress was removed.

Point **P** is the **limit of proportionality** — after this, the graph is no longer a straight line but starts to **bend**. At this point, the material **stops** obeying **Hooke's law**, but would still **return** to its **original shape** if the stress was removed.

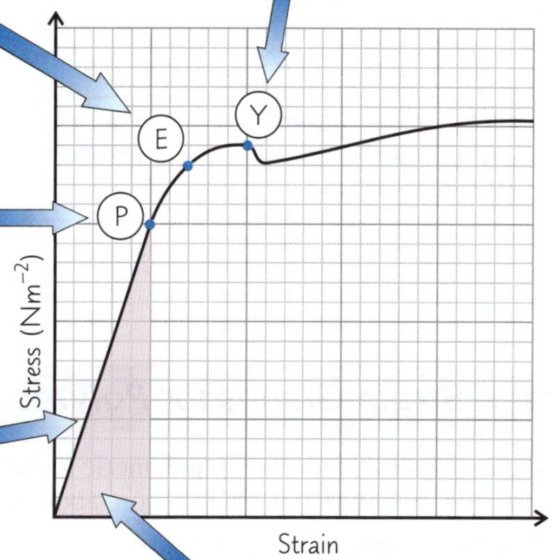

Before point **P**, the graph is a **straight line** through the **origin**. This shows that the material is obeying **Hooke's law** (page 52). The **gradient** of the line is constant — it's the **Young modulus** (see pages 56-57).

The **area** under the first part of the graph gives the **energy stored** in the **material per unit volume** (see page 57).

Stress-Strain Graphs for **Brittle** Materials **Don't Curve**

The graph shown below is typical of a **brittle** material.

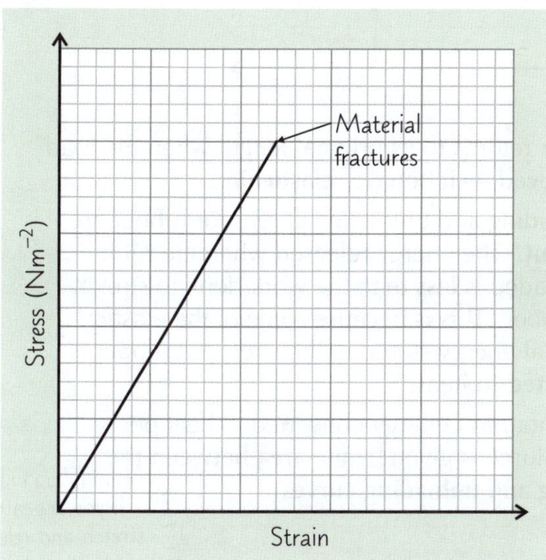

1) The graph starts the same as the one above — with a **straight line through the origin**. So brittle materials also obey **Hooke's law**.
2) However, when the **stress** reaches a certain point, the material suddenly **fractures** (breaks) — it **doesn't deform plastically**.
3) **Brittle materials** are described on p.58.

Melissa had no idea her brother was planning on testing what he had learnt about brittle materials.

Stress-Strain and Force-Extension Graphs

Stronger *Materials can Withstand* ***More Stress*** *Before they* ***Break***

The graph below shows the stress-strain curves for materials of different strengths and stiffnesses.

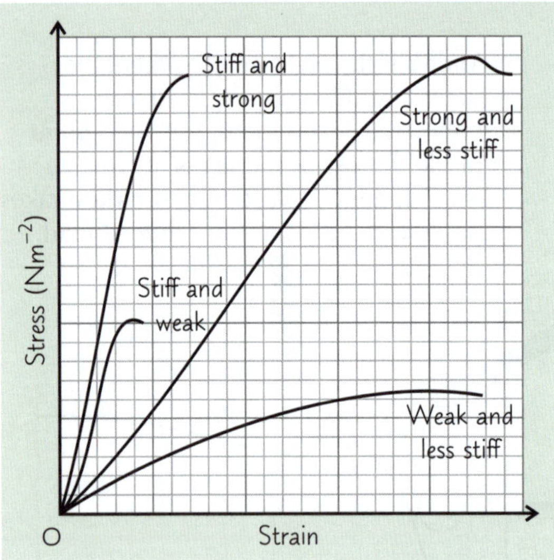

1) Different materials have different **breaking stresses** (p.54).
2) The **stronger** the material, the higher the breaking stress.
3) **Stiff** materials are difficult to stretch or compress. They have a **large** Young modulus.
4) For a given stress, a stiff material will have a lower strain (i.e. a smaller extension) than a less stiff material.
5) A stiff material **doesn't** have to be strong (and vice versa). Some stiff materials **break** under a low stress, and some strong materials **aren't** very stiff.

When a line on a stress-strain graph just stops, you can assume the material has reached its breaking stress and fractured (unless the question says otherwise).

Steve took a break after a stressful day at the office.

Stress-Strain *Graphs for* ***Polymeric*** *Materials* ***Vary***

1) The **molecules** that make up **polymeric** (or polymer) **materials** are arranged in **long chains** (there's more about these chains on page 59).
2) They have a **range** of properties, so different polymers have different **stress-strain graphs**.
3) The examples show the stress-strain graphs for two polymers — **polythene** and **rubber**.

Example: This is the stress-strain graph for **polythene**.

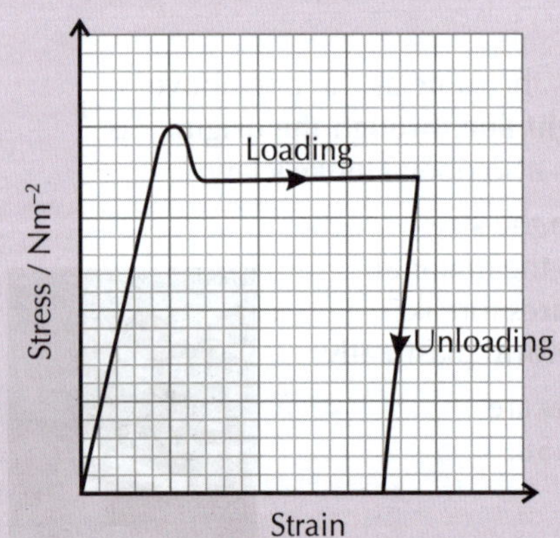

1) Polythene behaves **plastically** — applying a stress to it stretches it into a new shape.
2) Polythene is a **ductile** material.

Example: This is the stress-strain graph for **rubber**.

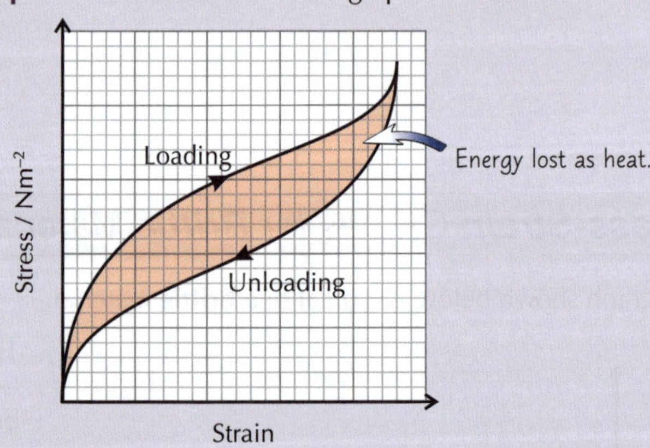

1) **Rubber** returns to its original length when the load is removed — it behaves **elastically**.
2) The loading and unloading curves for rubber are **different**. The energy released when the rubber is unloaded is **less** than the work done to stretch the rubber. This is because some of the elastic potential energy stored in the stretched rubber is **converted to heat**.
3) The amount of energy converted to heat per unit volume is given by the **area between the loading and unloading curves**.

If you repeatedly stretch and release a rubber band, it gets hotter. Give it a go.

Stress-Strain and Force-Extension Graphs

Force-Extension *Graphs Are Similar to* ***Stress-Strain*** *Graphs*

1) Force-extension graphs look a lot like **stress-strain** graphs, but they show slightly different things.
2) Force-extension graphs are **specific** for the tested **object** and **depend on its dimensions**. Stress-strain graphs describe the **general behaviour** of a **material**, they are **independent of the dimensions**.
3) You can plot a force-extension graph of what happens when you gradually **remove** a **force** from an object. The **unloading** line doesn't always match up with the **loading** line though.

You first met force-extension graphs on page 52.

A force-extension graph for a metal wire

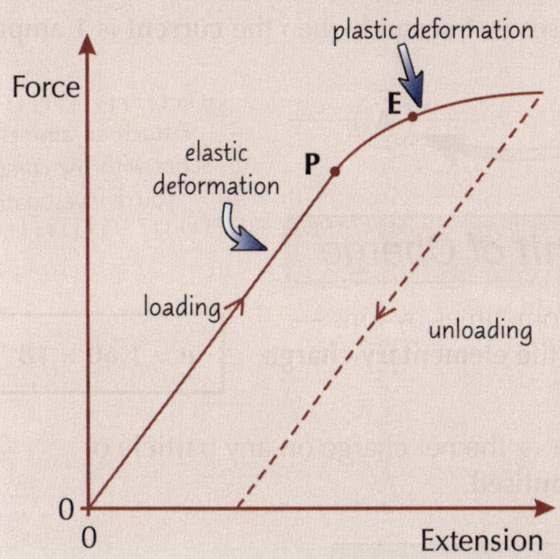

1) This graph is for a metal wire that has been stretched beyond its **limit of proportionality (P)** so the graph starts to **curve**.
2) When the load is **removed**, the **extension decreases**.
3) The unloading line is **parallel** to the loading line because the stiffness constant ***k*** is still the same (since the forces between the atoms are the same as they were during the loading).
4) But because the wire was stretched beyond its **elastic limit (E)** and deformed **plastically**, it has been **permanently stretched**. This means the unloading line doesn't go through the origin.
5) The **area** between the two lines is the **work done** to permanently deform the wire.

Of course, if you apply a big enough load to fracture the object, you can't draw an unloading line — you just get a force-extension graph like one of these graphs.

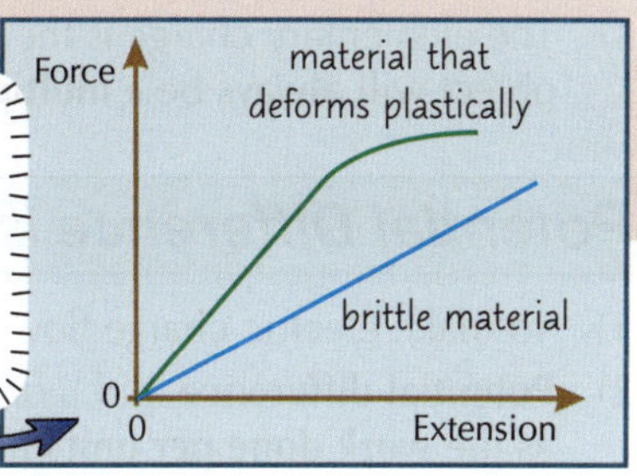

Practice Questions

Q1 What is the difference between the limit of proportionality and the elastic limit?

Q2 Describe what happens at the yield point.

Q3 What are polymeric materials?

Q4 A metal wire is stretched beyond its elastic limit. Why does the unloading line on the force-extension graph for the wire not go through the origin?

Exam Questions

Q1 Sketch a stress-strain graph to show (use the same axes):

a) A stiff and weak material being stretched to its breaking point. [1 mark]

b) A strong and less stiff material being stretched to its breaking point. [1 mark]

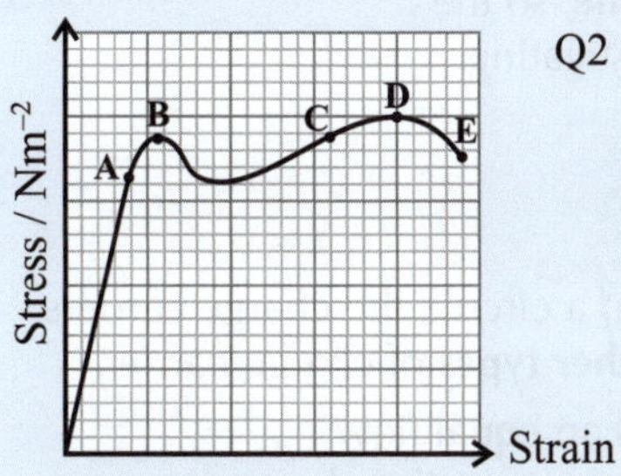

Q2 The graph on the left shows a stress-strain curve for mild steel.

a) State which letter, A-E, corresponds to the steel's limit of proportionality. [1 mark]

b) State which letter, A-E, corresponds to the steel's yield point. [1 mark]

My sister must be brittle — she's always snapping...

In case you were wondering, I haven't just drawn the graphs on these three pages for fun (though I did enjoy myself) — they're there for you to learn. I find the best way to remember each one is to understand why it has the shape it does — if that sounds too much like hard work, then at least make sure you can describe the shape of all of them.

Charge, Current and Potential Difference

Electricity's brilliant, I love it. It's what gets me out of bed in the morning. (Not literally of course, that'd be quite painful.)

Current is the *Rate* of *Flow* of *Charge*

1) The **current** in a **wire** is like **water** flowing in a **pipe**. The **amount** of water that flows depends on the **flow rate** and the **time**. It's the same with electricity — **current is the rate of flow of charge.**

$$I = \frac{\Delta Q}{\Delta t}$$ where I is the current in amperes, ΔQ is the charge in coulombs, and Δt is the time taken in seconds

Remember that conventional current flows from + to –, the opposite way from electron flow.

2) The **coulomb** is the **unit of charge**.

One coulomb (C) is defined as the **amount of charge** that passes in **1 second** when the **current** is **1 ampere**.

3) You can measure the current flowing through part of a circuit using an **ammeter**. This is the circuit symbol for an ammeter:

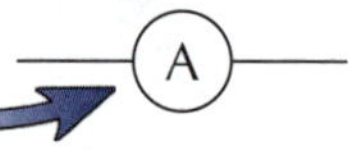

Attach an ammeter in series with the component you're investigating.

The Charge on an Electron *is the* Smallest Unit *of Charge*

1) In electrical circuits, charge is usually carried by **electrons** (or sometimes by ions — see below). Electrons all carry the **same charge**, **–e**, where **e** is the **elementary charge**:

e = 1.60 × 10^{-19} C

2) Protons carry an **opposite charge** of the **same magnitude**, **+e**.
3) The elementary charge is the **smallest unit** that charge comes in — the net charge on any particle or object will **always** be a **multiple of e**. We say that charge is **quantised**.

Potential Difference is the *Work Done* per *Unit Charge*

1) To make electric charge flow through a conductor, you need to do **work** on it.
2) **Potential difference** (p.d.), or **voltage**, is defined as the **work done per unit charge moved**:

$$V = \frac{W}{Q}$$ W is the work done in joules (see p.48). It's the energy transferred in moving the charge.

The **potential difference** across a component is **1 volt** (V) when you do **1 joule** of work moving **1 coulomb** of charge through the component. This **defines** the volt.

1 V = 1 J C^{-1}

Back to the 'water analogy' again. The p.d. is like the pressure that's forcing water along the pipe.

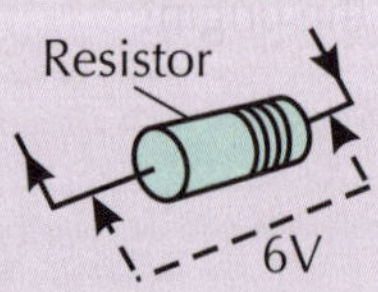

Here you do 6 J of work moving each coulomb of charge through the resistor, so the p.d. across it is 6 V. The energy gets converted to heat.

3) You can measure the potential difference across a component using a **voltmeter**. This is the circuit symbol for a voltmeter:

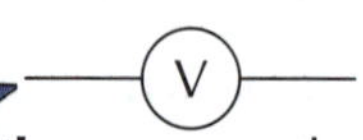

4) Remember, the potential difference across components in parallel is **the same**, so the **voltmeter** should be connected in **parallel** with the component you're investigating.

The maximum value that a voltmeter or ammeter can measure is called the full scale deflection.

Charge Carriers in Liquids and Gases are *Ions*

1) A current is the **rate of flow of charged particles** (called **charge carriers**). In a circuit, the charge carriers are free electrons (sometimes called **conduction electrons**), but there are **other types** of charge carrier.
2) A flow of **positively-charged** particles produces **exactly** the **same current** as an **equal** flow of negatively-charged particles in the **opposite direction**. This is why we use **conventional current**, defined as 'in the same direction as a flow of positive charges'.
3) **Ionic crystals** like sodium chloride are **insulators**. Once **molten**, though, the liquid **conducts**. Positive and negative **ions** are the **charge carriers**. The **same thing** happens in an **ionic solution** like copper sulfate solution.
4) **Gases** are **insulators**, but if you apply a **high enough voltage** electrons get **ripped out** of **atoms**, giving you **ions** along a path. You get a **spark**.

Charge, Current and Potential Difference

Work Done on a Charge Equals the Kinetic Energy it Gains

1) When a **charged particle** is accelerated by a **potential difference**, the energy transferred to it is equal to the work done on the particle, $W = VQ$ (see previous page). For an electron (with charge of size e), this can be written as $W = Ve$.
2) The energy transferred is equal to the **kinetic energy** gained by the electron.
3) **Kinetic energy** = $\frac{1}{2}mv^2$ (see page 50), so:

$$eV = \frac{1}{2}mv^2$$

where m is the electron's mass and v is its velocity

This gives you the velocity of a single electron accelerated through a potential difference — don't get it confused with mean drift velocity (coming up below).

The Mean Drift Velocity is the Average Velocity of the Charge Carriers

When **current** flows through a wire, you might imagine the **electrons** all moving uniformly in the **same direction**. In fact, they move **randomly** in **all directions**, but tend to **drift** one way. The **mean drift velocity** is just the **average velocity** and it's **much, much less** than the electrons' **actual speed**. (Their actual speed is about 10^6 ms^{-1}.)

The Current Depends on the Mean Drift Velocity:

The **current** is given by the equation:

$$I = nqvA$$

where: I = electrical current (A)
n = number density of charge carriers (m^{-3}) (number per unit volume)
q = charge on each charge carrier (C) (for electrons, $q = -e = -1.60 \times 10^{-19}$ C.)
v = mean drift velocity (ms^{-1})
A = cross-sectional area (m^2)

So...

- If you double the **number of charge carriers**, the **current doubles**.
- If you **double the area**, the **current doubles**, like this:

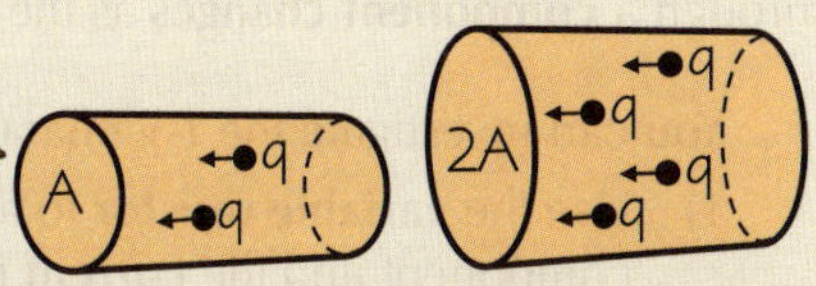

- If the carriers move **twice as fast** you get twice the charge in the same time — **twice the current**.
- If you **double the charge** of each carrier, you get twice the charge in the same time — **twice the current**.

Practice Questions

Q1 Describe in words and symbols how current and charge are related.

Q2 Explain what is meant by the sentence "charge is quantised".

Q3 Define potential difference.

Q4 Write an equation for the velocity of an electron accelerated by a p.d. of V, in terms of its charge and mass.

Q5 What happens to the current in a wire if the mean drift velocity of the electrons is halved?

Exam Questions

Q1 A battery delivers 4500 C of electric charge to a circuit in 10 minutes. Calculate the average current. [1 mark]

Q2 A kettle runs off the mains supply (230 V) and has an overall efficiency of 88%. Calculate how much electric charge will pass through the kettle if it transfers 308 J of energy to the water it contains. [2 marks]

Q3 Copper has 1.0×10^{29} free electrons per m^3. Calculate the mean drift velocity of the electrons in a copper wire of cross-sectional area 5.0×10^{-6} m^2 when it is carrying a current of 13 A. [2 marks]

I can't even be bothered to make the current joke...

Talking of currant jokes, I saw this bottle of wine the other day called 'raisin d'être' — 'raison d'être' meaning 'reason for living', but spelled slightly differently to make 'raisin', meaning 'grape'. Ho ho. Chuckled all the way home.

Resistance and Conductance

Resistance is what causes components to heat up. It's how your toaster works, and why computers need cooling fans...

Everything has Resistance

1) If you put a **potential difference** (p.d.) across an **electrical component**, a **current** will flow.
2) **How much** current you get for a particular **p.d.** depends on the **resistance** of the component.
3) You can think of a component's **resistance** as a **measure** of how **difficult** it is to get a **current** to **flow** through it.
4) **Resistance** is measured in **ohms** (Ω). A component has a resistance of 1 Ω if a potential difference of 1 V makes a current of 1 A flow through it.

Mathematically, **resistance** is: $R = \frac{V}{I}$
This equation **defines** resistance.

You also need to know the formula for the **inverse** of resistance — **conductance**, ***G***. ⇒ $G = \frac{I}{V}$

This is a measure of how good an electrical conductor a component is. It's measured in Ω^{-1} or siemens, S.

5) The equation for the power of a circuit is given by $P = VI$ (you'll see this on p.70). The resistance equation can be substituted into this to give you a whole new way of calculating electrical power. ⇒ $V = IR$, so $P = VI = I^2R$
6) This power is the rate at which a component converts electrical energy into other types of energy, e.g. heat. This is known as **power dissipation**. This can be useful, e.g. the dissipation of power as light from a bulb. In other situations, power dissipation causes problems we need to work round. E.g. computers have cooling fans to get rid of some of the heat that builds up in their circuits, and mains electricity is transmitted at a high voltage (so low current) to minimise the power dissipated during transmission.

I-V Graphs Show How Resistance Varies

The term '***I-V* characteristic**' refers to a **graph** which shows how the **current** (*I*) flowing through a **component changes** as the **potential difference** (*V*) across it is increased.

You can investigate the *I-V* characteristic of a component using a **test circuit** like this one:

1) Use the **variable resistor** to alter the **potential difference** across the component and the **current** flowing through it, and record *V* and *I*.
2) **Repeat** your measurements and take **averages** to reduce the effect of random errors (see p.12) on your results.
3) **Plot a graph** of current against potential difference from your results. This graph is the ***I-V* characteristic** of the component and you can use it to see how the **resistance** changes.

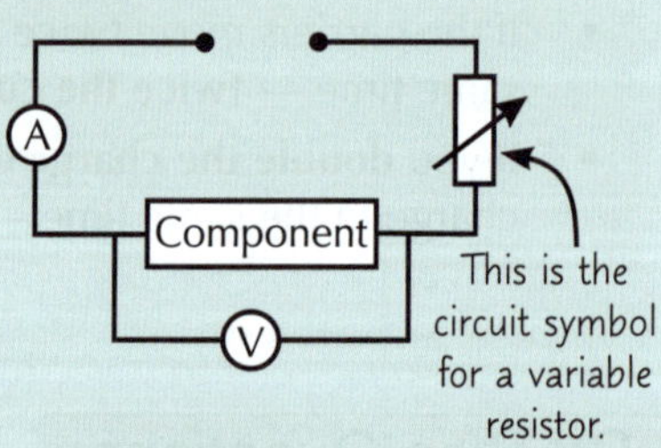

If you have access to a computer, you could enter your data into a spreadsheet and use this to plot the graph.

For an Ohmic Conductor, *R* is a Constant

Metal wires and resistors are ohmic conductors

Conductors that **obey** Ohm's law (mostly metals) are called **ohmic conductors**. Ohm's law states that:

Provided external factors such as **temperature** are **constant**, the **current** through an ohmic conductor is **directly proportional** to the **potential difference** across it (that's $V = IR$).

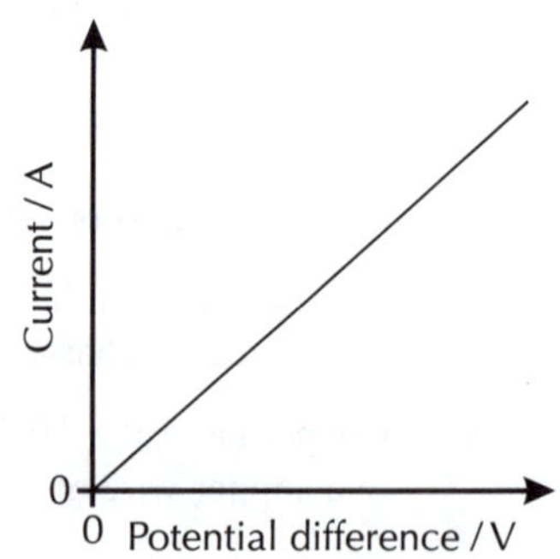

1) As you can see from the graph, **doubling** the **p.d. doubles** the **current**.
2) The gradient is **constant**, which means that **resistance** is **constant**.
3) The **shallower** the **gradient** of the characteristic ***I-V*** graph, the **greater** the **resistance** of the component.
4) Remember, Ohm's law is **only** true for **ohmic conductors** where external factors like **temperature** are **constant**.
5) **Non-ohmic conductors** don't have this relation between current and p.d. There are **examples** of non-ohmic conductors on the next page.

Resistance and Conductance

The I-V Characteristic for a Filament Lamp is Curved

Filament lamp circuit symbol:

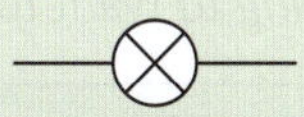

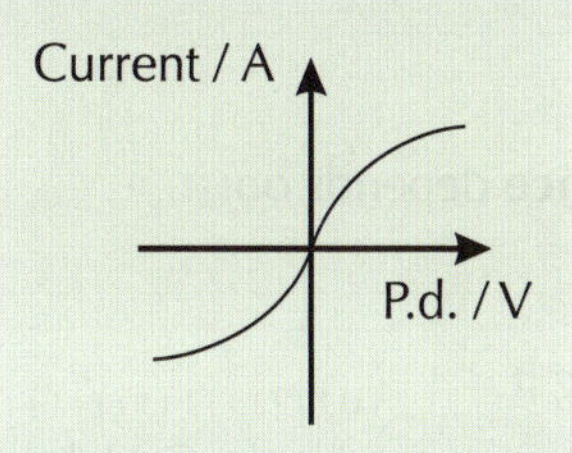

1) The characteristic graph for a **filament lamp** is a **curve**, which starts **steep** but gets **shallower** as the **potential difference rises.**
2) The **filament** in a lamp is just a **coiled up** length of **metal wire**, so you might think it should have the **same characteristic graph** as a **metallic conductor**.
3) However, **current** flowing through the lamp **increases** its **temperature**, so its **resistance increases** (see page 69).

The Resistance of a Thermistor Depends on Temperature

Thermistor circuit symbol:

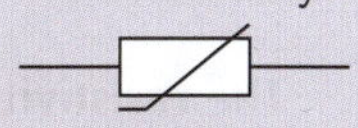

A **thermistor** is a **resistor** with a **resistance** that depends on its **temperature**, so you can use them as **temperature sensors**.

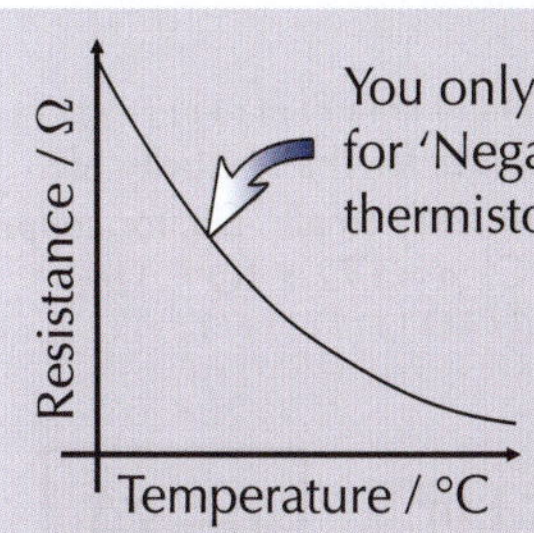

You only need to know about **NTC thermistors** — NTC stands for 'Negative Temperature Coefficient'. The resistance of an NTC thermistor **decreases** with **temperature** (so its conductance increases).

The characteristic ***I-V* graph** for an NTC thermistor looks like this: As the voltage **increases**, the current **increases**. More current leads to an **increase in temperature** and so a **decrease in resistance**. This in turn means more current can flow, so the graph **curves upwards.**

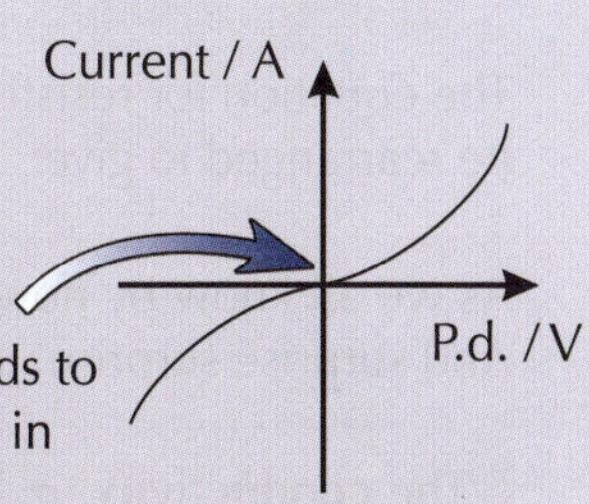

An **LDR** (**light dependent resistor**), is similar to a thermistor, but is sensitive to **light**, not heat — the more light falls on it, the lower its resistance.

Light dependent resistor symbol:

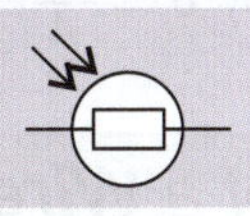

Filament lamps, thermistors, LDRs and diodes are all examples of non-ohmic conductors. You can tell by looking at their *I-V* graphs.

Diodes Only Let Current Flow in One Direction

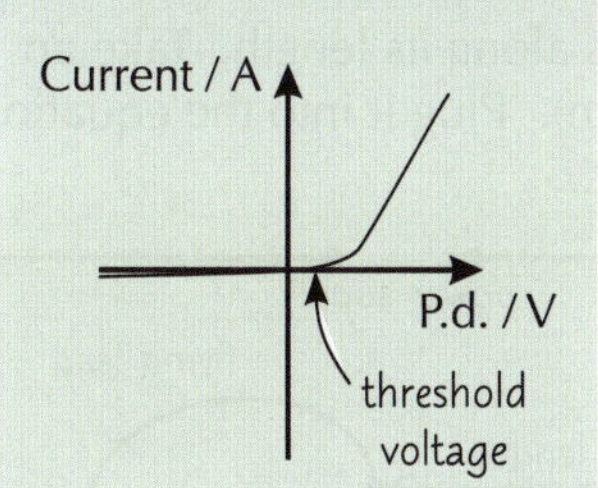

Diodes (including light emitting diodes (LEDs)) are designed to let **current flow** in **one direction** only. You don't need to be able to explain how they work, just what they do.

1) **Forward bias** is the **direction** in which the **current** is **allowed to flow.**
2) **Most** diodes require a **threshold voltage** of about **0.6 V** in the **forward direction** before they will conduct.
3) In **reverse bias**, the **resistance** of the diode is **very high** and the current that flows is **very tiny**.

Diode circuit symbol:

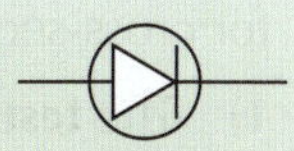

LED circuit symbol:

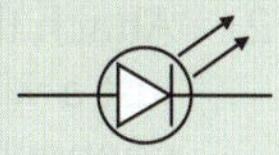

Practice Questions

Q1 State the formulas for resistance and conductance.

Q2 Sketch the test circuit used to investigate the *I-V* characteristic of a component, and explain how it is used.

Q3 State Ohm's law. Give an example of an ohmic conductor.

Q4 Draw an *I-V* characteristic graph for a diode. Label the areas of forward bias and reverse bias.

Exam Question

Q1 a) A current of 2.8 A is passed through a wire from a 1.5 V power supply. Find the resistance of the wire. [1 mark]

b) When the same power supply is connected to a different wire, the current is 0.15 A. Calculate the conductance of this wire. [1 mark]

c) Explain why the wires become warm after a period of operation. [2 marks]

You light up my world like an LED — with One-Directional current...

The examiners like testing this kind of stuff as if you've really done the experiment, so your results aren't perfect and you have fun errors to deal with. If errors aren't your thing, have a quick flick to page 12 to calm your nerves.

Electrical Properties of Solids

From a remote-controlled car to a supercomputer... if what you're building involves electricity, you're going to want to know about resistivity and conductivity...

Three Things Determine Resistance

If you think about a nice, **simple electrical component**, like a **length of wire**, its **resistance** depends on:

1) **Length (L).** The **longer** the wire, the **more difficult** it is to make a **current flow**.
2) **Area (A).** The **wider** the wire, the **easier** it will be for the electrons to pass along it.
3) **Resistivity (ρ).** This **depends** on the **material** the wire's made from, as the **structure** of the material may make it easy or difficult for charge to flow. In general, resistivity depends on **external factors** as well, like **temperature**.

ρ is the Greek letter rho, the symbol for resistivity.

The **resistivity** of a material is defined as the **resistance** of a **1 m length** with a **1 m²** **cross-sectional area** — it's given by $\rho = \frac{RA}{L}$. Resistivity is measured in **ohm metres** (Ωm).

The equation for resistivity can be rearranged to give:

$$R = \frac{\rho L}{A}$$

where R = resistance in Ω, A = cross-sectional area in m^2, and L = length in m

Typical values for the resistivity of conductors are really small. E.g. for copper (at 25 °C) $\rho = 1.72 \times 10^{-8}$ Ωm.

As we all know by now, conductance is the inverse of resistance (p.66). And surprise surprise... the inverse of resistivity is **conductivity**, σ.

The **conductivity**, σ, of a material is defined as the **conductance** of a **1 m length** with a **1 m²** **cross-sectional area** — it's given by $\sigma = \frac{GL}{A}$. It's measured in **siemens per metre** (S m^{-1}).

$$G = \frac{\sigma A}{L}$$

To Find the Resistivity of a Wire you Need to Find its Resistance

You'll need to **measure** the test wire's cross-sectional area before you start. Assume that the wire is **cylindrical**, and so its cross-section is **circular**. Then you can find the cross-sectional area using: **area of a circle = πr^2**.

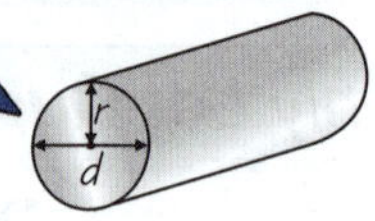

Using a **micrometer**, measure the **diameter** of the wire for at least **three** different points along its length. Take an **average** value of the diameter and divide by **two** to get the **radius** (make sure this is in m). Plug it into the equation for cross-sectional area and... **ta da**. Now you can get your teeth into the electricity bit...

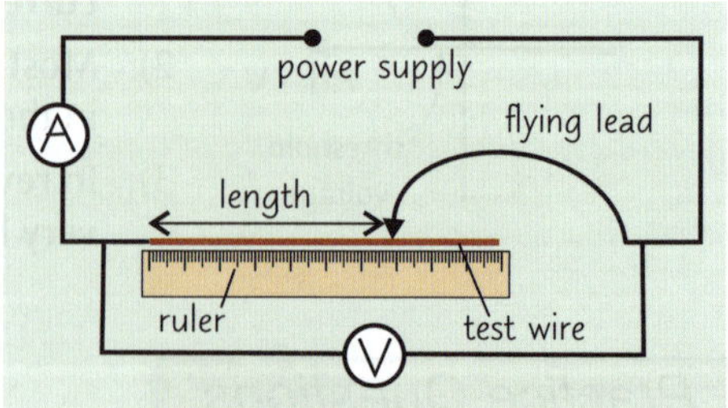

1) The **test wire** should be **clamped** to a ruler with the circuit attached to the wire where the ruler reads zero.
2) Attach the **flying lead** to the test wire — the lead is just a wire with a crocodile clip at the end to allow connection to any point along the test wire.
3) Record the **length** of the test wire **connected** in the circuit, the **voltmeter reading** and the **ammeter reading**.
4) Use your readings to calculate the **resistance** of the length of wire, using: $R = \frac{V}{I}$
5) Repeat this measurement and calculate an average resistance for the length.
6) Repeat for several **different** lengths, for example between 0.10 and 1.00 m.
7) Plot your results on a graph of R against L, and draw a **line of best fit**.

The **gradient** of the line of best fit is equal to $\frac{R}{L} = \frac{\rho}{A}$. So **multiply** the **gradient** of the line of best fit by the **cross-sectional area** of the wire to find the resistivity of the wire material.

You could find ρ for a single length using the resistivity formula, but this will give you an overestimate as it'll include the resistance of the ammeter and all the wires and connections.

8) The other components of the circuit also have a resistance, but the gradient of the line of best fit isn't affected by the resistance within the rest of the circuit.
9) The **resistivity** of a material depends on its **temperature**, so you can only find the resistivity of a material **at a certain temperature**. Current flowing in the test wire can cause its temperature to increase, so you need to try to keep the temperature of the test wire **constant**, e.g. by only having small currents flow through the wire.
10) To find the conductivity of a wire instead, do the same experiment. As $\sigma = 1/\rho$, you can use your final result to calculate the metal's conductivity if that's what you need.

Electrical Properties of Solids

Different Materials have Different Numbers of Charge Carriers

How conductive a material is depends on its **number density of mobile charge carriers** — the number of free electrons (or ions that are free to move) there are per cubic metre of the material. The more **mobile charge carriers** a material has per unit volume, the **better** a **conductor** it will be.

Metals

In a **metal**, the **charge carriers** are **free electrons** (see p.64). Metals are **good conductors** because they have absolutely shedloads of them — the **number density of mobile charge carriers** is **high**.

If you **increase** the **temperature** of a metal, the **number** of mobile charge carriers **stays about the same**. As the electrons move, they scatter from the metallic lattice. As the temperature increases, the lattice **vibrates** more, increasing the electron scattering, so the electrons are slightly less free to move. This means that as the **temperature increases**, the **conductivity** of a metal will slightly **decrease**.

Semiconductors

Semiconductors have a much **lower** charge carrier number density (fewer free electrons) than metals, so they have a **lower conductivity**.

As you **increase** the **temperature** of a semiconductor, more electrons are freed to conduct. This means that as the temperature **increases**, the **conductivity** of a semiconductor **rapidly increases**.

Thermistors are made up of semiconductors whose conductivity changes with **temperature**, as described above. **Light dependent resistors** (LDRs) are made up of semiconductors whose conductivity is mostly controlled by **light** rather than heat — their conductivity **increases** with increasing light levels.

Insulators

A **perfect insulator** wouldn't have **any mobile charge carriers**, so it wouldn't be able to conduct at all.

Superconductors Have Zero Resistivity

1) If you **cool** some materials (e.g. mercury) down to below a '**transition temperature**', their **resistivity disappears entirely** and they become a **superconductor**.
2) Without any resistance, **none** of the electrical energy is turned into heat, so **none** of it's wasted. That means you can start a current flowing in a circuit using a magnetic field, take away the magnet and the current would carry on flowing **forever**. Neat.
3) There's a catch, though. Most 'normal' conductors, e.g. metals, have transition temperatures below **10 kelvin (–263 °C)**. Getting things that cold is **hard**, and **really expensive**.
4) Superconducting wires could be used to make **power cables** that transmit electricity without any **loss** of power, really **strong electromagnets** that **don't** need a constant power source or **electronic circuits** that work really **fast**, because there's no resistance to slow them down.

Practice Questions

Q1 What three factors does the resistance of a length of wire depend on?

Q2 Write down the units of resistivity and conductivity.

Q3 Why are semiconductors poorer conductors than metals at lower temperatures?

Q4 Explain why insulators do not conduct electricity.

Q5 What is a superconductor? Describe how mercury can be made to be a superconductor.

Exam Question

Q1 a) Describe the equipment and method you would use to measure the resistivity of copper using a copper wire. You should include a labelled circuit diagram as part of your answer. [6 marks]

b) Explain why the conductivity of a copper wire differs from the conductivity of a NTC thermistor. [2 marks]

Insulator Airlines — the no-charge carriers...

That resistivity experiment's a popular one to come up in exams — so make sure you learn it. Try to think about where errors are creeping into your measurements too, and how you might be able to reduce them... see p.12 for ideas...

Power, Work and Electrical Energy

Power and energy are pretty familiar concepts — and here they are again. Same principles, just different equations.

Power** is the **Rate** of **Transfer** of **Energy

Power (*P*) is **defined** as the **rate** of **transfer** of **energy**. It's measured in **watts (W)**, where **1 watt** is equivalent to **1 joule per second**.

in symbols: $P = \frac{E}{t}$

You might sometimes see power defined as the rate of doing work. Work done, *W*, is the same as the transfer of energy.

There's a really simple formula for **power** in **electrical circuits**:

$$P = VI$$

This makes sense, since:

1) **Potential difference** (*V*) is defined as the **energy transferred** per **coulomb** (or the **work done** per **unit charge**, see p.64).
2) **Current** (*I*) is defined as the **number** of **coulombs** transferred per **second**.
3) So **p.d.** × **current** is **energy transferred per second**, i.e. **power**.

Arnold had a pretty high resistance to doing work.

You also know (from the definition of **resistance**) that $V = IR$ (see p.66). **Combining** this with the equation above gives you loads of **different ways** to **calculate power**.

$$P = VI \qquad P = \frac{V^2}{R} \qquad P = I^2R$$

Obviously, which equation you should use depends on what **quantities** you're given in the **question**.

Example: A robotic mutant Santa from the future converts 750 J of electrical energy into heat every second.
a) What is the operating power of the robotic mutant Santa?
b) All of the robotic mutant Santa's components are connected in series, with a total resistance of 30 Ω. What current flows through his wire veins?

a) Power $= \frac{E}{t} = \frac{750}{1} = \mathbf{750\ W}$ b) $P = I^2R$ so $I = \sqrt{\frac{P}{R}} = \sqrt{\frac{750}{30}} = \sqrt{25} = \mathbf{5\ A}$

Energy** is Easy to **Calculate** if you Know the **Power

Sometimes it's the **total energy** transferred that you're interested in. In this case you simply need to **multiply** the **power** by the **time**. So: $E = VIt$ (or $E = \frac{V^2}{R}t$ or $E = I^2Rt$)

Make sure that the time is in seconds before you use these equations.

Example: The circuit diagram on the right is part of an electric kettle. A current of 4.0 A flows through the kettle's heating element once it is connected to the mains (230 V).
The kettle takes 4.5 minutes to boil the water it contains. How much energy does the kettle's heating element transfer to the water in the time it takes to boil?

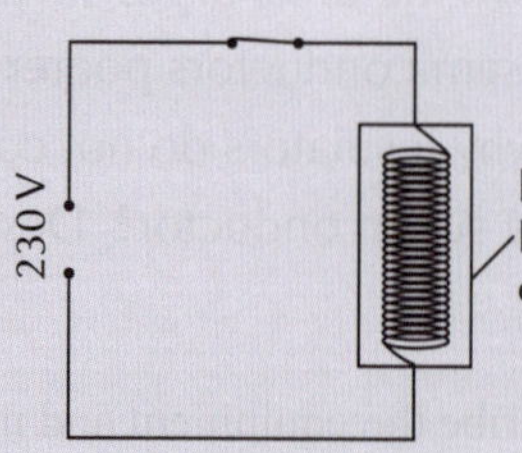

Time the kettle takes to boil in seconds = 4.5 × 60 = 270 seconds.

Use the equation $E = VIt = 230 \times 4.0 \times 270 = 248\ 400$ J = **250 kJ (to 2 s.f.)**

Remember, this is the circuit symbol for an open switch:

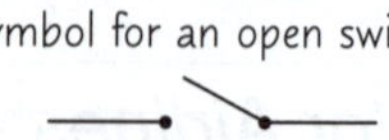

This is a closed switch:

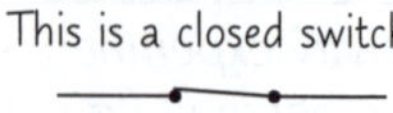

Power, Work and Electrical Energy

Electricity Companies Don't Use Joules and Watts

Electricity companies charge customers for '**units**' of electricity. Another name for a unit is a **kilowatt-hour (kWh)**. If you know the **power** of an **appliance** and the **length of time** it's used for you can calculate the **work** it does in kWh.

Work Done (kWh) = Power (kW) × Time (h)

1 kWh = 3.6 million joules

1 kW = 1000 W
1 hour = 3600 seconds

The **joule** is such a **small amount** of energy compared with the amount a typical household uses every month that it's **impractical**.

Cost of Electricity is the Price per Unit Times the Number of Units Used

To work out the **cost of electricity** you need to know **how much you've used** (in **units**) and the **price of each unit**. Then it's a simple matter of **multiplying** these two numbers together:

Cost = No. of units × Price per unit

Watch out for the units — you need power in kW (not watts), time in hours (not minutes or seconds) and money in either pence or pounds.

It's Important to Save Energy

Electricity doesn't just cost a packet — producing it has **environmental impacts** too:

- Burning **fossil fuels** (**coal**, **oil** and **gas**) releases **carbon dioxide** into the atmosphere, which adds to the **greenhouse effect** and contributes to **climate change**. Sulfur dioxide released by burning oil and coal can also cause **acid rain**.
- **Nuclear** power stations produce **radioactive waste**, which can be **dangerous** to the environment and human health if it isn't **disposed of safely**.
- Even **renewable** energy sources like **biofuels** have an impact because lots of **land** is needed to produce them.

Using **energy saving devices** can not only save you **money**, but helps to **reduce these impacts**.

1) **Energy saving** (fluorescent) **light bulbs** are more expensive than **traditional filament bulbs**, but are about four times more **efficient**. They also last for about **ten years** (as opposed to a year for filament bulbs) so reduce the number of bulbs needing to be **disposed of** and **replaced**.
2) **LED light bulbs** are even **more efficient** than energy saving bulbs, and last even longer, but they **cost more**.

Practice Questions

Q1 Power is measured in watts. What is 1 watt equivalent to?

Q2 What equation links power, potential difference and resistance?

Q3 Write down the equation linking power, current and resistance.

Q4 What equation would you use to find the cost of using an electrical appliance for a given amount of time?

Q5 Compare filament, energy saving and LED bulbs in terms of their environmental impact.

Exam Questions

Q1 A 12 V car battery supplies a current of 48 A for 2.0 seconds to the car's starter motor. The total resistance of the connecting wires is 0.010 Ω.

a) Calculate the energy transferred by the battery in this time. [1 mark]

b) Calculate the energy wasted as heat in the wires. [1 mark]

Q2 A customer is buying a washing machine. Model A has an average power of 470 W on a typical cycle, and model B has an average power of 410 W. Model A takes 135 minutes to wash a load of laundry, and model B takes 125 minutes.

a) Calculate how many kilowatt-hours each model uses in a typical cycle. [2 marks]

b) The customer does two loads of laundry a week on a typical cycle. Electricity costs 16.2p per unit. Calculate how much each model of washing machine would cost him a year to run. [4 marks]

Ultimate cosmic powers...

Many equations for you to get a handle on here, but you know what they say — practice makes perfect. Get practising.

Internal Resistance and E.m.f.

There's resistance everywhere — inside batteries, in all the wires (although it's very small) and in the components themselves. I'm assuming the resistance of the wires is zero on the next two pages, but you can't always do this.

Batteries have **Resistance**

Remember, I'm assuming that the resistance of the wires in the circuit is zero.

Resistance comes from **electrons colliding** with **atoms** and **losing energy** to other forms.

In a **battery**, **chemical energy** is used to make **electrons move**. As they move, they collide with atoms inside the battery — so batteries **must** have resistance. This is called **internal resistance**.

Internal resistance is what makes **batteries** and **cells warm up** when they're used.

Chemical reactions in the battery produce electrical energy.

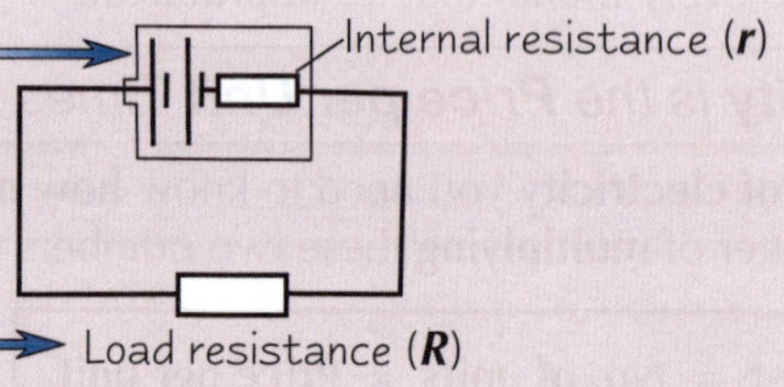

Load resistance is the total resistance of all the components in the external circuit. You might see it called 'external resistance'.

You might also see this equation written in terms of work done, W.

$$\varepsilon = \frac{E}{Q}$$

1) The amount of **electrical energy** the battery produces for each **coulomb** of charge is called its **electromotive force** or **e.m.f.** (ε). Be careful — e.m.f. **isn't** actually a force. It's measured in **volts**.
2) The **potential difference** across the **load resistance** (***R***) is the **energy transferred** when **one coulomb** of charge flows through the **load resistance**. This potential difference is called the **terminal p.d.** (*V*).
3) If there was **no internal resistance**, the **terminal p.d**. would be the **same** as the **e.m.f.** However, in **real** power supplies, there's **always some energy lost** overcoming the internal resistance.
4) The **energy wasted per coulomb** overcoming the internal resistance is called the **lost volts** (***v***).

Conservation of energy tells us:

energy per coulomb supplied by the source = **energy per coulomb transferred in load resistance** + **energy per coulomb wasted in internal resistance**

There are Loads of **Calculations** with **E.m.f.** and **Internal Resistance**

Examiners can ask you to do **calculations** with **e.m.f.** and **internal resistance** in loads of **different** ways. You've got to be ready for whatever they throw at you.

$$\varepsilon = V + v \qquad \varepsilon = I(R + r)$$
$$V = \varepsilon - v \qquad V = \varepsilon - Ir$$

Learn these equations for the exam. They might not be on your formula sheet.

These are all basically the **same equation**, just written differently. If you're given enough information you can calculate the e.m.f. (ε), terminal p.d. (*V*), lost volts (***v***), current (***I***), load resistance (***R***) or internal resistance (***r***). Which equation you should use depends on what information you've got, and what you need to calculate.

You Can Work Out the **E.m.f.** of **Multiple** Cells in **Series** or **Parallel**

For cells **in series** in a circuit, you can calculate the **total e.m.f.** of the cells by **adding** their individual e.m.f.s.

$$\varepsilon_{total} = \varepsilon_1 + \varepsilon_2 + \varepsilon_3 + ...$$

This makes sense if you think about it, because each charge goes through each of the cells and so gains e.m.f. (electrical energy) from each one.

See p.74 for all the rules for parallel and series circuits.

For identical cells **in parallel** in a circuit, the **total e.m.f.** of the combination of cells is the **same size** as the e.m.f. of each of the individual cells.

$$\varepsilon_{total} = \varepsilon_1 = \varepsilon_2 = \varepsilon_3 + ...$$

This is because the current will split equally between identical cells. The charge only gains e.m.f. from the cells it travels through — so the overall e.m.f. in the circuit doesn't increase.

Internal Resistance and E.m.f.

Time for an Example E.m.f. Calculation Question...

Example Three identical cells each with an e.m.f. of 2.0 V and an internal resistance of 0.20 Ω are connected in parallel in the circuit shown to the right. A current of 0.90 A is flowing through the circuit. Calculate the total p.d. across the cells.

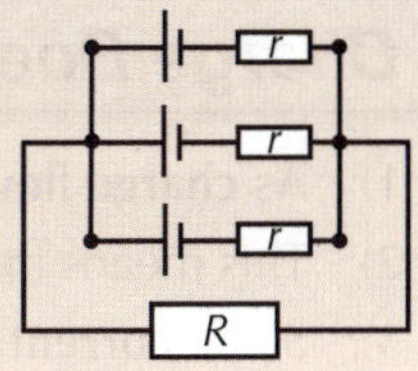

First calculate the lost volts, v, for 1 cell using $v = Ir$.

Since the current flowing through the circuit is split equally between each of the three cells, the current through one cell is $I/3$. So for 1 cell: $v = I/3 \times r = 0.90/3 \times 0.20 = 0.30 \times 0.20 = 0.06$ V

Then find the terminal p.d. across 1 cell using the equation: $V = \varepsilon - v = 2 - 0.06 = 1.94$

So the total p.d. across the cells combined = **1.9 V (to 2 s.f.)**

Investigate Internal Resistance and E.m.f. With This Circuit

1) **Vary** the **current** in the circuit by changing the value of the **load resistance** (***R***) using the variable resistor. **Measure** the **p.d.** (***V***) for several different values of **current** (***I***).
2) Record your data for V and I in a table, and **plot the results** in a graph of V against I.

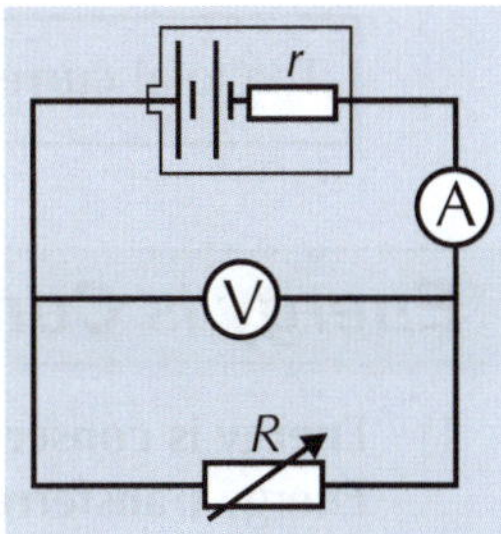

To find the **e.m.f.** and **internal resistance** of the cell, start with the equation: $\varepsilon = V + Ir$

1) Rearrange to give $V = -rI + \varepsilon$
2) Since ε and r are constants, that's just the equation of a **straight line**:

Equation of a straight line
$y = mx + c$
gradient — y-intercept

3) So the intercept on the vertical axis is ε.
4) And the gradient is $-r$.

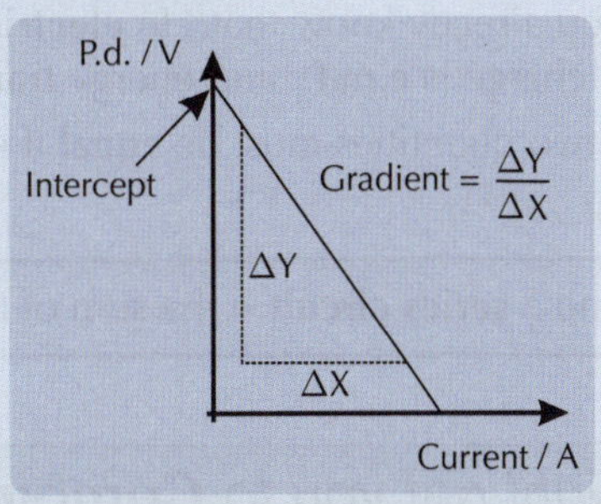

You probably don't need to take repeated readings in an experiment like this one, but make sure you get plenty of data points to draw your line.

An **easier** way to **measure** the **e.m.f.** of a **power source** to just connect a **voltmeter** across its **terminals**. Voltmeters have a very **high resistance**, but a **small current** will still flow through them. This means there must be some **lost volts**, which means you measure a value **very slightly less** than the **e.m.f.** (Although in practice the difference isn't usually significant.)

Practice Questions

Q1 What causes internal resistance? Write down the equation linking energy transferred, e.m.f. and charge.

Q2 What is the difference between e.m.f. and terminal p.d.?

Q3 What is meant by 'lost volts'?

Q4 Give an example of a source of e.m.f and describe an experiment to find the value of its e.m.f.

Exam Questions

Q1 A battery with an internal resistance of 0.8 Ω and an e.m.f. of 24 V powers a dentist's drill with resistance 4.0 Ω.

a) Calculate the current in the circuit when the drill is connected to the power supply. [1 mark]

b) Calculate the voltage across the drill while it is being used. [1 mark]

Q2 A bulb of resistance R is powered by two cells connected in series, each with internal resistance r and e.m.f. ε. Which expression represents the current flowing through each cell? [1 mark]

A $\frac{\varepsilon}{R+r}$ **B** $\frac{\varepsilon}{2(R+2r)}$ **C** $\frac{2\varepsilon}{R+2r}$ **D** $\frac{\varepsilon}{R+2r}$

Why'd the physicist swallow a multimeter? To find his internal resistance...

Thank you, thank you, I'm here all week. A jam-packed pair of pages here, but it's all important stuff. Make sure you know the difference between terminal p.d. and e.m.f., and that you've got a handle on all those equations.

Conservation of Energy & Charge in Circuits

There are some things in Physics that are so fundamental that you just have to accept them. Like the fact that there's loads of Maths in it. And that energy is conserved. And that Physicists get more homework than everyone else.

Charge Doesn't 'Leak Away' Anywhere — it's Conserved

1) As **charge flows** through a circuit, it **doesn't** get **used up** or **lost**.
2) This means that whatever **charge flows into** a junction will **flow out** again.
3) Since **current** is **rate of flow of charge**, it follows that whatever **current flows into** a junction is the same as the current **flowing out** of it.

Example: *CHARGE FLOWING IN 1 SECOND*

$Q_1 = 6\text{ C} \Rightarrow I_1 = 6\text{ A}$ → $Q_2 = 2\text{ C} \Rightarrow I_2 = 2\text{ A}$ and $Q_3 = 4\text{ C} \Rightarrow I_3 = 4\text{ A}$

$I_1 = I_2 + I_3$

Kirchhoff's first law says:

> The total **current entering a junction** = the total **current leaving it.**

Energy conservation is vital.

Energy is Conserved too

1) **Energy is conserved.** You already know that. In **electrical circuits**, **energy** is **transferred round** the circuit. Energy **transferred to** a charge is **e.m.f.**, and energy **transferred from** a charge is **potential difference**.
2) In a **closed loop**, these two quantities must be **equal** if energy is conserved (which it is).

Kirchhoff's second law says:

> The **total e.m.f.** around a **series circuit** = the **sum** of the **p.d.s** across each component.

(or $\varepsilon = \Sigma IR$ in symbols)

Exam Questions might get you to Combine Resistors in Series and Parallel

A **typical exam question** could give you a **circuit** with bits of information missing, leaving you to fill in the gaps. Not the most fun... but on the plus side you get to ignore any internal resistance stuff (unless the question tells you otherwise)... hurrah. Here are some **important rules**:

Series Circuits:

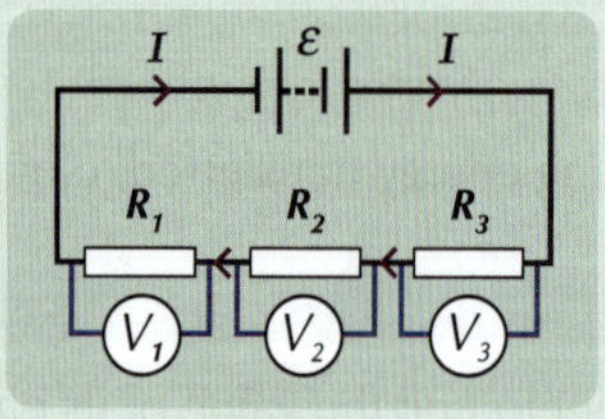

1) **same current** at **all points** of the circuit (since there are no junctions)
2) **e.m.f. split** between **components** (by Kirchhoff's 2nd law), so: $\varepsilon = V_1 + V_2 + V_3$
3) $V = IR$, so if I is constant: $IR_{total} = IR_1 + IR_2 + IR_3$
4) cancelling the Is gives:

$$R_{total} = R_1 + R_2 + R_3$$

5) As $R = \frac{1}{G}$, you can write this in terms of conductance:

$$\frac{1}{G_{total}} = \frac{1}{G_1} + \frac{1}{G_2} + \frac{1}{G_3}$$

Parallel Circuits:

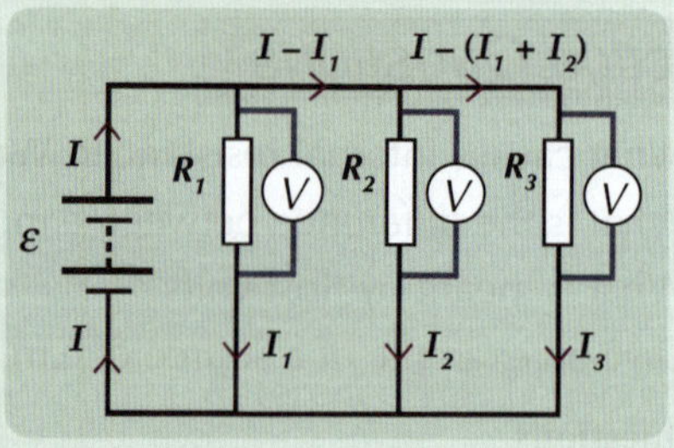

1) **current** is **split** at each **junction**, so: $I = I_1 + I_2 + I_3$
2) **same p.d.** across **all components** (three separate loops — within each loop the e.m.f. equals sum of individual p.d.s)
3) so, $V/R_{total} = V/R_1 + V/R_2 + V/R_3$
4) cancelling the Vs gives:

$$\frac{1}{R_{total}} = \frac{1}{R_1} + \frac{1}{R_2} + \frac{1}{R_3}$$

$\varepsilon = V$ in this case, as we're ignoring internal resistance.

5) As $R = \frac{1}{G}$, you can write this in terms of conductance:

$$G_{total} = G_1 + G_2 + G_3$$

Conservation of Energy & Charge in Circuits

Worked Exam Question

Example:

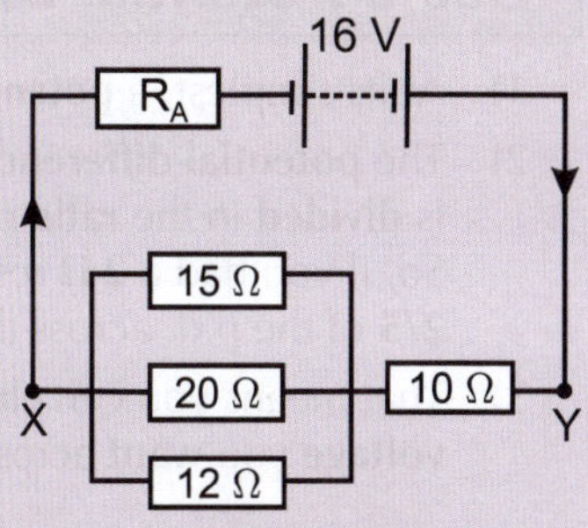

A battery of e.m.f. 16 V and negligible internal resistance is connected in a circuit as shown on the right. All values in the circuit are given to 2 s.f.

a) Show that the group of resistors between X and Y could be replaced by a single resistor of resistance 15 Ω.

You can find the combined resistance of the 15 Ω, 20 Ω and 12 Ω resistors using:
$1/R = 1/R_1 + 1/R_2 + 1/R_3 = 1/15 + 1/20 + 1/12 = 1/5 \quad \Rightarrow R = 5\ \Omega$
So overall resistance between X and Y can be found by $R = R_1 + R_2 = 5 + 10 =$ **15 Ω**

b) If $R_A = 20\ \Omega$:
i) calculate the potential difference across R_A.

Careful — there are a few steps here. You need the p.d. across R_A, but you don't know the current through it. So start there: total resistance in circuit = 20 + 15 = 35 Ω,
so current through R_A can be found using $I = V_{total}/R_{total}$: $I = 16/35$ A
then you can use $V = IR_A$ to find the p.d. across R_A: $V = 16/35 \times 20 = 9.142...=$ **9.1 V (to 2 s.f.)**

ii) calculate the current in the 15 Ω resistor.

You know the current flowing into the group of three resistors and out of it, but not through the individual branches. But you know that their combined resistance is 5 Ω (from part a) so you can work out the p.d. across the group:
$V = IR = 16/35 \times 5 = 16/7$ V
The p.d. across the whole group is the same as the p.d. across each individual resistor, so you can use this to find the current through the 15 Ω resistor:
$I = V/R = (16/7) / 15 =$ **0.15 A (to 2 s.f.)**

Practice Questions

Q1 Write Kirchhoff's first and second laws.

Q2 State the formulas used to combine the resistance of resistors in series and in parallel.

Q3 Find the current through and potential difference across each of two resistors, each with a conductance of 0.2 S, when they are placed in a circuit containing a 5 V battery, and are wired: a) in series, b) in parallel.

Exam Question

Q1 For the circuit on the right:

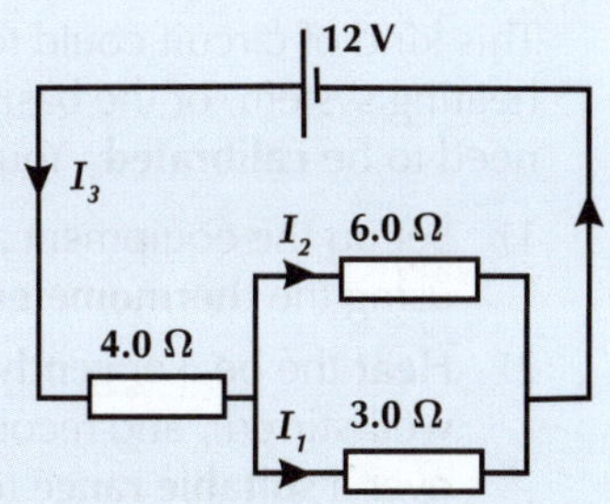

a) Calculate the total effective resistance of the three resistors in this combination. [2 marks]

b) Calculate the main current, I_3. [1 mark]

c) Calculate the potential difference across the 4.0 Ω resistor. [1 mark]

d) Calculate the potential difference across the parallel pair of resistors. [1 mark]

e) Using your answer from part d), calculate the currents I_1 and I_2. [2 marks]

This is a very purple page — needs a bit of yellow I think...

V = IR is the formula you'll use most often in these questions. Make sure you know whether you're using it on the overall circuit, or just one specific component. It's amazingly easy to get muddled up — you've been warned.

The Potential Divider

It's probably best not to bring up this topic when you're having dinner with your in-laws — it's potentially quite divisive.

Use a Potential Divider to Get a Fraction of an Input Voltage

1) At its simplest, a **potential divider** is a circuit with a **voltage source** and a couple of **resistors** in series.
2) The **potential difference** of the voltage source (e.g. a power supply) is **divided** in the **ratio** of the **resistances**. As an equation: $$\frac{V_1}{V_2} = \frac{R_1}{R_2}$$ So, if you had a **2 Ω** resistor and a **3 Ω** resistor, you'd get **2/5** of the p.d. across the **2 Ω** resistor and **3/5** across the **3 Ω**.
3) That means you can **choose** the **resistances** to get the **voltage** you **want** across one of them.

This rearranges to give $V_1/R_1 = V_2/R_2$. As $I = V/R$ this just means the current is the same through both resistors, which you know from page 74.

In the circuit shown, R_2 has $\frac{R_2}{R_1 + R_2}$ of the total resistance.

So: $$V_{out} = \frac{R_2}{R_1 + R_2} V_{in}$$

E.g. if $V_{in} = 9$ V and you want V_{out} to be 6 V,

then you need: $\frac{R_2}{R_1 + R_2} = \frac{6}{9}$ which gives $R_2 = 2R_1$.

So you could have, say, $\boldsymbol{R_1 = 100\ \Omega}$, $\boldsymbol{R_2 = 200\ \Omega}$

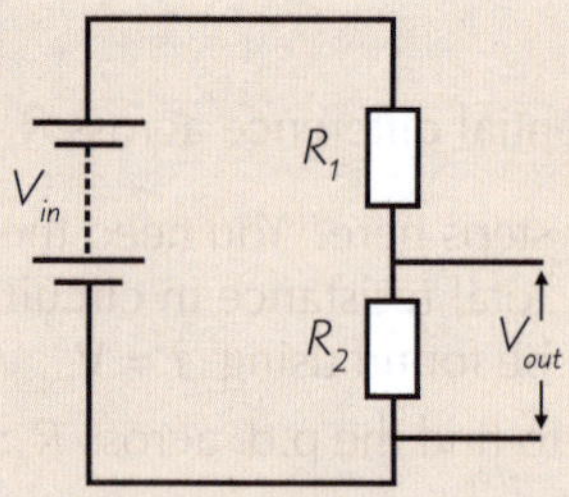

This is the circuit symbol for a cell...

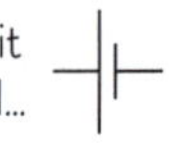

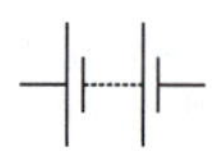

...and this is the symbol for a battery.

4) This circuit can be used for **calibrating voltmeters**, which have a **very high resistance**.
5) If you put something with a **relatively low resistance** across R_2 though, you start to run into **problems**. You've **effectively** got **two resistors** in **parallel**, which will **always** have a **total** resistance **less** than R_2. That means that V_{out} will be **less** than you've calculated, and will depend on what's connected across R_2. Hrrumph.

Add an LDR or Thermistor for a Light or Temperature Sensor

1) Potential dividers can be made into sensors by including components whose resistance changes with external factors, for example light dependent resistors and thermistors (p.69). This means $\boldsymbol{V_{out}}$ **varies** with light or heat, so you can make a potential divider that works as a light or heat **sensor**.
2) The **circuit** needs to be **calibrated** so you know how the voltage across the component and V_{out} varies as external conditions change — e.g. knowing the voltage across a thermistor at a given temperature.
3) You can use an experiment like the one below to plot a calibration curve of voltage against an external factor:

Example: Calibration of an electronic thermometer.

Here's a potential divider using an **NTC thermistor**.

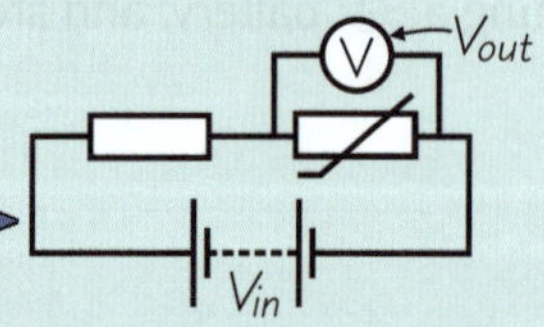

Think about safety before you start — keep the rest of the circuit as far away from the Bunsen burner and the water bath as possible.

This kind of circuit could form part of the circuit for a **thermostat** in a central heating system, or the basis of an **electronic thermometer**. First, the circuit would need to be **calibrated**. You can do this using the equipment shown on the right:

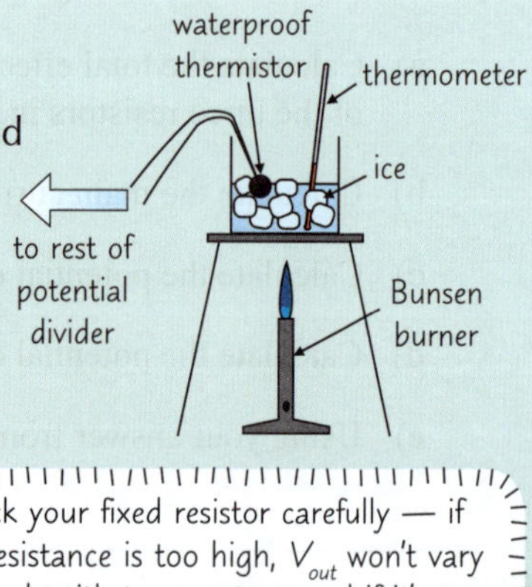

1) Set up the equipment as shown, then measure the **temperature** of the water using the **thermometer**, and record the **voltage** across the thermistor.
2) **Heat** the beaker **gently** using the Bunsen burner (make sure the water is well-stirred), and record the temperature and the voltage at **regular intervals** over a **suitable range** (e.g. at 5 °C intervals over a range of 0-100 °C).
3) Plot a **graph** of temperature against voltage from your results. This graph is the thermistor's **calibration curve**. You can use it to find the temperature of the thermistor from the voltage across it, without needing the thermometer — the thermistor and the calibration curve together are effectively **another thermometer**.

Pick your fixed resistor carefully — if its resistance is too high, V_{out} won't vary enough with temperature, and if it's too low, V_{out} might vary over a bigger range than your voltmeter can handle.

The Potential Divider

A Potentiometer uses a Variable Resistor to give a Variable Voltage

1) Imagine you have a long length of wire connected to a power supply. If the wire is **uniform** (i.e. same cross-sectional area and material throughout), then its **resistance** is **proportional** to its **length**.
2) This means that if you were to connect a voltmeter across different lengths of the wire, the **potential difference** you'd record would be **proportional** to the **length** you'd connected it over — you're measuring across a bigger share of the total resistance so you get a bigger potential difference.

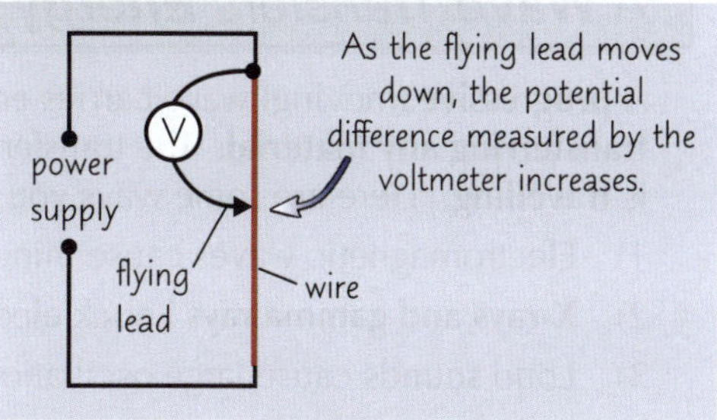

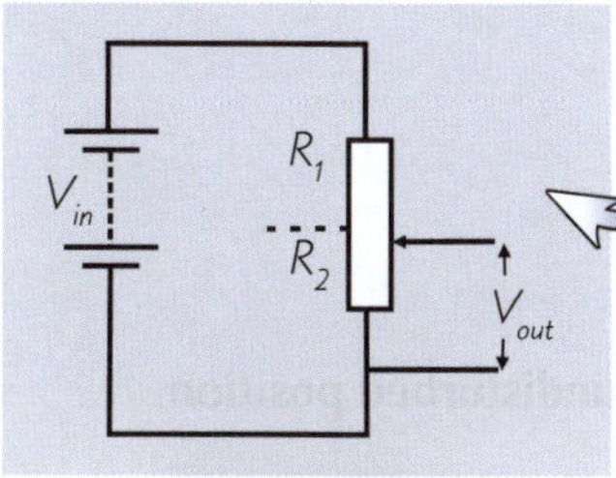

3) This is the basis of how a **potentiometer** works. A **potentiometer** is basically a potential divider, with a variable resistor replacing R_1 and R_2 (it's even sometimes **called** a potential divider just to confuse things).
4) You move a **slider** or turn a knob to **adjust** the **relative sizes** of R_1 and R_2. That way you can vary V_{out} from **0 V** up to the source voltage.
5) This is dead handy when you want to be able to **change** a **voltage continuously**, like in the **volume control** of a stereo:

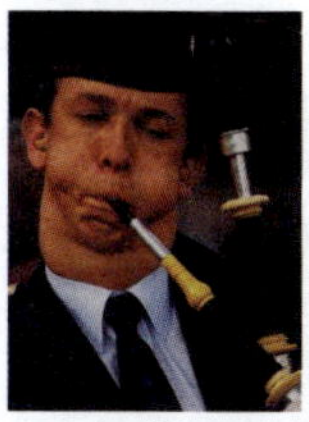

I've often wished bagpipes had a volume control. Or just an off switch.

Example:
Here, V_{in} is replaced by the input signal (e.g. from a CD player) and V_{out} is the output to the amplifier and loudspeaker.

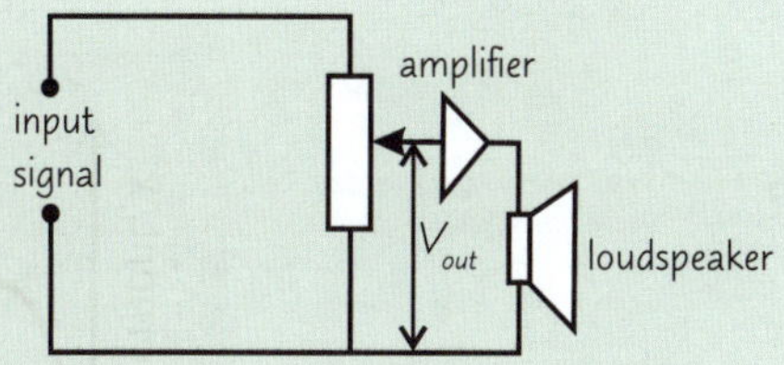

Practice Questions

Q1 Look at the electronic thermometer circuit on page 76.
How could you change the circuit so that it could be used to detect changes in light intensity?

Q2 The NTC thermistor in the circuit on page 76 has a resistance of 300 Ω at a temperature T_1, and 900 Ω at a temperature T_2. The fixed resistor has a value of 100 Ω. Show that $V_{out}(T_1) = 1.5$ V and $V_{out}(T_2) = 0.6$ V.

Exam Questions

Q1 In the circuit on the right, all the resistors have the same value. Calculate the p.d. between:

a) A and B. [1 mark]

b) A and C. [1 mark]

c) B and C. [1 mark]

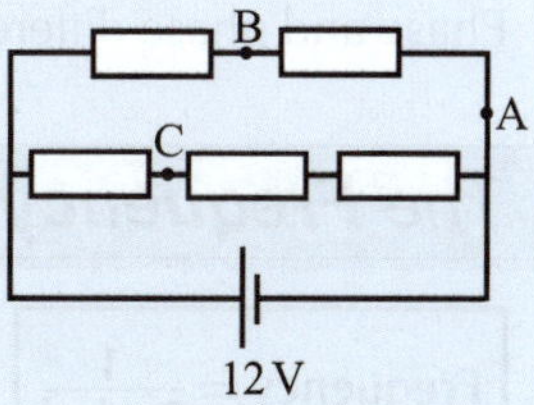

Q2 Look at the circuit on the right. All the resistances are given to 2 significant figures.

a) Calculate the p.d. between A and B as shown by a high resistance voltmeter placed between the two points. [1 mark]

b) A 40.0 Ω resistor is now placed between points A and B. Calculate the p.d. across AB and the current flowing through the 40.0 Ω resistor. [4 marks]

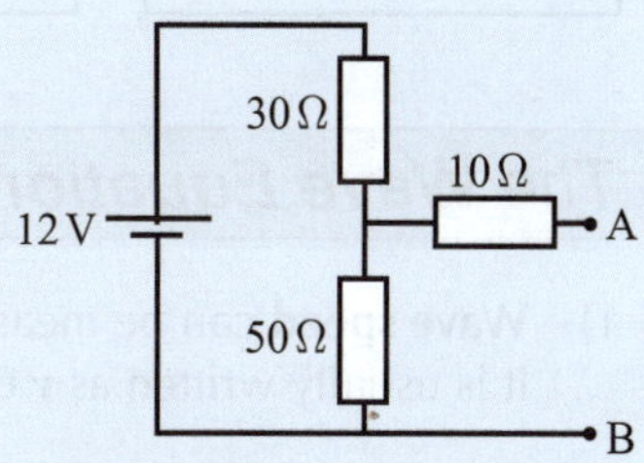

OI... YOU... [bang bang bang]... turn that potentiometer down...

You'll probably have to use a potentiometer in every experiment you do with electricity from now on in, so you'd better get used to them. I can't stand the things myself, but then lab and me don't mix — far too technical.

Progressive Waves

Aaaah... waggling ropes about. It's all good clean fun as my mate Richard used to say...

A Wave Transfers Energy Away From Its Source

A **progressive** (moving) wave carries **energy** from one place to another **without transferring any material**. The transfer of energy is in the **same direction** as the wave is **travelling**. Here are some ways you can tell waves carry energy:

1) Electromagnetic waves cause things to **heat up**.
2) **X-rays** and **gamma rays** knock electrons out of their orbits, causing **ionisation**.
3) Loud **sounds** cause large oscillations in air particles which can make things **vibrate**.
4) **Wave power** can be used to **generate electricity**.

Since waves carry energy away, the **source** of the wave **loses energy**.

You Need to Know These Bits of a Wave

1) **Displacement**, x, metres — how far a **point** on the wave has **moved** from its **undisturbed position**.
2) **Amplitude**, A, metres — the **maximum magnitude** of the **displacement**.
3) **Wavelength**, λ, metres — the **length** of **one whole wave cycle**, e.g. from **crest** to **crest** or **trough** to **trough**.

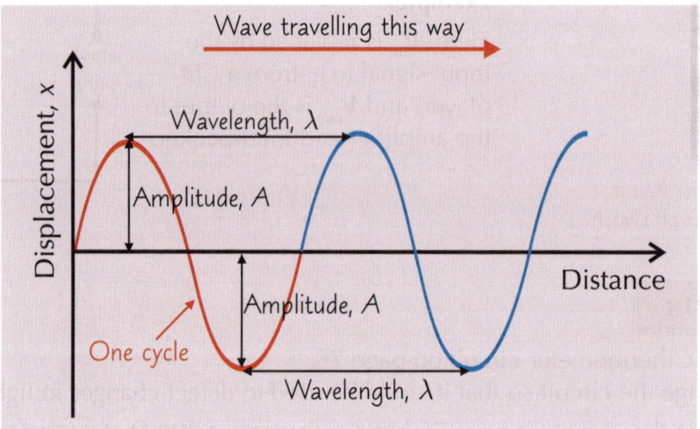

4) **Period**, T, seconds — the **time taken** for a **whole cycle** (vibration) to complete.
5) **Frequency**, f, hertz — the **number** of **cycles** (vibrations) **per second** passing a given **point**.
6) **Phase** — a measurement of the **position** of a certain **point** along the wave cycle.
7) **Phase difference** — the amount one wave lags behind another.

Phase and phase difference are measured in angles (in degrees or radians). See p.90.

The Frequency is the Inverse of the Period

$$\text{Frequency} = \frac{1}{\text{Period}} \qquad f = \frac{1}{T}$$

It's that simple.
Get the **units** straight: **1 Hz = 1 s^{-1}**.

The Wave Equation Links Wave Speed, Frequency and Wavelength

1) **Wave speed** can be measured just like the speed of anything else. It is usually written as v or c.

$$\text{Wave speed } (v) = \frac{\text{Distance } (d)}{\text{Time } (t)}$$

2) From this equation, you can write another for the speed of the wave, called the **wave equation**:

$$\text{Speed of wave } (v) = \text{frequency } (f) \times \text{wavelength } (\lambda) \qquad v = f\lambda$$

Remember, you're not measuring how fast a physical point (like one molecule of rope) moves. You're measuring how fast a point on the **wave pattern** moves.

Progressive Waves

Oscilloscopes Display Waves

1) A cathode ray **oscilloscope** (CRO) measures **voltage**. It **displays** waves from a **signal generator** as a function of **voltage** over **time**.
2) The displayed wave is called a **trace**.
3) The screen is split into squares called **divisions**.
4) The vertical axis is in **volts**. The **volts per division** shown on this axis is controlled by the **gain dial**.
5) The horizontal axis is in **seconds** — also called the **timebase**. The **seconds per division** shown on this axis is controlled by the **timebase** dial.
6) You can alter the gain and timebase to make it **easy to read** off measurements.

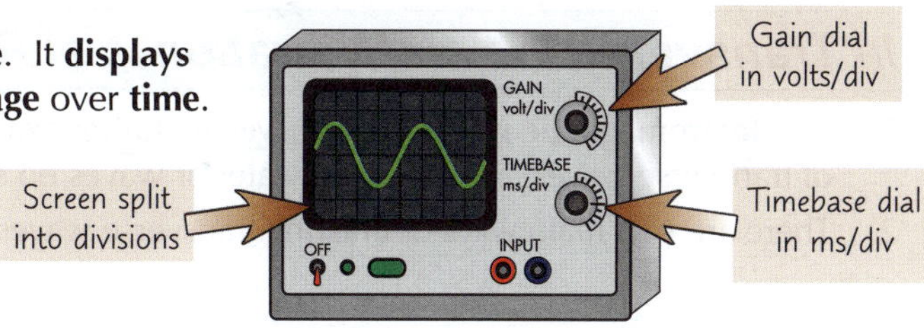

The period of a wave is found from: the no. of divisions in one complete wave × the timebase setting.

If you plug an **AC (alternating current) supply** into an oscilloscope, you get a trace that goes up and down in a regular pattern — some of the time it's positive and some of the time it's negative.

A **microphone** converts **sound waves** into **electrical signals** which can be seen on an **oscilloscope**.

An *Oscilloscope* Can be Used to Find the *Speed of Sound*

1) Set up the experiment as shown in the diagram below and set the **frequency** of the signal generator to around 2-6 kHz. A **2-beam oscilloscope** must be used as this can display **two waves** — one for the **signal generator** generating a **sound wave** and one for the **same sound wave** being **received** by the **microphone**.
2) A 2-beam oscilloscope has **three dials** — two are the gain dials for **each** input signal and the other dial is the timebase dial for **both** input signals. Adjust the **dials** so that you can see at least one **complete cycle** of each wave.
3) **Change** the **distance** between the microphone and loudspeaker so that the **peaks** of one wave **line up** with the **troughs** of the other wave (i.e. the waves are **out of phase** — see p.90) and then **measure** the distance between the microphone and loudspeaker.
4) Calculate the **frequency**, *f*, of the wave by measuring the **period**, *T*, of the wave from the oscilloscope display and using it in $f = \frac{1}{T}$ (see p.78). The **resolution** of the oscilloscope will be **better** than the resolution of the signal generator, so measuring and calculating the frequency this way will give you a **smaller uncertainty** than just reading the frequency from the signal generator.

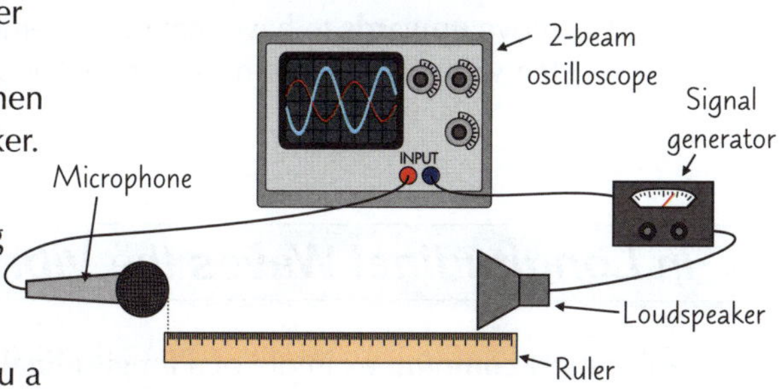

5) Move the microphone **away from** (or **towards**) the loudspeaker so that the microphone's corresponding wave on the oscilloscope **moves** one **full wavelength** along the signal generator's wave — the peaks of one wave and the troughs of the other wave will **line up** again. Measure the **new** distance between the microphone and loudspeaker.
6) The **difference** in your two recorded distances is equal to the **wavelength** of the sound wave.
7) Repeat steps 5-6 and use your results to find a **mean value** for the distance moved by the microphone each time (i.e. the **wavelength**). Then use the equation $v = f\lambda$ to find the value for the **speed of sound**.

Practice Questions

Q1 Define and give the units of the frequency, displacement, amplitude and period of a wave.

Q2 Write down the wave equation.

Q3 Describe an experiment involving a 2-beam oscilloscope, a loudspeaker and a microphone to measure the speed of sound.

Exam Question

Q1 An oscilloscope has the gain set to 2.0 volts/div and a timebase set to 3.0 ms/div. It is displaying the trace of a wave that has a wave speed of 280 ms^{-1}.

a) State the maximum voltage of the trace. [1 mark]

b) Calculate the frequency and wavelength of the wave. [5 marks]

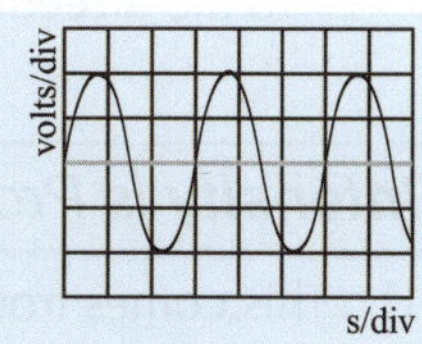

I don't care about expensive things — all I care about is wave speed...

Make sure you know what all the waves terms (amplitude, frequency etc.) mean — or this section could get confusing...

Longitudinal and Transverse Waves

Get a long spring and have a go at making different waves. Or sit there beeping pretending to be a microwave.

*In **Transverse Waves**, **Vibration** is at Right Angles to the **Direction** of Travel*

1) All **electromagnetic waves** are **transverse**. Other examples of transverse waves are ripples on water or waves on strings.
2) There are **two** main ways of **drawing** transverse waves:

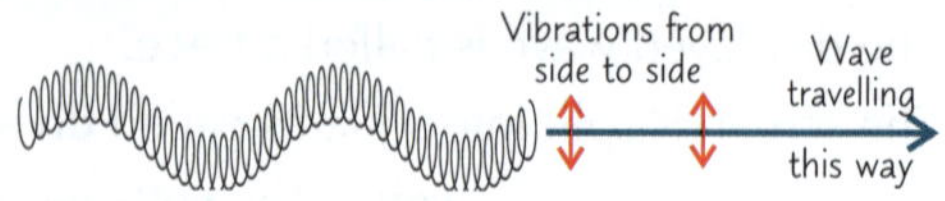

They can be shown as **graphs** of **displacement** against **distance along the path of the wave.**

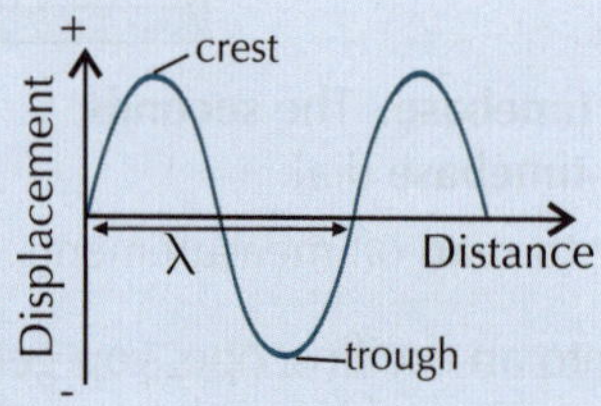

Or, they can be shown as graphs of **displacement against time** for a point as the wave passes.

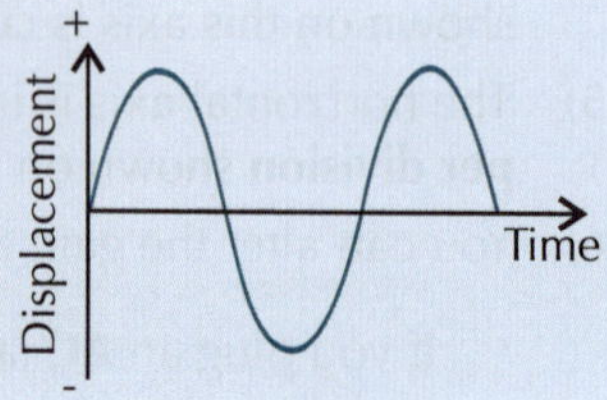

3) Both sorts of graph often give the **same shape**, so make sure you check out the label on the ***x*-axis**. Displacements **upwards** from the centre line are given a **+ sign**. Displacements downwards are given a **– sign**.
4) You can work out what **direction** a point on a wave is moving in when given a snapshot of the wave.

Example: Look at the snapshot of the wave on the right. Which direction is point A on the wave moving in?

1) Look at which **direction** the wave is **travelling** in — here the wave is moving from **left** to **right**.
2) The displacement of the wave **just to the left** of point A is **greater** than point A's. So as the wave travels along, point A will need to move **upwards** to have that displacement. (If the displacement to the left was less than point A's, point A would need to move down.)

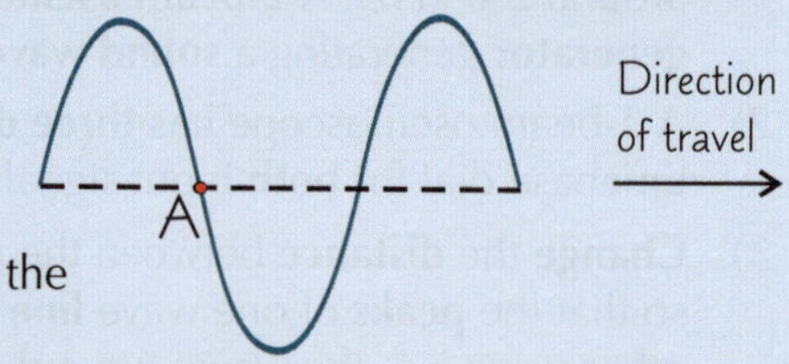

*In **Longitudinal Waves** the **Vibrations** are **Along** the Direction of Travel*

The most **common** example of a **longitudinal wave** is **sound**. A sound wave consists of alternate **compressions** and **rarefactions** of the **medium** it's travelling through. (That's why sound can't go through a vacuum.) Some types of **earthquake shock waves** are also longitudinal.

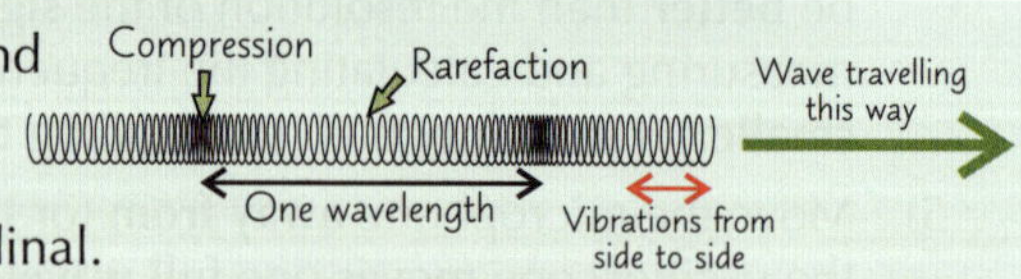

It's hard to **represent** longitudinal waves **graphically**. You'll usually see them plotted as **displacement** against **time**. These can be **confusing** though, because they look like a **transverse wave**.

Intensity** is a Measure of How Much **Energy** a Wave is **Carrying

1) When you talk about "**brightness**" for light or "**loudness**" for sound, what you really mean is **how much light** or **sound** energy hits your eyes or your ears **per second**.
2) The scientific measure of this is **intensity**.

Intensity is the **rate of flow** of **energy** per **unit area** at **right angles** to the **direction of travel** of the wave. It's measured in $\mathbf{Wm^{-2}}$.

$$\text{Intensity} = \frac{\text{Power}}{\text{Area}} \qquad I = \frac{P}{A}$$

Intensity** is Proportional to **Amplitude Squared

1) This comes from the fact that **intensity** is **proportional** to **energy**, and the energy of a wave depends on the square of the **amplitude**.
2) From this you can tell that for a **vibrating source** it takes four times as much energy to double the size of the vibrations.

$$\text{Intensity} \propto (\text{Amplitude})^2$$

Longitudinal and Transverse Waves

All ***Electromagnetic (EM) Waves*** *Have Some* ***Properties*** *In Common*

1) All EM waves travel in a **vacuum** at a **speed** of **3.00×10^8 ms^{-1}** (to 3 s.f.), and at **slower** speeds in other media.
2) They are **transverse** waves consisting of **vibrating electric** and **magnetic fields**.
   The **electric** and **magnetic** fields are at **right angles** to each other and to the **direction of travel**.
3) Like all waves, EM waves can be **refracted** (p.87), **reflected** and **diffracted** (p.84-85) and can undergo **interference** (p.90). They also obey $v = f\lambda$ (v = velocity, f = frequency, λ = wavelength).
4) Like all progressive waves, progressive EM waves **carry energy**.
5) EM waves are transverse so, like all transverse waves, they can be **polarised** (see page 82).

Some ***Properties Vary*** *Across the* ***EM Spectrum***

EM waves with different wavelengths behave differently in some respects. The spectrum is split into seven categories: **radio waves**, **microwaves**, **infrared**, **visible light**, **ultraviolet**, **X-rays** and **gamma rays**.

1) The longer the wavelength, the more **obvious** the wave characteristics — long radio waves diffract round hills.
2) **Energy** is directly proportional to **frequency**. **Gamma rays** have the **highest energy**; **radio waves** the **lowest**.
3) In general, the **higher** the **energy**, the more **dangerous** the wave — some can even cause **ionisation**.

Type	Approximate Wavelength / m	Penetration	Uses
Radio waves	10^{-1} — 10^{6}	Pass through matter.	Radio transmissions.
Microwaves	10^{-3} — 10^{-1}	Mostly pass through matter, but cause some heating.	Radar. Microwave cooking. TV transmissions.
Infrared (IR)	7×10^{-7} — 10^{-3}	Mostly absorbed by matter, causing it to heat up.	Heat detectors. Night vision cameras. Remote controls. Optical fibres.
Visible light	4×10^{-7} — 7×10^{-7}	Absorbed by matter, causing some heating.	Human sight. Optical fibres.
Ultraviolet (UV)	10^{-8} — 4×10^{-7}	Absorbed by matter. Slight ionisation.	Sunbeds. Security marks that show up under UV.
X-rays	10^{-13} — 10^{-8}	Mostly pass through matter, but cause ionisation as they pass.	To see damage to bones and teeth. Airport security scanners. To kill cancer cells.
Gamma rays	10^{-16} — 10^{-10}	Mostly pass through matter, but cause ionisation as they pass.	Irradiation of food. Sterilisation of medical instruments. To kill cancer cells.

Practice Questions

Q1 Draw a displacement-time graph for a point on a wave as the wave passes.

Q2 Describe the difference between the vibrations in a transverse wave and a longitudinal wave.

Q3 State the two types of field that make up an EM wave and describe the direction of their vibrations.

Q4 Name four properties that all electromagnetic waves have.

Q5 Name the seven types of EM wave that make up the electromagnetic spectrum.

Q6 Which types of electromagnetic radiation have the highest and lowest energies?

Q7 What are the approximate wavelengths for: a) radio waves b) microwaves?

Exam Question

Q1 a) A 10.0 W light beam is shone onto a screen with an area of 0.002 m^2.
Calculate the intensity of the light beam on the screen. [1 mark]

b) The intensity of the light on the screen is increased until it is exactly triple the original beam intensity. Which of the following describes the amplitude of the light waves in the beam compared to their original amplitude?

A It is 3 times larger. B It has halved. C It is 9 times larger. D It is $\sqrt{3}$ larger. [1 mark]

I've got UV hair...

No really I have, you just can't see it. Aaanyway... moving swiftly on. Lots of facts to learn on these pages — and that's just what you have to do I'm afraid, sit and learn it then make sure you really know it. Not much fun, but there you go.

Polarisation of Waves

Polarisation is all about making a wave move up and down in only one direction. No, not that one direction.

A ***Polarised Wave*** *Only* ***Oscillates*** *in One Direction*

1) If you **shake a rope** to make a **wave**, you can move your hand **up and down** or **side to side** or in a **mixture** of directions — it still makes a **transverse wave**.
2) But if you try to pass **waves in a rope** through a **vertical fence**, the wave will only get through if the **vibrations** are **vertical**. The fence filters out vibration in other directions. This is called **polarising** the wave.

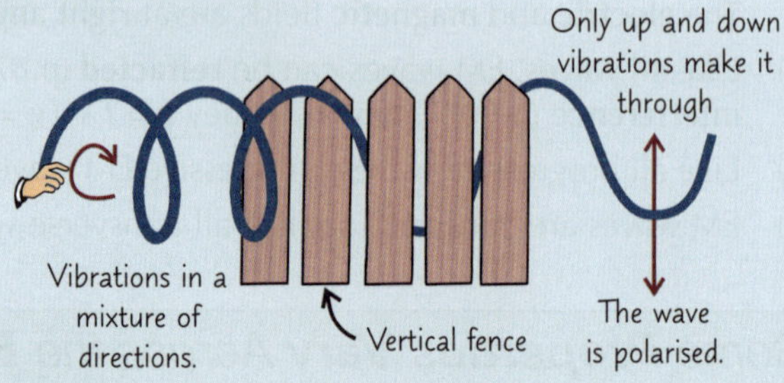

Electromagnetic Radiation *can be Polarised*

1) **Electromagnetic radiation** (e.g. light) is made up of two transverse waves vibrating in different directions. (The things vibrating are actually electric and magnetic fields.)
2) A **polarising filter** acts a bit like the fence. Light that has passed through the polarising filter will only be vibrating in one direction.
3) If you have two polarising filters at **right angles** to each other, then **no** light will get through as all directions of vibration will be blocked.
4) Polarisation **can only happen** for **transverse** waves. You **can't** polarise **longitudinal waves** like sound. The fact that you can polarise light is one **indication** that it's a transverse wave.

Try moving around a TV or radio **aerial**. The signal **comes** and **goes** because the incoming waves are **polarised**, so you need your aerial to **align** with them for a **strong** signal.

Investigate ***Polarisation*** *of* ***Light*** *Using Two* ***Polarising Filters***

You can observe polarisation by shining unpolarised white light through two polarising filters.

1) Align the transmission axes of two **polarising filters** so they are both **vertical**. Shine unpolarised light on the first filter. Keep the position of the **first filter fixed** and **rotate** the second one.
2) Light that passes through the first filter will always be **vertically polarised**.
3) When the transmission axes of the two filters are **aligned**, **all** of the light that passes through the first filter also passes through the second.
4) As you rotate the second filter, the amount of light that passes through the second filter **varies**.

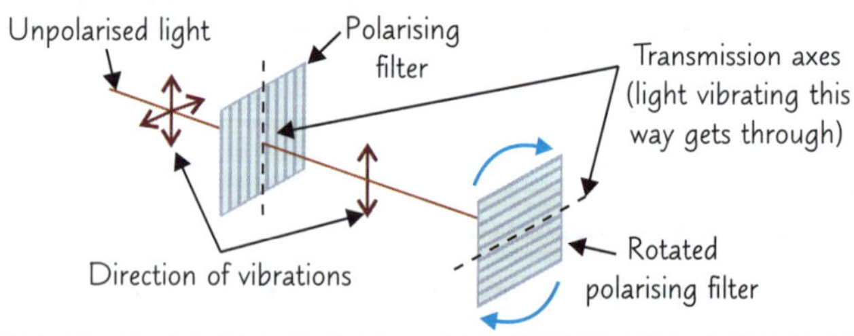

Just like vectors, you can think of the transmission axis of the rotating filter as having a **vertical** and **horizontal component**. The **larger** the **vertical component**, the **more** vertically polarised light will pass through the filter.

5) As the second filter is rotated, **less** light will get through it as the **vertical** component of the second filter's transmission axis **decreases**. This means the **intensity** of the light getting through the second filter will gradually **decrease**.
6) When the two transmission axes are at **45°** to each other, the intensity will be **half** that getting through the first filter. When they're at **right angles** to each other **no** light will pass through — the **intensity** is **0**.
7) As you continue turning, the intensity should then begin to **increase** once again.
8) When the two axes **realign** (after a 180° rotation), **all** the light will be able to pass through the second filter again.

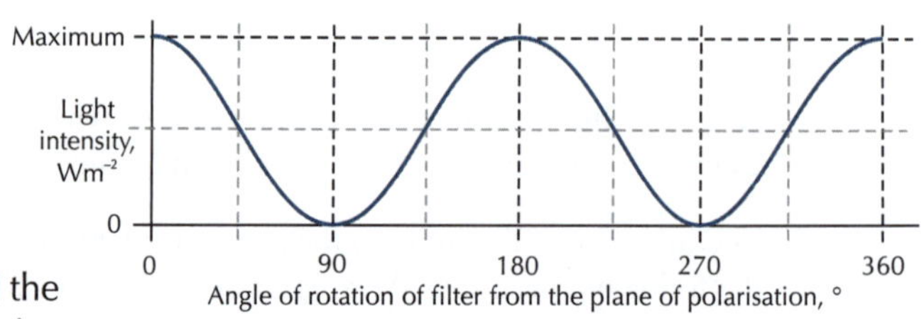

You come across polarising filters more often than you'd think. For example, **3D films** use polarised light to create depth — the filters in each lens are at right angles to each other so each eye gets a slightly different picture. **Polaroid sunglasses** also use polarising filters — reflected light is partially polarised so the sunglasses block this out to help prevent glare.

Polarisation of Waves

You Can **Polarise Microwaves** Using a **Metal Grille**

Polarising filters don't work on **microwaves** — their **wavelength** is too long. Instead, **metal grilles** (squares full of metal wires which are all aligned) are used to polarise them. You can investigate the polarisation of microwaves using a **microwave transmitter** and a **microwave receiver** linked to a **voltmeter**.

1) Place a metal **grille** between the microwave **transmitter** and **receiver** as shown below. (Handily, microwave transmitters transmit **polarised** microwaves, so you only need one metal grille.)
2) The intensity of microwaves passing through the grille is at a **maximum** when the direction of the vibration of the microwaves and the wires on the grille are at **right angles** to each other.
3) As you rotate the grille, the **intensity** of polarised microwaves able to pass through the grille **decreases**, so the reading on the voltmeter **decreases**.
4) When the wires of the metal grille are **aligned** with the direction of the polarised waves, **no signal** will be shown on the voltmeter.

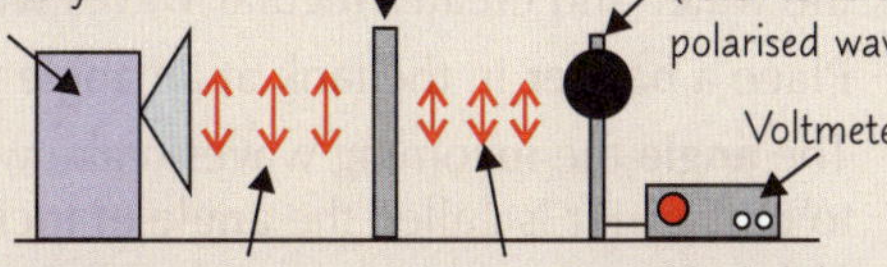

The **intensity** drops to **zero** when the wires are **aligned** with the **direction of polarisation** of the microwaves, because the grille is **absorbing their energy** (see below).

Be careful here — this is the **opposite effect** to polarising filters on the previous page, where aligning the polarising filter with the direction of vibrations gave the maximum intensity of waves passing through the filter.

> Make sure all of your electrical equipment is safely connected before you turn it on — microwave transmitters operate at very high voltages.

1) The vibrating electric field of the microwave **excites** electrons in the metal grille.
2) The energy of the incoming microwaves is **absorbed** by the grille and **re-emitted** in **all directions**.
3) Only a few of those re-emitted waves are vibrating in the **direction** of the microwave receiver.
4) The microwave **receiver** only receives microwaves in **one plane**, so even if the **re-emitted** wave travels towards the receiver, it might not be picked up.
5) When the wires and vibrations of the waves are **aligned**, **more** electrons are excited than when they're at right angles to each other — all the energy is absorbed and the **intensity** reading drops to **zero**.
6) When the wires and vibrations are at **right angles** to each other, some electrons in the grille are still excited and so there is still a **small drop** in **intensity**.

Practice Questions

Q1 What is meant by a polarised wave? Why can't you polarise sound waves?

Q2 Describe an experiment that shows visible light can be polarised.

Q3 Explain why the intensity of vertically polarised microwaves passing through a metal grille will drop to zero when the grille is aligned with the direction of polarisation.

Exam Question

Q1 Two polarising filters are placed on top of each other and held in front of a source of white unpolarised light.

a) No light can be seen through the filters.
State the angle between the transmission axes of the two filters. [1 mark]

b) The filters are rotated so that the angle between their transmission axes is 45°.
Describe the difference in the intensity of the light once it has passed through both filters compared to the light once it has only passed through the first filter. [1 mark]

c) Give one use of polarising filters. [1 mark]

Forget polarisation, I need a mental filter...

...to stop me talking rubbish all the time. Anyway, the important thing to remember about polarisation — it's a case of filtering out different directions of wave vibrations. And make sure you really know the experiment for polarising light.

Reflection and Diffraction

All waves share some properties — they spread out (diffract), reflect off stuff and refract (more on that on p.87-89).

Waves Can Be **Reflected**

Reflection means the wave is **bounced back** when it **hits a boundary**. The **angle of incidence** always **equals** the **angle of reflection**.

You can show the reflection of water waves in a **ripple tank**.

1) **Ripple tanks** are shallow tanks of water that you can generate a wave in.
2) This is done by an **oscillating paddle**, which continually dips into the water and creates regular waves with straight, parallel wave fronts.
3) Place a **barrier** in the tank at an **angle** to the wave fronts.
4) The **angle** the incoming waves make with the **normal** to the barrier is called the **angle of incidence**, ***i***.
5) You should see the waves **reflecting** off the barrier and travelling in a different direction to the way they arrived.
6) The angle between the **direction** of the **reflected waves** and the **normal** to the barrier is called the **angle of reflection**, ***r***.
7) You can **change** the **angle** of **incidence** to see that the angle of **reflection** changes by the same amount. They are always **equal** to each other.

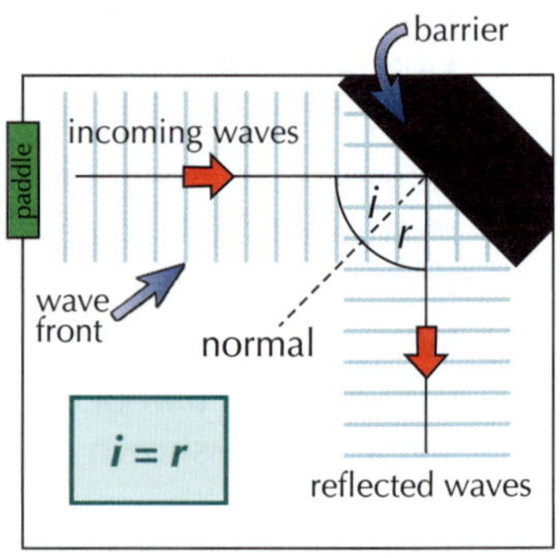

Waves Go **Round Corners** and **Spread out** of **Gaps**

The way that **waves spread out** as they come through a **narrow gap** or go round obstacles is called **diffraction**. **All** waves diffract, but it's not always easy to observe. The amount of diffraction depends on the **size of the gap** in comparison to the **wavelength** of the wave.

You Can Use a **Ripple Tank** to Investigate **Diffraction**

1) Objects are placed into a ripple tank to create a **barrier** with a **gap** in the middle of it.
2) This gap can be varied to see the effects this has on how the waves spread through the tank.

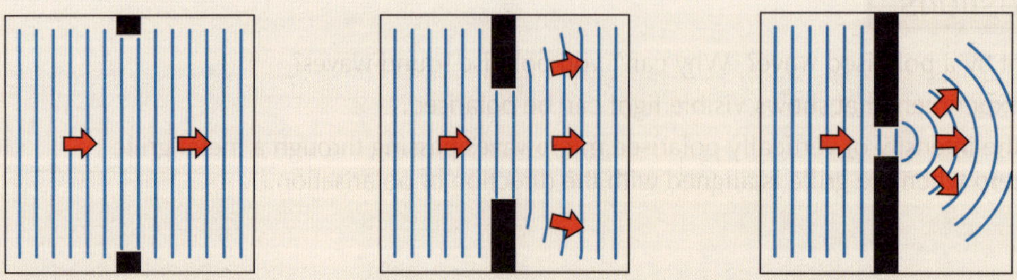

When the gap is **a lot bigger** than the **wavelength**, diffraction is **unnoticeable**.

You get **noticeable diffraction** through a gap **several** wavelengths wide.

You get the **most** diffraction when the gap is **the same** size as the **wavelength**.

As the gap decreases, the diffraction becomes more noticeable until the gap becomes too small and the water waves cannot pass through it anymore. The waves are then **reflected** back on themselves.

When **sound** passes through a **doorway**, the **size of gap** and the **wavelength** are usually roughly **equal**, so **a lot** of **diffraction** occurs. That's why you have no trouble **hearing** someone through an **open door** to the next room, even if the other person is out of your **line of sight**. The reason that you can't **see** him or her is that when **light** passes through the doorway, it is passing through a **gap** around a **hundred million times bigger** than its wavelength — the amount of diffraction is **tiny**.

Reflection and Diffraction

Diffraction can be **Explained** Using **Huygens' Construction**

1) Huygens developed a **general model** of the propagation of **waves** in what is now known as **Huygens' construction**:

> **HUYGENS' CONSTRUCTION**: Every point on a wavefront may be considered to be a **point source** of **secondary wavelets** that spread out in the forward direction at the speed of the wave. The **new wavefront** is the surface that is **tangential** to all of these **secondary wavelets**.

This diagram shows how this works:

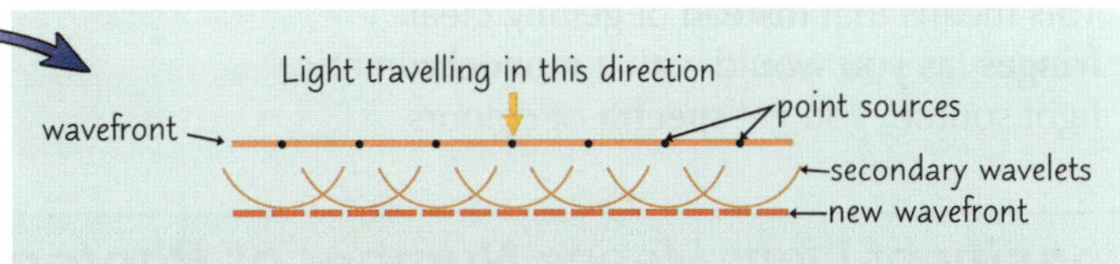

2) Huygens' construction can be used to explain the shape of the wavefronts as light travels through a **slit**. The **secondary wavelets** that pass through the slit are what produce the curve of the **new wavefront** emerging from the slit.

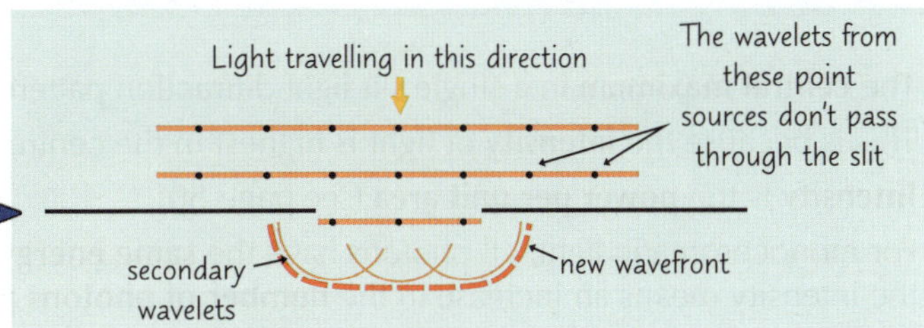

You Get **Diffraction** Around an **Obstacle** Too

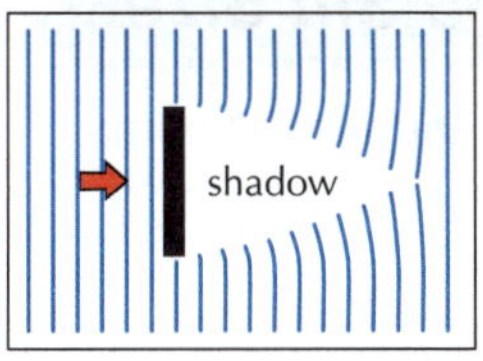

When a wave meets an **obstacle**, you get diffraction around the edges. Behind the obstacle is a '**shadow**', where the wave is blocked. The **wider** the obstacle compared with the wavelength of the wave, the less diffraction you get, and so the **longer** the shadow.

This can also be shown using Huygens' construction:

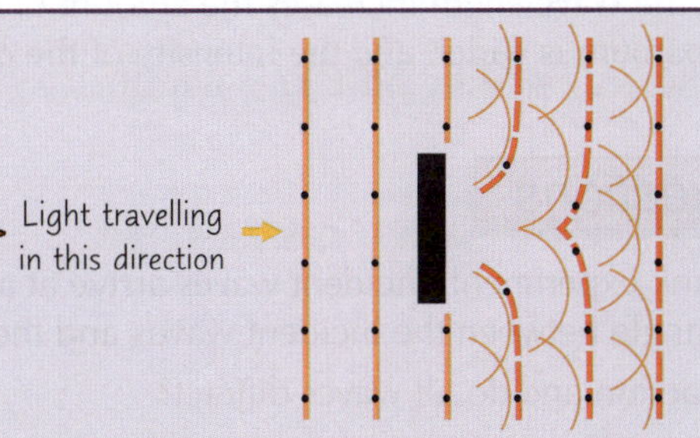

Exactly the same principles apply here as they did for the diagrams above.

Light Shone Through a **Narrow Slit** Can Form a **Diffraction Pattern**

To observe a clear **diffraction pattern** for light, you need to use a **monochromatic**, **coherent** (see p.90) light source. Monochromatic just means all the light has the same **wavelength** (and frequency) and so is the same **colour**. **Lasers** are a monochromatic and coherent light source.

Demonstrating Light Diffraction Patterns with a Laser

You need to be very careful when using lasers — see page 92.

If the wavelength of light is about the **same size** as the aperture, you get a diffraction pattern.

You'll see a **central bright fringe** (central maximum), with dark and bright fringes **alternating** on either side. The dark and bright fringes are caused by **destructive** and **constructive interference** of light waves (see p.90).

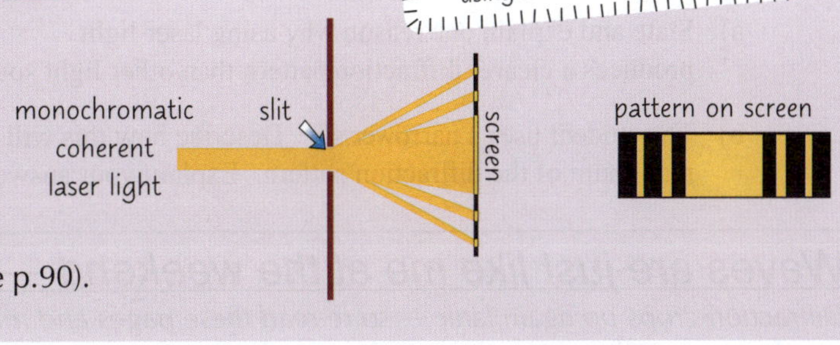

Reflection and Diffraction

Diffracted *White Light* Creates *Spectra* of Colours

1) **White light** is actually a **mixture** of different colours, each with different **wavelengths.**
2) When white light is shone through a single narrow slit, all of the different wavelengths are **diffracted** by different amounts.
3) This means that **instead** of getting clear **fringes** (as you would with a **monochromatic** light source) you get **spectra** of colours.

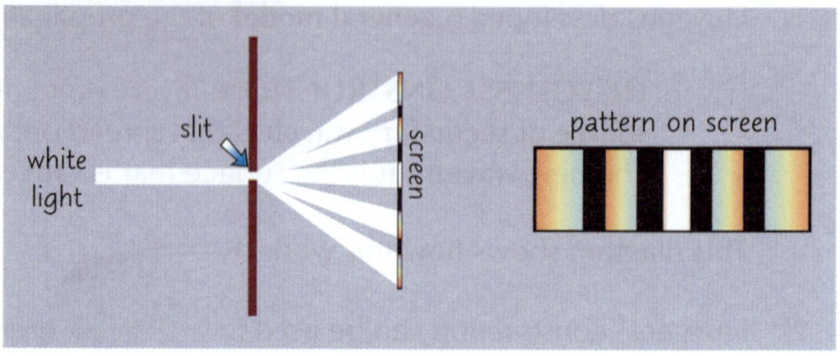

Intensity of Light Means *Number of Photons*

1) The **central maximum** in a single slit light diffraction pattern is the **brightest** part of the pattern.
2) This is because the **intensity** of light is highest in the centre.
3) **Intensity** is the **power per unit area** (see page 80).
4) For monochromatic light, all photons have the **same energy**, so an increase in the intensity means an increase in the **number of photons per second**.
5) So there are **more photons** per **unit area** hitting the central maximum per second than the other bright fringes.

See p.126 for the photon model of light.

The *Width* of the *Central Maximum* Varies with *Wavelength* and *Slit Size*

When light is shone through a single slit, there are **two** things which affect the **width** of the central maximum:

1) **Increasing** the **slit width** decreases the amount of diffraction. This means the central maximum is narrower, and the **intensity** of the central maximum is **higher**.
2) **Increasing** the **wavelength** increases the amount of diffraction. This means the central maximum is wider, and the intensity of the central maximum is **lower**.

Practice Questions

Q1 In a ripple tank experiment, incident waves arrive at a barrier's surface at an angle of 30° to the normal. What is the angle between the incident waves and the reflected waves?

Q2 What is diffraction and do all waves diffract?

Q3 a) Sketch the pattern produced when monochromatic light is shone through a narrow slit onto a screen.
b) If the wavelength of the monochromatic light was decreased, what would happen to the central maximum?

Q4 What would you expect to see when white light is shone through a thin slit onto a screen?

Exam Questions

Q1 A mountain lies directly between you and a radio transmitter. Explain, with the use of a diagram, why you can pick up long-wave radio broadcasts from the transmitter but not short-wave radio broadcasts. [3 marks]

Q2 A student shines a laser beam through a narrow slit onto a screen.

a) State and explain one reason why using laser light produces a clearer diffraction pattern than other light sources. [2 marks]

b) The student uses a narrower slit. Describe how this will affect the central maximum of the diffraction pattern. Explain your answer. [3 marks]

Waves are just like me at the weekend — they like to spread out...

Diffraction crops up again later — so re-read these pages and make sure you really understand it.

Refraction

The stuff on the next few pages explains why your legs look short in a swimming pool.

Refraction Occurs When the Medium a Wave is Travelling in Changes

Refraction is the way a wave **changes direction** as it enters a **different medium**. The change in direction is a result of the wave **slowing down** or **speeding up**. You can tell if the wave is speeding up or slowing down by the way it **bends towards** or **away** from the normal.

1) If a light ray bends **towards** the normal — it is **slowing** down. The ray is going from a **less** optically dense material to a **more** optically dense material.
2) If the ray bends **away** from the normal — the wave is **speeding up**. It is going from an optically **denser** material to a **less** optically dense material.
3) The speed changes because the **wavelength** of the wave is changing and the **frequency** stays **constant** ($v = f\lambda$).

The more optically dense material will have a higher refractive index (see below).

If light travels from a **less** optically dense material to a **more** optically dense material, the wave **slows down**, the **wavelength decreases** and the **frequency** stays the **same**.

The Refractive Index of a Material Measures How Much It Slows Down Light

Light goes fastest in a **vacuum**. It **slows down** in other materials, because it **interacts** with the particles in them. The more **optically dense** a material is, the more light **slows down** when it enters it.

The **refractive index** of a material, n, is the **ratio** between the **speed of light** in a **vacuum**, c, and the speed of light in that **material**, c_s.

$$n = \frac{c}{c_s}$$

$c = 3.00 \times 10^8$ ms^{-1}.

The speed of light in air is only a tiny bit smaller than c. So you can assume the refractive index of air is 1.

Snell's Law uses Angles to Calculate the Refractive Index

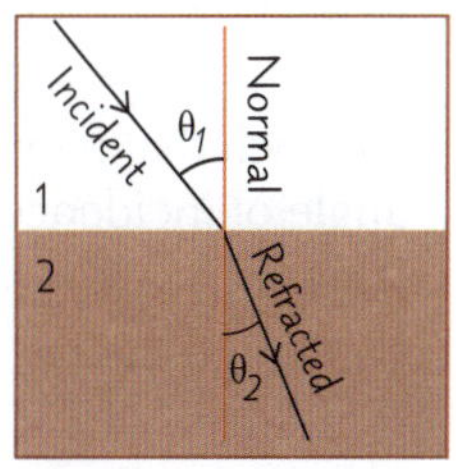

1) The **angle** the **incoming light** makes to the **normal** is called the **angle of incidence**, θ_1. The **angle** the **refracted ray** makes with the **normal** is the **angle of refraction**, θ_2.
2) The light is crossing a **boundary**, going from a medium with **refractive index** n_1 to a medium with refractive index n_2.
3) When light enters an **optically denser** medium it is refracted **towards** the normal.
4) n_1, n_2, θ_1 and θ_2 are related by **Snell's law**: $n_1 \sin\theta_1 = n_2 \sin\theta_2$

You can do this experiment to find the refractive index of any solid, transparent material.

You can use a **ray box** to find the **refractive index** of a glass block:

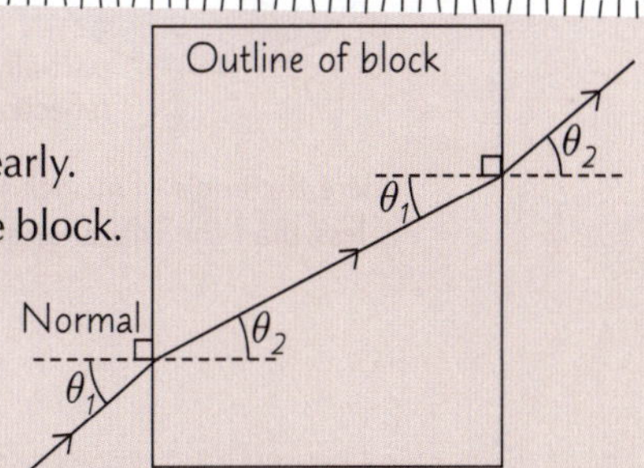

1) Place a glass block on a piece of paper and draw around it.
2) Use the ray box to shine a beam of light into the glass block. Turn off any other lights so you can see the path of the light beam through the block clearly.
3) **Trace** the path of the **incoming** and **outgoing** beams of light either side of the block.
4) Remove the block and join up the two paths you've drawn with a **straight line** that follows the path the light beam took through the glass block. You should be able to see from your drawing how the path of the ray **bent** when entering and leaving the block.
5) Measure the angles of incidence (θ_1) and refraction (θ_2) where the light enters and exits the block. Air is **less** optically dense than glass, so as the light **enters** the glass block it **bends towards** the normal ($\theta_1 > \theta_2$) as it **slows down**. The beam should **bend away** from the normal as it **exits** the block ($\theta_2 > \theta_1$) and **speeds up**.
6) Rearrange **Snell's law** to make the refractive index of the **material** the subject, and substitute in $n = 1$ for **air** and the values you found for θ_1 and θ_2 to calculate a value for n.
7) The **percentage uncertainty** (see p.10) in your measurements for θ_1 and θ_2 will be **smaller** for **larger angles**, so it's better to do the experiment at large angles and then **repeat** the experiment to find an **average** of your results.
8) Your result should be more **precise** (see p.12) if you use a **narrower** beam of light as the **uncertainty** in the **position** of the beam will be lower.

Alternatively, you could plot a graph of $\sin\theta_1$ against $\sin\theta_2$. The gradient of the graph will be equal to the refractive index of the material.

You might see the angles of incident and refraction written as θ_i and θ_r.

Refraction

*Light Leaving an **Optically Denser Material** is Refracted **Away** from the **Normal***

When light **goes from** an optically denser material into an optically **less dense** material (e.g. glass to air), interesting things can happen.

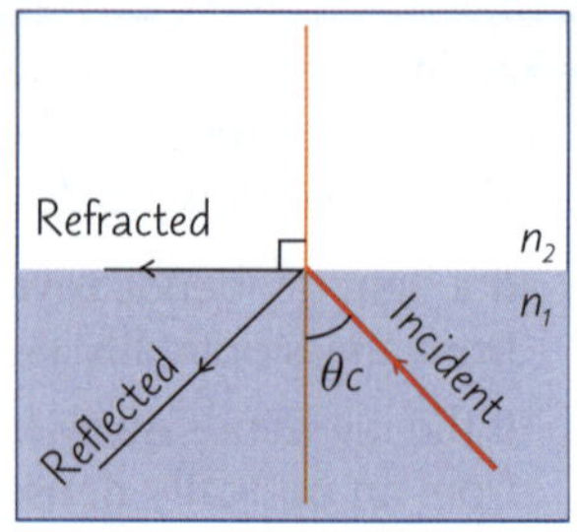

1) Shine a ray of light at a boundary going from refractive index n_1 to n_2, then gradually **increase** the angle of incidence.
2) The light is refracted away from the normal, so as you increase the angle of incidence, the angle of **refraction** gets closer and closer to **90°**.
3) Eventually θ_1 reaches a **critical angle** θ_c for which $\theta_2 = 90°$. The light is **refracted** along the **boundary**.
4) As sin 90° = 1, Snell's law, $n_1 \sin\theta_1 = n_2 \sin\theta_2$, becomes $n_1 \sin\theta_c = n_2 \times 1$ so:

$$\sin\theta_c = \frac{n_2}{n_1} = {}_1n_2$$

where ${}_1n_2$ is the relative refractive index between material 1 and material 2

5) For a **material-to-air** boundary, this simplifies to: $\sin\theta_c = \frac{1}{n}$ where n is the refractive index of the material
6) As θ_1 becomes **greater** than the **critical angle**, refraction is impossible. All the light is **reflected** back into the material — this is called **total internal reflection**.

Since sin can only take values between -1 and 1, total internal reflection can only happen if $\sin\theta_c < 1$ so ${}_1n_2 < 1$.

*You can Investigate **Critical Angles** and **Total Internal Reflection** with **Glass Blocks***

1) Shine a light ray into the **curved face** of a semi-circular glass block so that it always enters at **right angles** to the edge — this means the ray won't **refract** as it enters the block, just when it leaves from the straight edge.
2) Vary the angle of **incidence**, θ_1, until the light beam refracts so much that it exits the block along the **straight edge**. This angle of incidence is the **critical angle**, θ_c, for glass-air boundary.
3) If you increase the angle of incidence so it's **greater** than θ_c, you'll find the ray is reflected from the straight edge of the block.

You can rearrange the formula for the critical angle above and put in your value for θ_c to find the refractive index of the block.

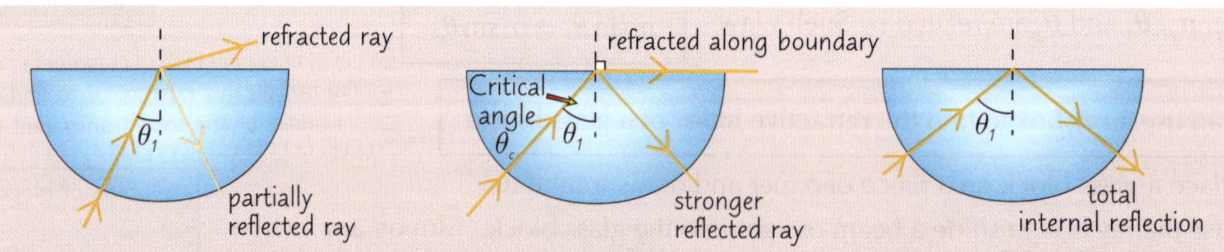

Here the angle of incidence is **less** than the critical angle

Here the angle of incidence is **equal** to the critical angle

Here the angle of incidence is **greater** than the critical angle.

Optical Fibres** Use **Total Internal Reflection

An optical fibre is a very **thin flexible tube** of **glass** or **plastic** fibre that can carry **light signals** over long distances and round corners. One type is a **step-index** optical fibre.

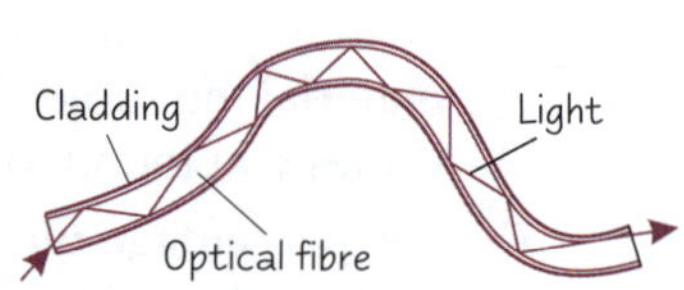

1) Step-index optical fibres themselves have a **high refractive index** but are surrounded by **cladding** with a lower refractive index to allow **total internal reflection**. Cladding also protects the fibre from **scratches** which could let **light escape**.
2) Light is shone in at **one end** of the fibre. The fibre is so **narrow** that the light always **hits the boundary** between the fibre and cladding at an **angle bigger** than the **critical angle**.
3) So all the light is **totally internally reflected** from boundary to boundary until it reaches the other end.

Refraction

Dispersion and Absorption Cause Signal Degradation

A **signal** (a stream of pulses of light) travelling down an optical fibre can be **degraded** by **absorption** or by **dispersion**. **Signal degradation** can cause **information** to be **lost**.

Absorption Causes Loss in Amplitude

As the signal travels, some of its energy is lost through **absorption** by the **material** the fibre is made from. This energy loss results in the **amplitude** of the signal being **reduced**.

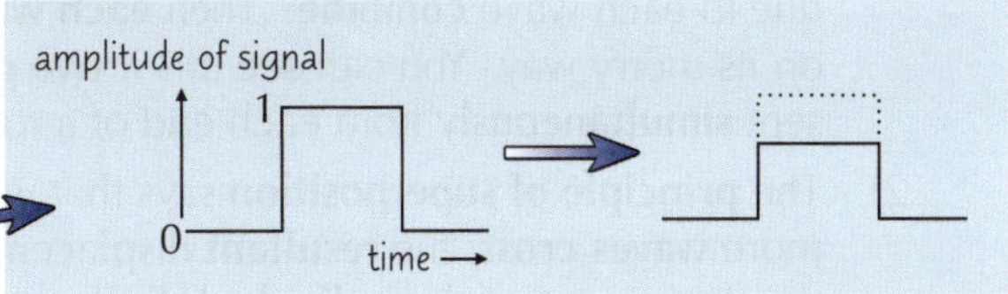

Dispersion Causes Pulse Broadening

There are two types of dispersion that can degrade a signal.

1) **Modal dispersion** — light rays enter the fibre at different angles, and so take different **paths**. The rays which take a **longer path** take longer to reach the other end than those that travel down the **middle** of the fibre. A **single-mode** fibre only lets light take **one path**, so it stops modal dispersion.
2) **Material dispersion** — light consists of different **wavelengths** that travel at different speeds in the fibre — this causes some light wavelengths to reach the end of the fibre faster than others. Using **monochromatic light** can stop material dispersion.

Both types of dispersion lead to **pulse broadening**. The signal sent down the fibre is broader at the other end. Broadened pulses can **overlap** each other and confuse the signal.

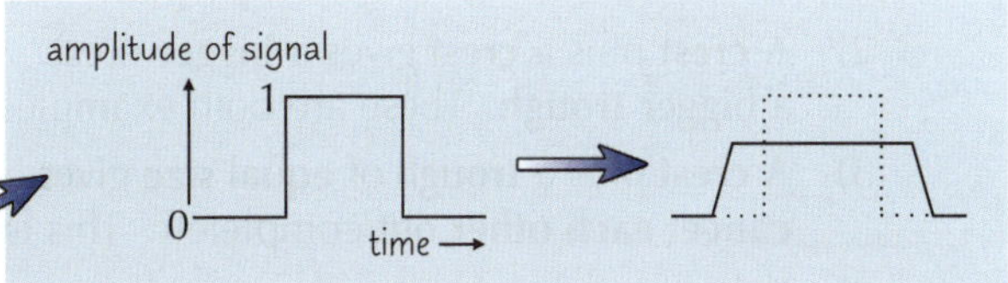

An **optical fibre repeater** can be used to **boost** and regenerate the signal every so often, which can **reduce** signal **degradation** caused by both **absorption** and **dispersion**.

Practice Questions

Q1 Why does light go fastest in a vacuum and slow down in other media?

Q2 What is the formula for the critical angle for a ray of light at a material/air boundary?

Exam Questions

Q1 a) Light travels in diamond at 1.24×10^8 ms^{-1}. Calculate the refractive index of diamond. [1 mark]

b) Calculate the angle of refraction if light strikes a facet of a diamond ring at an angle of 50° (to 2 s.f.) to the normal of the air/diamond boundary. [2 marks]

Q2 An adjustable underwater spotlight is placed on the floor of an aquarium tank. When the light points upwards at a steep angle a beam comes through the surface of the water into the air, and the tank is dimly lit. When the spotlight is placed at a shallower angle, no light comes up through the water surface, and the tank is brightly lit.

a) Explain what is happening. [2 marks]

b) It is found that the beam into the air disappears when the spotlight is pointed at any angle of less than 41.25° to the floor. Calculate the refractive index of water. [2 marks]

Q3 a) Explain the ways in which the cladding is designed to keep transmitted light inside an optical fibre. [2 marks]

b) The cladding functions as expected, but there is still some information loss when a step-index optical fibre is used to transmit light signals over long distances.
Discuss the potential causes of this loss of information and how the design and operation of the optical fibre could be altered to reduce information loss over long transmission distances. [6 marks]

I don't care about expensive things — all I care about is wave speed...

Physics examiners are always saying how candidates do worst in the waves bit of the exam. You'd think they'd have something more important to worry about — third world poverty, war, Posh & Becks... But no.

Superposition and Coherence

When two waves get together, it can be either really impressive or really disappointing.

Superposition Happens When Two or More Waves Pass Through Each Other

1) At the **instant** the waves **cross**, the **displacements** due to each wave **combine**. Then **each wave** goes on its merry way. You can **see** this if **two pulses** are sent **simultaneously** from each end of a rope.
2) The **principle of superposition** says that when two or more **waves cross**, the **resultant** displacement equals the **vector sum** of the **individual** displacements.

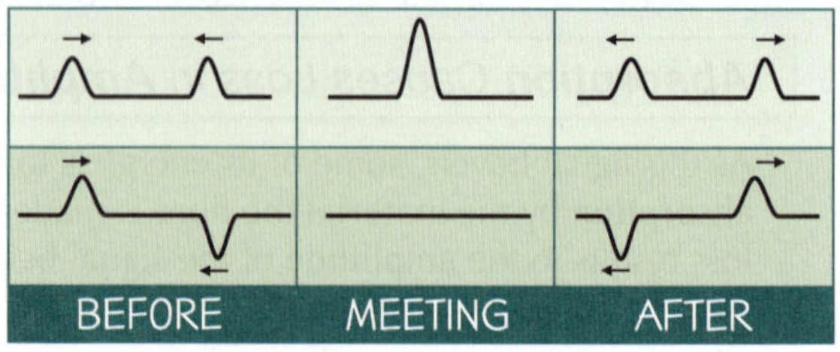

"**Superposition**" means "one thing on top of another thing". You can use the same idea in **reverse** — a **complex wave** can be separated out mathematically into **several simple** sine waves of various sizes.

Interference can be Constructive or Destructive

1) When two or more waves **superpose** with each other, the effect is called **interference**.
2) A **crest** plus a **crest** gives a **bigger crest**. A **trough** plus a **trough** gives a **bigger trough**. These are both examples of **constructive interference**.
3) A **crest** plus a **trough** of **equal size** gives... **nothing**. The two displacements **cancel each other out** completely. This is called **destructive interference**.
4) If the **crest** and the **trough** aren't the **same size**, then the destructive interference **isn't total**. For the interference to be **noticeable**, the two **amplitudes** should be **nearly equal**.

Graphically, you can superimpose waves by adding the individual displacements at each point along the *x*-axis, and then plotting them.

In Phase Means In Step — Two Waves in Phase Have the Same Wavelength

1) Two points on a wave are **in phase** if they are both at the **same point** in the **wave cycle**. Points in phase have the **same displacement** and **velocity**.
2) On the graph on the right, points **A** and **B** are **in phase**; points **A** and **C** are **out of phase**.
3) It's mathematically **handy** to show one **complete cycle** of a wave as an **angle of 360° (2π radians)**.
4) **Two points** with a **phase difference** of **zero** or a **multiple of 360°** are **in phase**.
5) **Points** with a **phase difference** of **odd-number multiples** of **180° (π radians)** are **exactly out of phase**.

To convert from degrees to radians, multiply by π/180°. To convert from radians to degrees, multiply by 180°/π.

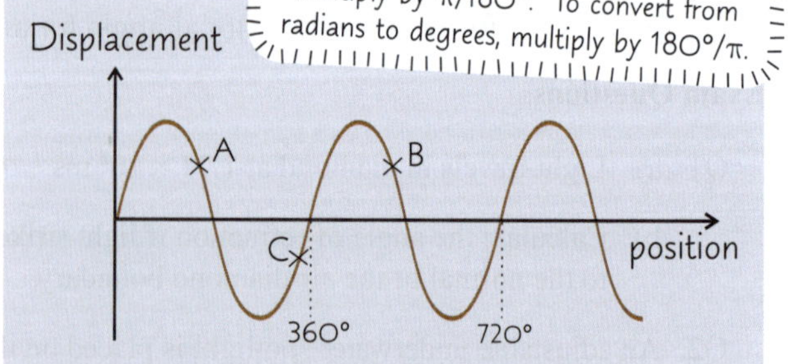

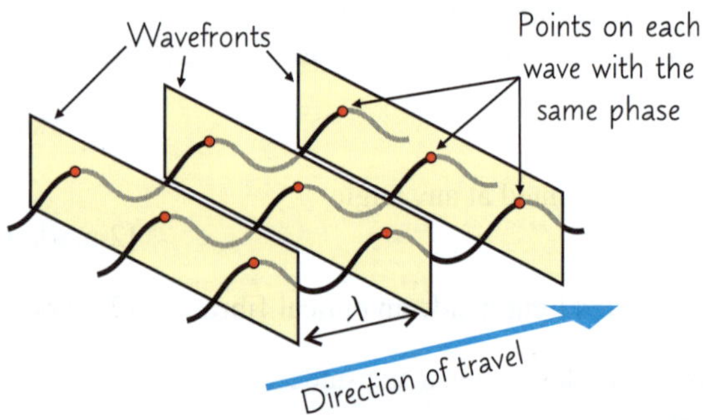

6) You can also talk about two **different waves** being **in phase**. **In practice** this usually happens because **both** waves came from the **same oscillator**. In **other** situations there will nearly always be a **phase difference** between two waves.
7) Two or more waves that are **coherent** (see below), **in phase** and travelling in the **same direction** will have **wavefronts**. These are imaginary planes that cut **across** all the waves, joining up all the **points** that are **in phase** with each other. The **distance** between each wavefront is equal to one **wavelength**, i.e. each wavefront is at the same point in the **cycle**.

To Get Interference Patterns the Two Sources Must Be Coherent

Interference **still happens** when you're observing waves of **different wavelength** and **frequency** — but it happens in a **jumble**. In order to get clear **interference patterns** (see p.92), the two or more sources must be **coherent**.

Coherent sources — they have the **same wavelength** and **frequency** and a **fixed phase difference** between them.

Superposition and Coherence

Constructive or Destructive Interference Depends on the Path Difference

1) Whether you get **constructive** or **destructive** interference at a **point** depends on how **much further one wave** has travelled than the **other wave** to get to that point (assuming the sources are coherent and in phase).
2) The **amount** by which the path travelled by one wave is **longer** than the path travelled by the other wave is called the **path difference**.
3) At **any point an equal distance** from both sources you will get **constructive interference**. You also get constructive interference at any point where the **path difference** is a **whole number of wavelengths**, because the waves arrive at the same point **in phase**. At points where the path difference is an odd number of **half wavelengths**, the waves arrive **out of phase** and you get **destructive interference**.

Constructive interference occurs when: **path difference = $n\lambda$ (where n is an integer)**

Destructive interference occurs when: **path difference = $\frac{(2n+1)\lambda}{2} = (n + \frac{1}{2})\lambda$**

If the sources are not in phase but are coherent, there will be constructive and destructive interference, but they will not occur at these path differences.

You Can Observe Interference With Sound Waves

1) Connect two **speakers** to the same oscillator (so the sound waves from them are **coherent** and **in phase**) and place them in line with each other.
2) Walk slowly across the room in front of them.
3) You will hear varying volumes of sound. At the points where the sound is **loudest**, the **path difference** is a **whole** wavelength.
4) The sound will be quietest at points where the path difference is an **odd** number of **half wavelengths**.

You may still hear some sound at the quietest points due to sound being reflected off walls and around the room.

Practice Questions

Q1 What is the principle of superposition?

Q2 If two points on a wave have a phase difference of 1440°, are they in phase?

Q3 What is a wavefront?

Q4 If there was a path difference of 5λ between two coherent, in phase waves, what kind of interference would occur?

Exam Questions

Q1 a) Two wave sources are coherent. Explain what this means. [2 marks]

b) Explain why you might have difficulty in observing interference patterns in an area affected by two waves from two sources even though the two sources are coherent. [1 mark]

Q2 Two waves from coherent sources meet and interfere. Which row of the table shows the correct type of interference that would occur with the stated phase and path difference? [1 mark]

	Phase Difference	Path Difference	Type of Interference
A	180°	λ	Constructive
B	180°	λ/2	Constructive
C	360°	λ	Destructive
D	360°	λ/2	Constructive

Learn this and you'll be in a super position to pass your exam...

...I'll get my coat.

A few crucial concepts here: a) interference can be constructive or destructive, b) you get constructive interference when the path difference is a whole number of wavelengths (for sources in phase), c) the sources must be coherent.

Two-Source Interference

You know what happens with one light source (p.85-86), so now it's time to see what happens with two. I can't wait...

*Demonstrating **Two-Source** Interference in **Water** and **Sound** is Easy*

1) It's **easy** to demonstrate **two-source interference** for either **sound** or **water** because they've got **wavelengths** of a handy **size** that you can **measure**.
2) You need **coherent** sources, which means the **wavelength** and **frequency** have to be the **same**. The trick is to use the **same oscillator** to drive **both sources**. For **water**, one **vibrator drives two dippers**. For sound, **one oscillator** is connected to **two loudspeakers**. (See diagram on page 91.)

*Demonstrating **Two-Source** Interference for **Light** is Harder*

Young's Double-Slit Experiment

1) To see **two-source** interference with light, you can either use two separate, **coherent** light sources or you can shine a **laser** through **two slits**. Laser light is **coherent** and **monochromatic**.
2) Young's double-slit experiment shines a laser through two slits onto a **screen**.
3) The slits have to be about the **same size** as the **wavelength** of the laser light so that it is diffracted — then the light from the slits acts like two coherent point sources.

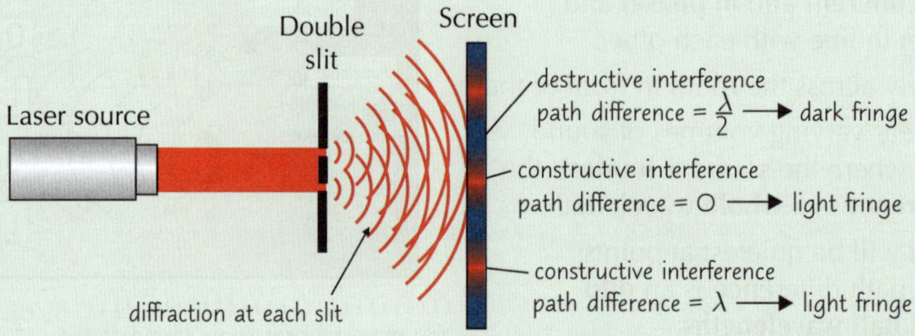

4) You get a pattern of light and dark **fringes**, depending on whether constructive or destructive **interference** is taking place. Thomas Young — the first person to do this experiment (with a lamp rather than a laser) — came up with an **equation** to **work out** the **wavelength** of the **light** from this experiment (see p.93).

Working with lasers is very **dangerous** because laser light is focused into a very direct, powerful beam of monochromatic light. If you looked at a laser beam **directly**, your eye's lens would focus it onto your retina, which would be **permanently damaged**.

To make sure you don't cause **damage** while using lasers, you should:

1) **Never** shine the laser **towards** a person.
2) Wear laser **safety goggles**.
3) Avoid shining the laser beam at a **reflective surface**.
4) Have a **warning sign** on display.
5) Turn the laser **off** when it's not needed.

You Can Do a Similar Experiment with Microwaves

To see interference patterns with **microwaves**, you can **replace** the laser and slits with two microwave **transmitter cones** attached to the **same** signal generator.

You also need to replace the screen with a microwave **receiver probe** (like the one used in the stationary waves experiment on page 98).

If you move the probe along the path of the green arrow, you'll get an **alternating pattern** of **strong** and **weak** signals — just like the light and dark fringes on the screen.

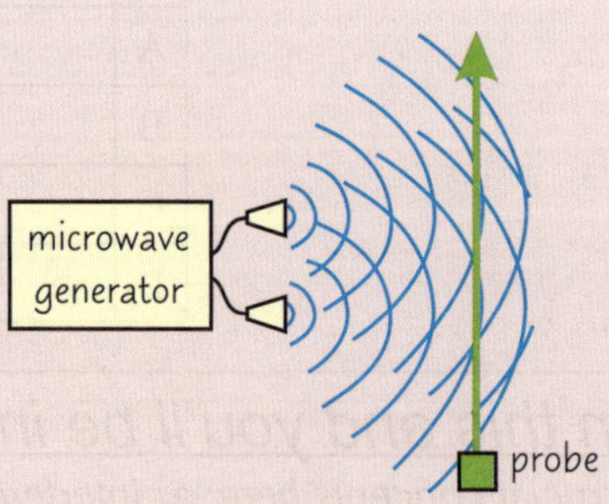

Two-Source Interference

Work Out the Wavelength with *Young's Double-Slit Formula*

1) The fringe spacing (***w***), wavelength (**λ**), spacing between slits (***s***) and the distance from slits to screen (***D***) are all related by **Young's double-slit formula**, which works for all waves.

Fringe spacing, $w = \frac{\lambda D}{s}$

'Fringe spacing' means the distance from the centre of one minimum to the centre of the next minimum or from the centre of one maximum to the centre of the next maximum.

Always check your fringe spacing.

2) Since the wavelength of light is so small you can see from the formula that a high ratio of ***D / s*** is needed to make the fringe spacing **big enough to see**.
3) Rearranging, you can use **λ = *ws / D*** to **calculate the wavelength** of light.
4) The fringes are **so tiny** that it's very hard to get an **accurate value of *w***. It's easier to measure across **several** fringes then **divide** by the number of **fringe widths** between them.

Some versions of this formula replace *w* with *x* or *d* and *s* with *a*.

Young's Experiment was *Evidence* for the *Wave Nature* of EM Radiation

1) Towards the end of the **17th century**, two important **theories of light** were published — one by Isaac Newton and the other by a chap called Huygens. **Newton's** theory suggested that light was made up of tiny particles, which he called "**corpuscles**". And **Huygens** put forward a theory using **waves**.
2) The **corpuscular theory** could explain **reflection** and **refraction**, but **diffraction** and **interference** are both **uniquely** wave properties. If it could be **shown** that light showed interference patterns, that would help settle the argument once and for all.
3) **Young's** double-slit experiment (over 100 years later) provided the necessary evidence. It showed that light could both **diffract** (through the narrow slits) and **interfere** (to form the interference pattern on the screen).

Of course, this being Physics, nothing's ever simple — give it another 100 years or so and the debate would be raging again, (p.126).

Practice Questions

Q1 In Young's experiment, why do you get a bright fringe at a point equidistant from both slits?

Q2 What does Young's experiment show about the nature of light?

Q3 Write down Young's double-slit formula.

Exam Questions

Q1 a) The diagram on the right shows two coherent light sources, S_1 and S_2, being shone on a screen. State the measurements you would need to take to calculate the wavelength of the light. [3 marks]

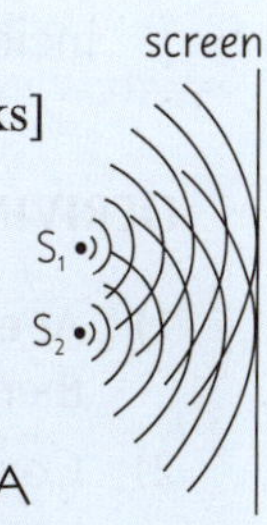

b) S_1 and S_2 may be slits in a screen behind which there is a source of laser light, instead of being two separate sources. State two safety precautions you should take when using this set-up. Explain your answer. [2 marks]

Q2 In an experiment to study sound interference, two loudspeakers are connected to an oscillator emitting sound at 1320 Hz and set up as shown in the diagram. They are 1.5 m apart and 7.3 m away from the line AC. A listener moving along the line hears minimum sound at A, maximum sound at B and minimum sound again at C. (You may assume that Young's double-slit formula can be used in this calculation).

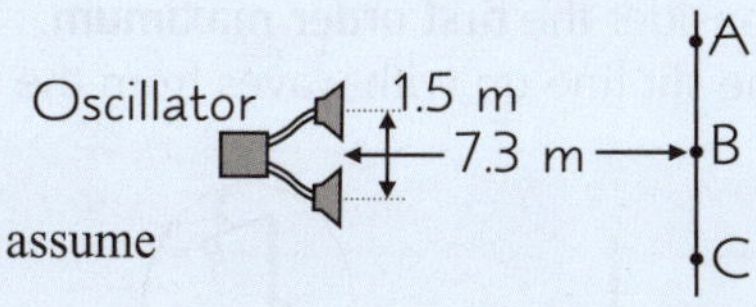

a) Calculate the wavelength of the sound waves if the speed of sound in air is taken to be 330 ms^{-1}. [1 mark]

b) Calculate the separation of points A and C. [1 mark]

Learn this stuff — or you'll be playing ketchup...

A few things to learn here — some diffraction experiments, a formula and a little bit of history. Be careful when you're calculating the fringe width by averaging over several fringes. Don't just divide by the number of bright lines. Ten bright lines will only have nine fringe widths between them, not ten. It's an easy mistake to make, but you have been warned.

Diffraction Gratings

What could possibly be more exciting than shining a laser through two slits? Shining a laser through more than two slits of course. Jeez, ask a stupid question...

Interference Patterns Get **Sharper** When You Diffract Through **More Slits**

1) You saw on pages 92-93 that **two sources** that produce waves that are **coherent** and **in phase** will result in an **interference pattern** (alternating bands of constructive and destructive interference). If you are using **light waves**, you can pass one **monochromatic** (one wavelength) **beam of light** through **two slits**. The slits are cut into the **same** piece of material, so that the wave passes through them at the **same time**. The light **diffracts** at both slits, producing two coherent sources of light. The interference pattern is made up of alternating **dark bands** and **light bands**.
2) You can repeat this experiment with **more than two equally spaced** slits. You get basically the **same shaped** pattern as for two slits — but the **bright bands** are **brighter** and **narrower** and the **dark areas** between are **darker**.
3) When **monochromatic light** is passed through a **grating**, which has **hundreds** of slits per millimetre, the interference pattern is **really sharp** because there are so **many beams reinforcing** the **pattern**.
4) Sharper fringes mean the distance between the maxima can be easily measured (**fringe width**).

You can **Measure** the **Wavelength** of Light using a **Diffraction Grating**

1) For **monochromatic** light, all the **maxima** are sharp lines. (It's different for white light — see page 95.)
2) There's a line of **maximum brightness** at the centre called the **zero order** line.
3) The lines just **either side** of the central one are called **first order lines**. The **next pair out** are called **second order** lines and so on.
4) Using the **fringe width**, x, and the distance to the screen, D, the angle the 1st order fringe makes with the zero order line can be calculated using **small angle approximations**.

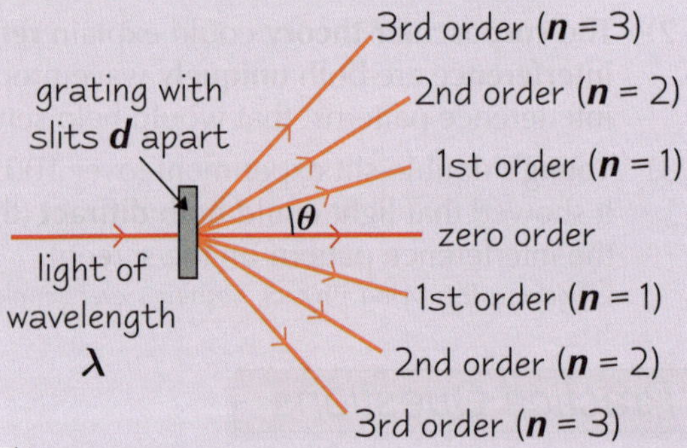

$$\tan\theta \approx \theta \text{ and } \tan\theta = \frac{x}{D}, \text{ so } \theta \approx \frac{x}{D}$$

If the grating has N slits per metre, then the slit spacing, d, is just $1/N$ metres.

5) If you know the slit separation, d, what order maximum you're observing, n, and the angle between this maximum and the incident light, θ, you can find the **wavelength** of the incident light.

$$d \sin\theta = n\lambda$$

DERIVING THE EQUATION:

1) At **each slit**, the incoming waves are **diffracted**. These diffracted waves then **interfere** with each other to produce an **interference pattern**.
2) Consider the **first order maximum**. This happens at the **angle** when the waves from one slit line up with waves from the **next slit** that are **exactly one wavelength** behind.

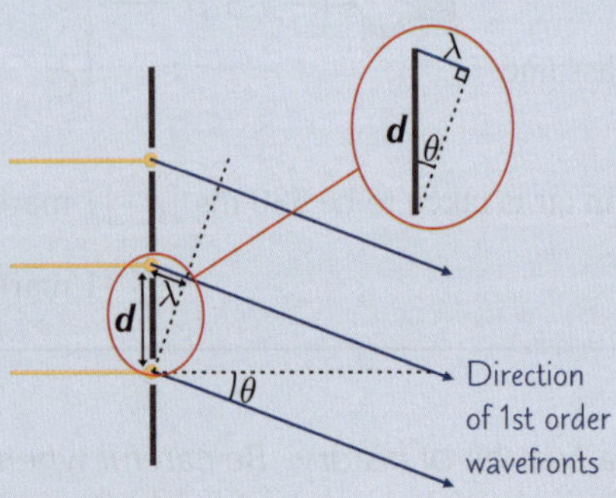

3) Call the **angle** between the **first order maximum** and the **incoming light** θ.
4) Now, look at the **triangle** highlighted in the diagram. The angle is θ (using basic geometry), d is the slit spacing and the **path difference** is λ.
5) So, for the first maximum, using trig:
$$d \sin\theta = \lambda$$
6) The other maxima occur when the path difference is 2λ, 3λ, 4λ, etc. So to make the equation **general**, just replace λ with $n\lambda$, where n is an integer — the **order** of the maximum.

Diffraction Gratings

You can Draw **General Conclusions** from **d sin θ = nλ**

1) If λ is **bigger**, **sin θ** is **bigger**, and so θ is **bigger**. This means that the larger the **wavelength**, the more the pattern will **spread out**.
2) If d is **bigger**, **sin θ is smaller**. This means that the **coarser** the **grating**, the **less** the pattern will **spread out**.
3) Values of **sin θ** greater than **1** are **impossible**. So if for a certain n you get a result of **more than 1** for **sin θ** you know that that order **doesn't exist**.

Diffraction Gratings Help to Identify **Elements** and Calculate **Atomic Spacing**

1) If you **diffract** white light through a **grating** then the patterns due to **different wavelengths** within the white light are **spread out** by **different** amounts (just like when white light is diffracted using a single slit (see page 86)).
2) Each **order** in the pattern becomes a **spectrum**, with **red** on the **outside** and **violet** on the **inside**. The **zero order maximum** stays **white** because all the wavelengths just pass straight through.
3) **Astronomers** and **chemists** often need to study spectra to help identify elements. They use diffraction gratings rather than prisms because they're **more accurate**.

4) The wavelength of **X-rays** is of a similar scale to the spacing between **atoms** in crystalline solids. This means that X-rays will form a **diffraction pattern** when directed at a thin crystal.
5) The crystal acts like a **diffraction grating** and the spacing between **atoms** (slit width) can be found from the diffraction pattern.
6) This is called **X-ray crystallography** — it was used to discover the structure of **DNA**.

Practice Questions

Q1 What equation is used to find the angle between the nth order maximum and the incident beam for a diffraction grating?

Q2 Derive the equation you quoted in Q1.

Q3 How is the diffraction grating pattern for white light different from the pattern for laser light?

Q4 What difference does it make to the pattern if you use a finer grating?

Exam Questions

Q1 Yellow laser light of wavelength 6.0×10^{-7} m is transmitted through a diffraction grating of 4.0×10^{5} lines per metre.

a) Calculate the angle to the normal at which the first and second order bright lines are seen. [3 marks]

b) State whether there is a fifth order line. Explain your answer. [1 mark]

Q2 Visible, monochromatic light is transmitted through a diffraction grating of 3.7×10^{5} lines per metre. The first order maximum is at an angle of 14.2° to the incident beam. Calculate the wavelength of the incident light. [2 marks]

Ooooooooooooo — pretty patterns...

Derivation — ouch. At least it's not a bad one though. As long as you learn the diagram, it's just geometry and a bit of trig from there. Make sure you learn the equation — that way, you know what you're aiming for. As for the rest of these pages, remember that the more slits you have, the sharper the image — and white light makes a pretty spectrum.

Stationary (Standing) Waves

Stationary waves are waves that... er... stand still... well, not still exactly... I mean, well... they don't go anywhere... um...

You get Stationary Waves When a Progressive Wave is Reflected at a Boundary

A stationary wave is the **superposition** of **two progressive waves** with the **same wavelength**, moving in **opposite directions**.

The progressive waves must also have the same speed and frequency.

1) Unlike progressive waves, **no energy** is transmitted by a stationary wave.
2) You can demonstrate stationary waves by attaching a **vibration transducer** at one end of a **stretched string** with the other end fixed. The transducer is given a wave frequency by a **signal generator** and creates that wave by vibrating the string.
3) The wave generated by the vibration transducer is **reflected** back and forth.
4) For most frequencies the resultant **pattern** is a **jumble**. However, if you alter the **signal generator** so the **transducer** produces an **exact number of waves** in the time it takes for a wave to get to the **end** and **back again**, then the **original** and **reflected** waves **reinforce** each other.
5) At these **"resonant frequencies"** you get a **stationary** (or **standing**) **wave** where the **pattern doesn't move** — it just sits there, bobbing up and down. Happy, at peace with the world...

A sitting wave.

Stationary Waves in Strings Form Oscillating "Loops" Separated by Nodes

1) Each particle vibrates at **right angles** to the string.
2) **Nodes** are where the **amplitude** of the vibration is **zero**.
3) **Antinodes** are points of **maximum amplitude**.
4) At resonant frequencies, an **exact number** of **half wavelengths** fits onto the string.

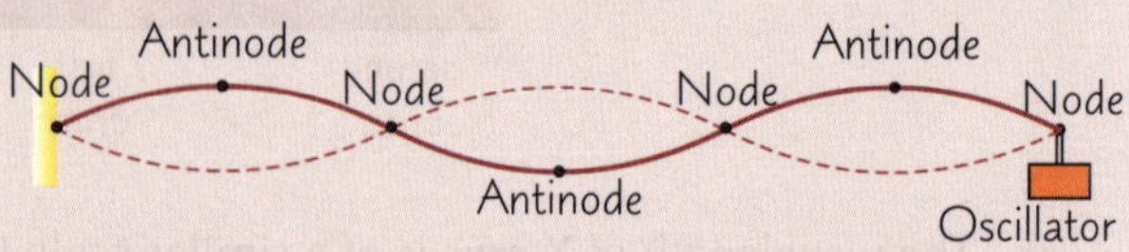

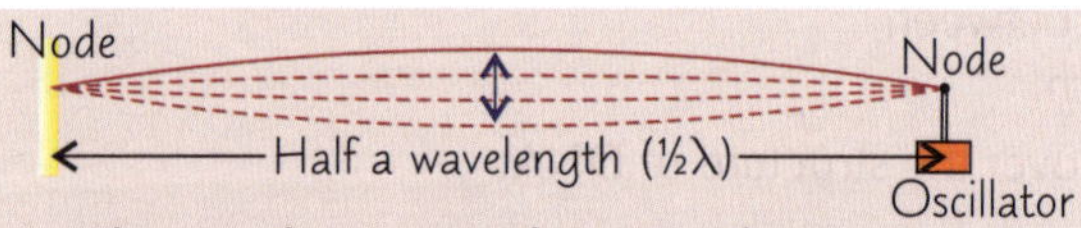

The standing wave above is vibrating at the **lowest possible** resonant frequency (the **fundamental mode of vibration** — also called the **first harmonic**). It has **one "loop"** with a **node at each end**.

This is the **second harmonic**. It is **twice** the **fundamental mode of vibration**. There are two **"loops"** with a **node** in the **middle** and **one at each end**.

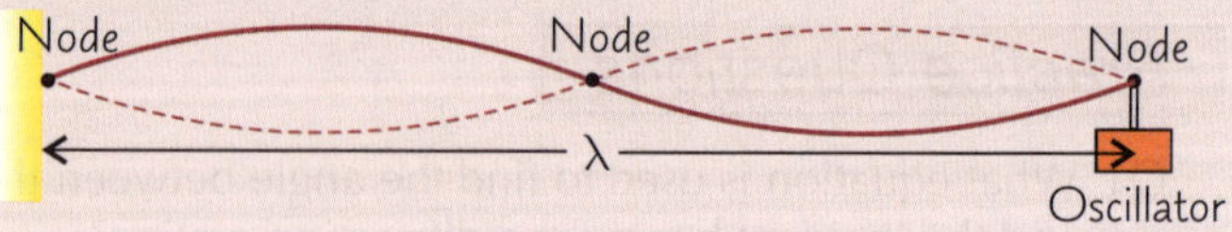

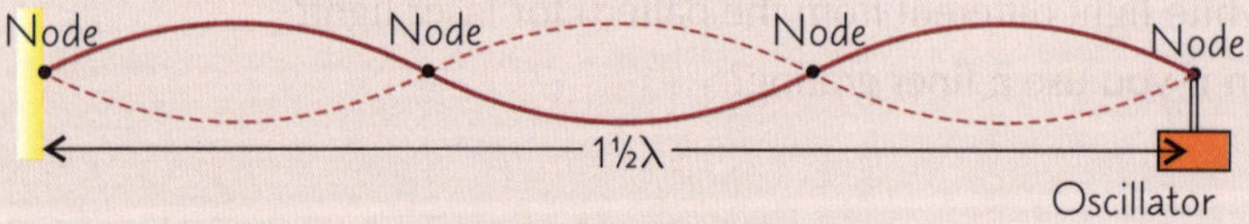

The **third harmonic** is **three times** the fundamental mode of vibration. **1½ wavelengths** fit on the string.

You can Demonstrate Stationary Waves with Sounds

Powder Can Show Stationary Waves in a Tube of Air

Stationary **sound** waves are produced in the **glass tube**.

The lycopodium **powder** (don't worry, you don't need to know what that is) laid along the bottom of the tube is **shaken away** from the **antinodes** but left **undisturbed** at the **nodes**.

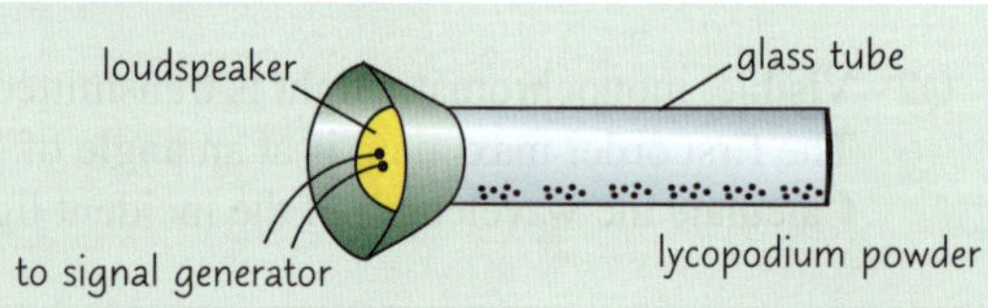

You can also demonstrate stationary waves with microwaves. This is covered on page 98.

Stationary (Standing) Waves

You Can Investigate **Factors** *Affecting the* **Resonant Frequencies** *of a String*

1) Start by measuring the **mass** (M) and **length** (L) of strings of different types using a **mass balance** and a ruler. Then find the **mass per unit length** of each string (μ) using: $\mu = \frac{M}{L}$ The units of μ are kgm^{-1}

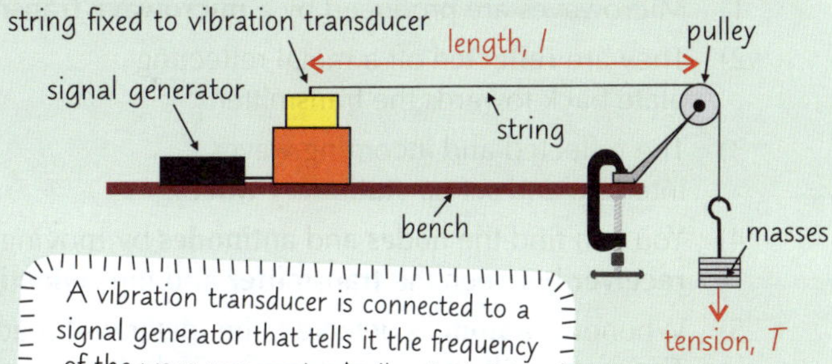

2) Set up the apparatus shown in the diagram with one of your strings. Record μ, measure and record the **length** (l) and work out the **tension** (T) using: $T = mg$ where m is the total mass of the masses in kg

A vibration transducer is connected to a signal generator that tells it the frequency of the wave you want. A vibrating plate on the transducer creates the wave.

3) Turn on the **signal generator** and vary the frequency until you find the **first harmonic** — i.e. a stationary wave that has a **node** at each end and a single **antinode**. This is the **frequency** of the first harmonic, f.
4) The **wavelength** of the wave, λ, is given by $\lambda = 2l$ for the **fundamental mode frequency**. The **frequency**, f, and the **velocity** of the wave, v, for the first harmonic are:

$$f = \frac{1}{2l}\sqrt{\frac{T}{\mu}} \qquad v = \sqrt{\frac{T}{\mu}}$$

The equation for v is true for <u>any</u> transverse wave on a string.

You can investigate how the **length**, **tension** or **mass per unit length** of the string affects the **resonant frequency** by:

- Keeping the string **type** (μ) and the **tension** (T) in it the same and altering the **length** (l). Do this by moving the **vibration transducer** towards or away from the pulley. Find the **first harmonic** again, and record f against l.
- Keeping the string **type** (μ) and **length** (l) the same and **adding** or **removing masses** to change the tension (T). Find the first harmonic again and record f against T.
- Keeping the **length** (l) and **tension** (T) the same, but using **different string** samples to vary μ. Find the first harmonic and record f against μ.

You can do this experiment with a different harmonic — just remember to use the same one throughout the experiment. You won't be able to use the equation for f though — this is just for the fundamental mode of vibration.

You should find the following from your investigation:

1) The **longer** the string, the **lower** the resonant frequency — because the **half wavelength** at the resonant frequency is longer.
2) The **looser** the string the **lower** the resonant frequency — because waves travel more **slowly** down a **loose** string.
3) The **heavier** (i.e. the more mass per unit length) the string, the **lower** the resonant frequency — because waves travel more **slowly** down the string. For a given **length** a **lower** wave speed, v, makes a **lower** frequency, f.

The **Notes** *Played by* **Stringed** *and* **Wind Instruments** *are Stationary Waves*

Transverse stationary waves form on the strings of **stringed instruments** like **violins** and **guitars**. Your finger or the bow sets the **string vibrating** at the point of contact. Waves are sent out in **both directions** and **reflected** back at both ends.

Longitudinal *Stationary Waves Form in a* **Wind Instrument** *or Other Air* **Column**

1) If a source of sound is placed at the open end of a flute, piccolo, oboe or other column of air, there will be some **frequencies** for which **resonance** occurs and a stationary wave is set up.

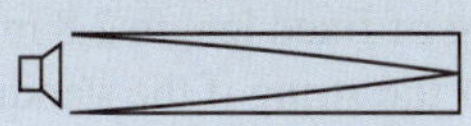

$l = \frac{\lambda}{4}$

2) If the instrument has a **closed end**, a **node** will form there. You get the lowest resonant frequency when the length, l, of the pipe is a **quarter wavelength**.

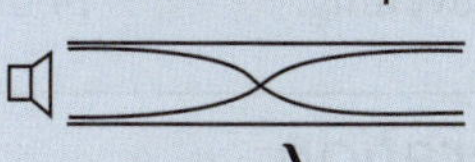

$l = \frac{\lambda}{2}$

3) **Antinodes** form at the **open ends** of pipes. If both ends are open, you get the lowest resonant frequency when the length, l, of the pipe is a **half wavelength**.

Remember, the sound waves in wind instruments are <u>longitudinal</u> — they <u>don't</u> actually look like these diagrams.

Stationary (Standing) Waves

You can Demonstrate Stationary Waves with Microwaves

1) Microwaves are produced by a **microwave transmitter**.
2) They are **reflected** off a metal reflecting plate back towards the transmitter.
3) The reflected and incoming waves interfere and set up **stationary waves**.
4) You can find the **nodes** and **antinodes** by moving the **microwave receiver** between the **transmitter** and the **reflecting** plate.
5) Whenever a **node** is detected, the meter will read a minimum value. The meter will show a **maximum** reading when an **antinode** is detected.

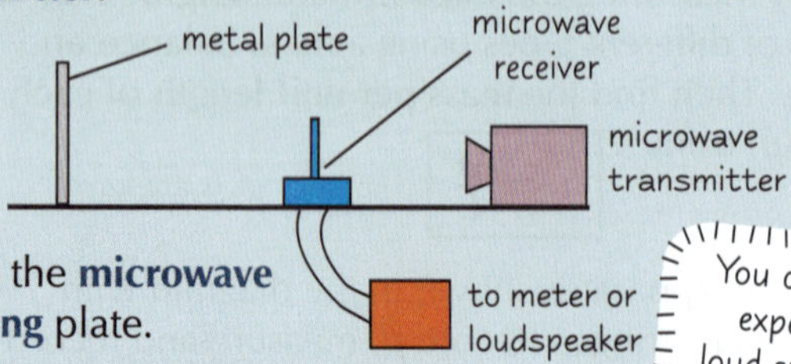

You could do a similar experiment using a loud speaker by moving around a microphone connected to a meter.

You can Use Stationary Waves to Measure the Speed of Sound

1) You can create a closed-end pipe by placing a **hollow tube** into a measuring cylinder of water.
2) Choose a tuning fork and note down the frequency of sound it produces (it'll be stamped on the side of it).
3) Gently tap the tuning fork and hold it just above the hollow tube. The sound waves produced by the fork travel down the tube and get reflected (and form a **node**) at the air/water surface.
4) Move the tube up and down until you find the **shortest distance** between the top of the tube and the water level that the sound from the fork **resonates** at (when the sound is at its loudest).
5) Just like with any closed pipe, this distance is a **quarter** of the **wavelength** of the stationary sound wave.
6) Once you know the frequency and wavelength of the stationary sound wave, you can work out the speed of sound (in air), v, using the equation $v = f\lambda$.

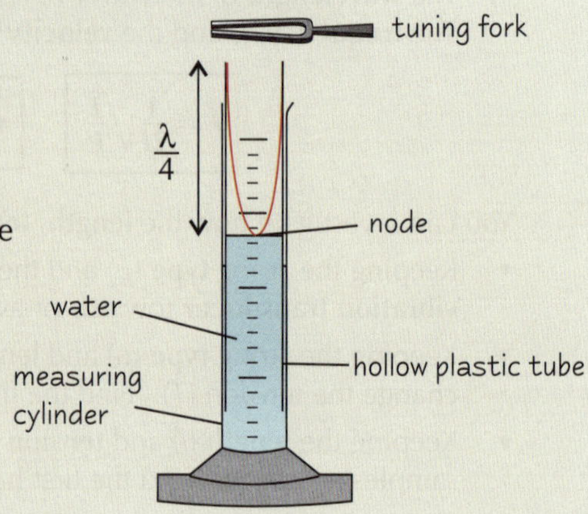

The antinode of the wave actually forms slightly above the top of the tube. You'd add an 'end correction value' on to take this into account, but you don't need to worry about it.

Practice Questions

Q1 How do stationary waves form?

Q2 Sketch the first three fundamental modes of vibration of a standing wave on three different diagrams.

Q3 At four times the first harmonic, how many half wavelengths fit on a violin string?

Q4 Describe an experiment to find the speed of sound in air using stationary waves.

Exam Questions

Q1 A stationary wave of three times the first harmonic is formed on a stretched string of length 1.2 m. Sketch a diagram showing the form of the wave. [2 marks]

Q2 A string of fixed length 2.8 m is oscillating at its fundamental mode of vibration. Calculate the mass of the string if the standing wave has velocity 16 ms^{-1} and the tension on the string is 3.2 N. [2 marks]

Q3 A stationary wave microwave experiment is set up, using microwaves of 1.00 GHz. At a certain position x, the meter connected to the microwave receiver reads 0.00. The receiver is moved 15.0 cm along the line of the stationary wave until the meter once again reads 0.00. Calculate the speed at which the microwaves are travelling. [4 marks]

CGP — putting the FUN back in FUNdamental mode of vibration...

Resonance was a big problem for the Millennium Bridge in London. The resonant frequency of the bridge was roughly normal walking pace, so as soon as people started using it they set up a huge stationary wave. An oversight, I feel...

Phasors

Phasors are a nifty way to help you imagine what's happening when waves interfere.

Phasors Represent the Phase of a Wave

You can use little rotating arrows to represent the phase (see p.90) of each point on a wave. These arrows are called **phasors**. The phasor **rotates anticlockwise** through one whole turn as the wave completes a full cycle.

1) One **complete cycle** of a wave is shown as one complete **revolution** of a phasor — the phasor will travel through an **angle of 360°**.
2) For **points** that are **in phase**, their phasors point in the **same direction**.
3) **Points** that are **exactly out of phase** (**antiphase**) have phasors which point in **opposite directions**. And there's not just phase and antiphase — points can have a **phase difference** of **any** angle.

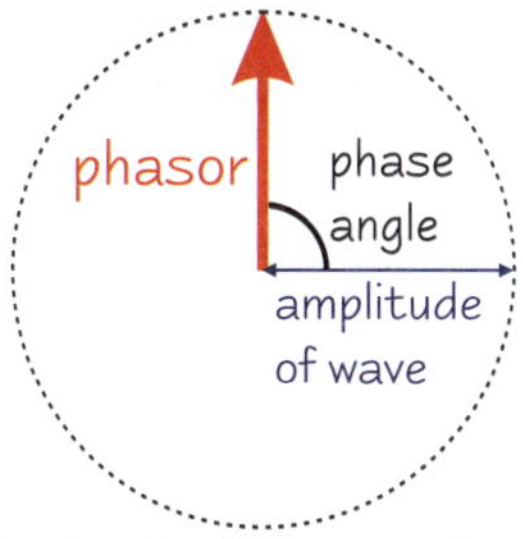

The length of the arrow shows the amplitude of the wave.

You Can Use Phasors to Show Superposition...

To superimpose waves (p.90) using phasors, just add the arrows tip to tail:

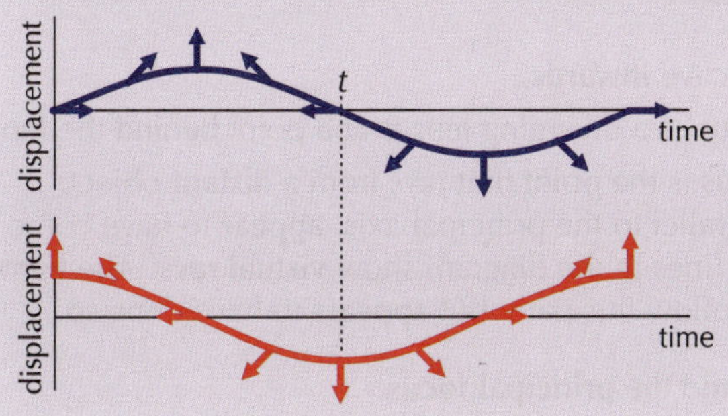

To find the resultant at time *t*, add the phasors tip to tail:

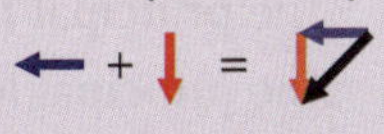

(So in this case, the resultant wave has a greater amplitude than the component waves and is 45° out of phase with both.)

...And Explain Diffraction Patterns

1) The **brightest** point of a diffraction pattern (p.85) is where light passes in a straight line from the slit to the screen. All the light waves that arrive there are **in phase**.
2) At all other bright points where light hits the screen, there is a **constant phase difference** between the waves arriving there, so the phasors point in slightly different directions and form a **smaller resultant**.
3) **Dark fringes** on the screen are where the phase difference between the light waves means their phasors add to form a **loop**, giving a **resultant of zero**.

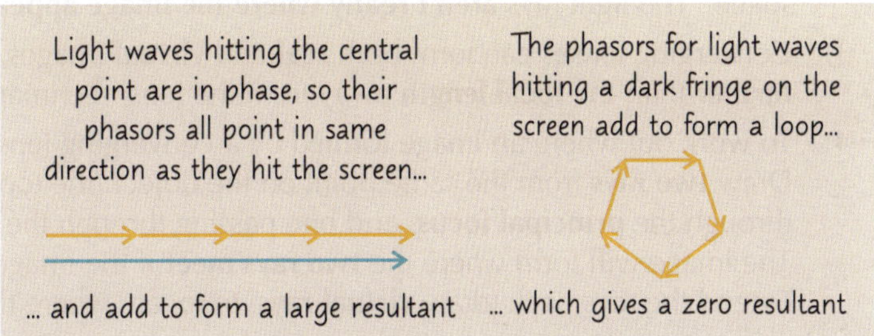

Practice Questions

Q1 Sketch a phasor with a phase angle of 135°.

Q2 Would a phasor with a phase angle of 225° be in or out of phase with the phasor in Q1?

Exam Question

Q1 Sound waves that are coherent and in phase, each with a wavelength of 0.6 m, are produced by two speakers. At a point of total destructive interference, the phasor describing one of the waves at that point is shown. Draw the phasor describing the second sound wave at that point and state whether its magnitude is bigger than, smaller than or the same as the first sound wave.

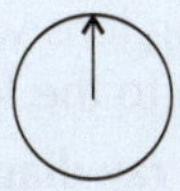

[2 marks]

Don't be phased by all this......

You may have to model photons using phasors — more coming up on that on pages 131-133.

Lenses

Lenses are pretty useful — they're used in glasses to help you see, they're used in telescopes to help you see, they're used in magnifying glasses to help you see, they're used in your eyes to help... you get the picture.

Converging Lenses Bring Light Rays Together

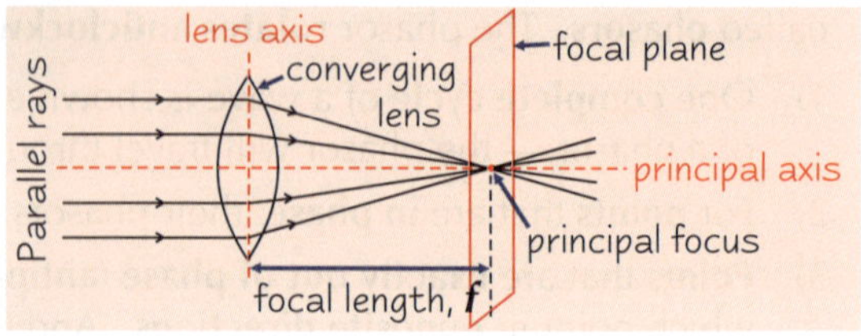

1) **Lenses** change the **direction** of light rays (and the **curvature of the light waves**, see next page) by **refraction** (see p.87-88).
2) **Converging** lenses bulge **outwards**.
3) Rays **parallel** to the **principal axis** of the lens converge onto a point called the **principal focus**. Parallel rays that **aren't** parallel to the principal axis converge somewhere else on the **focal plane** (see diagram).
4) The **focal length**, ***f***, is the distance between the **lens axis** and the **focal plane**.
5) *f* is **positive** for a **converging** lens because it is **in front of** the lens.

Diverging Lenses Spread Light Rays Out

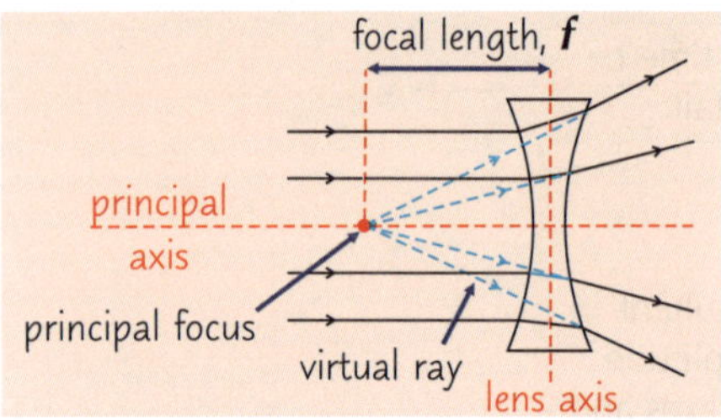

1) **Diverging** lenses cave **inwards**.
2) The **principal focus** of a diverging lens is at a point **behind** the lens.
3) The principal focus is the point that rays from a **distant** object, assumed to be parallel to the principal axis, **appear** to have come from. The dotted lines in the diagram show **virtual rays**. The light **doesn't** actually follow this path, but **appears** to have done so.
4) The **focal length**, *f*, is the distance between the **lens axis** and the **principal focus**.
5) *f* is **negative** for a **diverging** lens because it is **behind** the lens.

Images can be Real or Virtual

1) A **real image** is formed when light rays from a point on an object are made to **pass through** another point in space. The light rays are **actually there**, and the image can be **captured** on a **screen**.
2) A **virtual image** is formed when light rays from a point on an object **appear** to have come from another point in space. The light rays **aren't really where the image appears to be**, so the image **can't** be captured on a screen.
3) Converging lenses can form both **real** and **virtual** images, depending on where the object is. If the object is **further** than the **focal length** away from the lens, the image is **real**. If the object's **closer**, the image is **virtual**.
4) To work out where an image formed by a converging lens will appear, you can draw a **ray diagram**. Draw **two rays** from the same point on the object (the top is best) one **parallel** to the principal axis that passes through the **principal focus**, and one passing through the **centre** of the lens that **doesn't get refracted** (bent). The image will form where the **two rays meet** if the image is real. If the rays don't meet, the image is **virtual**. **Extend** the rays back (draw **virtual rays**) to locate where the two rays appear to have **come from**:

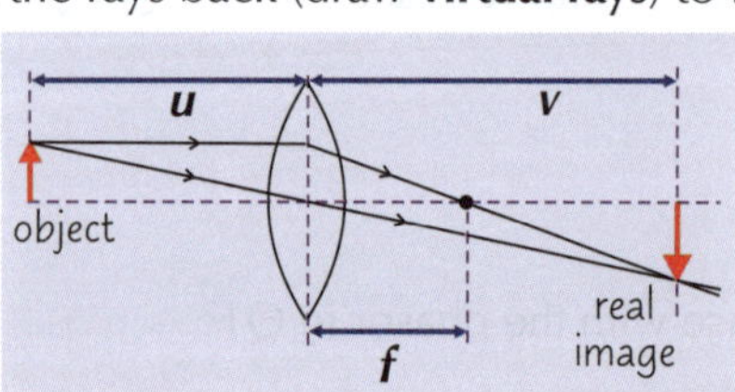

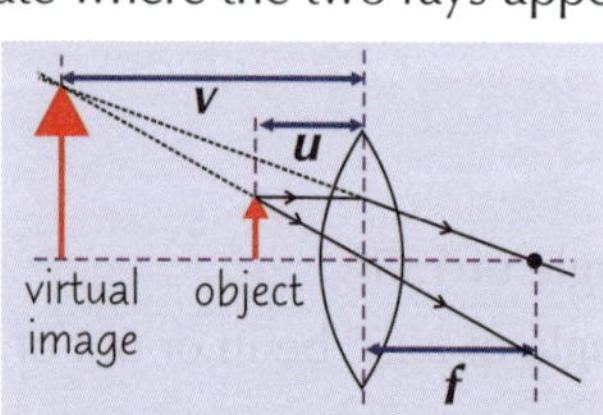

If an object sits on the principal axis, so will the image.

In the diagram, u = distance between object and lens axis, v = distance between image and lens axis (**positive** if image is **real**, **negative** if image is **virtual**), f = focal length.

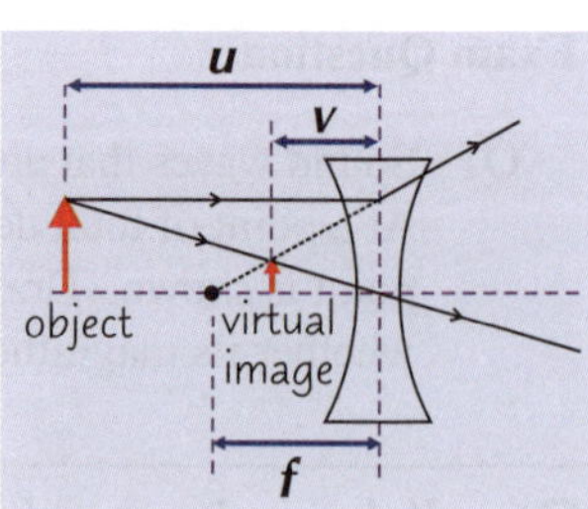

5) A diverging lens will **always** form a virtual image — the **position** of the **object** in relation to the **focal length** will not affect the **type** of image formed.
6) To draw a **ray diagram** for a diverging lens, draw **two rays** from the same point on the object. One must pass through the **centre** of the lens without getting refracted. The other travels **parallel** to the **principal axis** until it meets the centre of the lens. Then it **refracts** so that it **appears** to come from the **principal focus**. The image will form where the real and virtual rays **meet**.

Lenses

You can use the *Lens Equation* to Find *Where* an *Image* Will be *Formed*

1) The distances between a lens, the image and the source are related to each other by **the lens equation**:

$$\frac{1}{f} = \frac{1}{u} + \frac{1}{v}$$

f = focal length,
u = distance between object and lens axis,
v = distance between image and lens axis.

2) You can draw a **ray** or **wavefront diagram** to show these distances for a lens focussing light from a **distant** light source:

You might see a slightly different version of this equation, $\frac{1}{v} = \frac{1}{u} + \frac{1}{f}$, in which *u* is always negative.

This equation assumes that the lens is thin. Thin converging lenses can be used in glasses and contact lenses, magnifying glasses and microscopes.

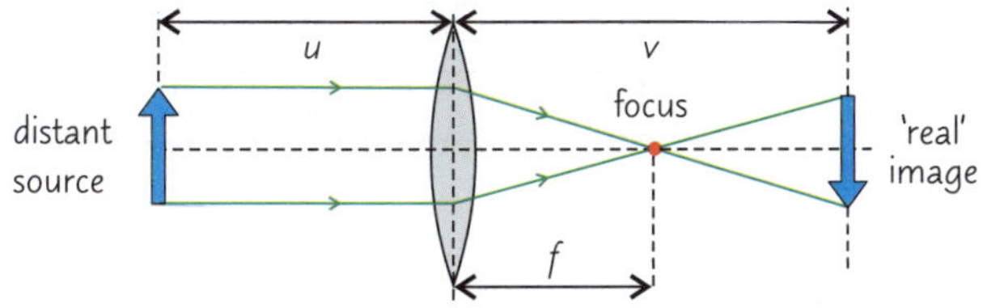

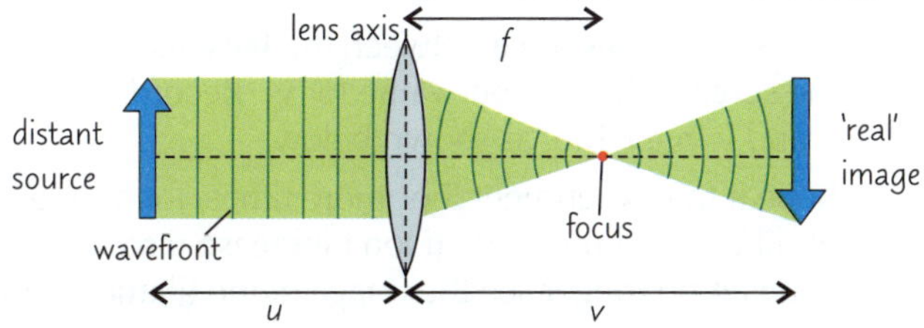

- You can calculate the **power** of a lens, which tells you the lens' ability to **bend light**. The higher the power, the more the lens will **refract** light.

$$P = \frac{1}{f}$$ where *P* is the power of the lens in dioptres (D).

- Imagine you have two or more **thin** lenses in a line, with their principal axes lined up. If the lenses are **touching** or are very **close together**, the **total power** of the lenses is found by just **adding** all the powers together — how simple.

$$P = P_1 + P_2 + P_3 + \ldots$$

A *Lens Changes* the *Curvature* of the *Light* Passing Through it

1) The **curvature** of a wave or a surface (such as the surface of a lens) is defined as:

$$\text{curvature} = \frac{1}{\text{radius of curvature}}$$

2) The **radius of curvature** is the radius a circle with the same curve as the wave or surface would have.
3) Waves are said to have a **negative curvature** when their curvature **decreases**, e.g. when they're diverging.
4) For a lens, the **radius of curvature** is just the distance from the **lens** to the **focus** — the **focal length**, ***f***.
5) The **amount of curvature** a lens **adds** to a wave passing through it is $\frac{1}{f}$ — which is just the **power** of the lens. The curvature is centred on the **principal focus**.

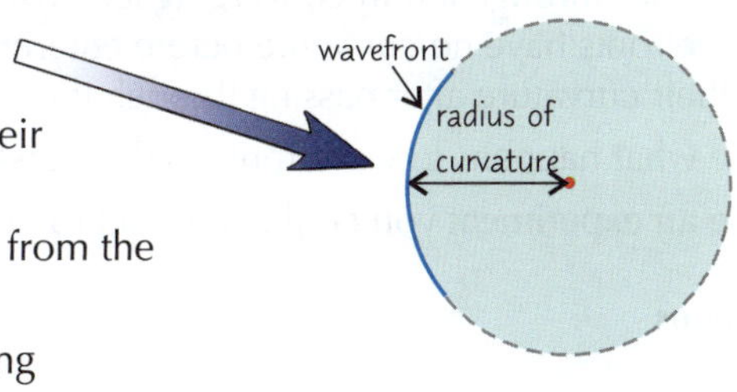

For example, the wavefronts for light from a **distant light source** approaching a converging lens will appear **flat**, i.e. the light will have **no curvature**. As the light passes through a converging lens, the wavefronts will gain a curvature of **1/*f***.

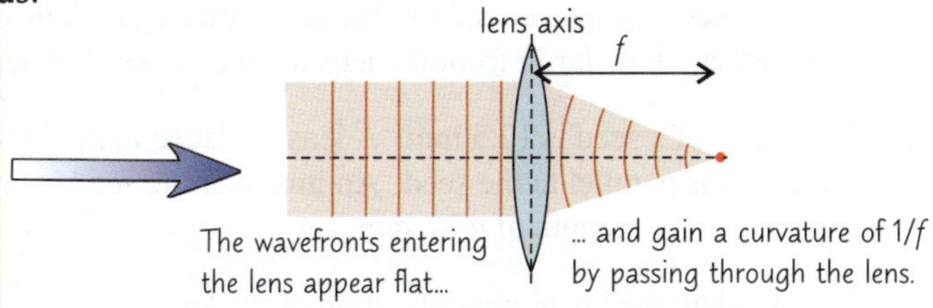

A Lens Can Produce a *Magnified Image*

Magnification is a ratio and so has no units.

The linear magnification, *m*, of an image can be calculated in two different ways. They are:

$$m = \frac{\text{image height}}{\text{object height}}$$

$$m = \frac{v}{u}$$

Lenses

Focus an Image on a Screen to Find the Power of a Converging Lens

You can determine the **focal length** of a thin converging lens by doing an experiment like this one...

1) Set up the equipment as shown, connecting the bulb to a low-voltage power supply.
2) Place the bulb exactly **0.200 m** away from the lens (i.e. $u = 0.200$ m) and turn on the power supply. Move the screen until you can see a clear picture of the filament on the screen.
3) Measure the **distance** between the **lens** and the **screen**. This is your value for v. Record u and v in a table, as shown below.
4) Repeat the experiment 5 or more times, for a range of different values of u. (Don't increase u so much that you can no longer see the image of the filament though.)
5) **Work out** $\frac{1}{u}$ and $\frac{1}{v}$ for each of your readings. **Add** them together to get $\frac{1}{f}$, which is the **power** of the lens.
6) Find the **average** of your values for $\frac{1}{f}$ and **divide** 1 by your answer to find the **focal length** f of your lens.

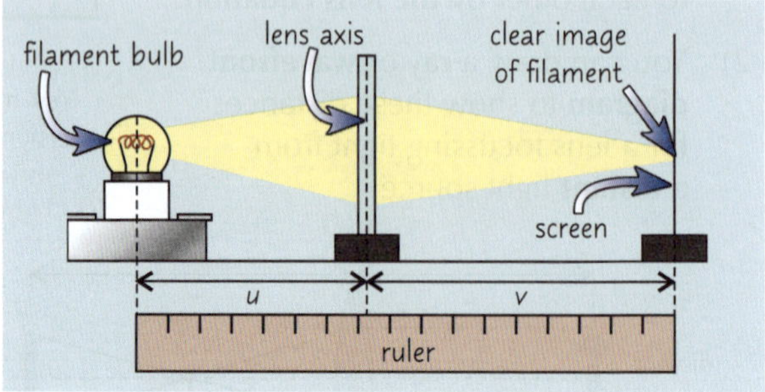

Distance from lamp to lens, u (metres)	Distance from lens to screen, v (metres)	$\frac{1}{u}$	$\frac{1}{v}$	Power, (dioptres) $\frac{1}{f}\left(=\frac{1}{u}+\frac{1}{v}\right)$
0.203	0.629	4.926...	1.589...	6.515...
...	...	...	...	...

Practice Questions

Q1 Define the focal length and power of a converging lens.

Q2 Write an equation to show how the object distance (u), image distance (v) and focal length of a thin lens (f) are related.

Q3 A wave passes through a thin converging lens with focal length f. If the wavefronts have no curvature before entering the lens, what is their curvature after passing through the lens?

Q4 Describe what happens to wavefronts as they pass through a thin converging lens.

Q5 Describe an experiment you could carry out to determine the focal length of a thin converging lens.

Exam Questions

Q1 a) Define the principal focus (focal point) of a converging lens. [1 mark]

b) An object was placed 0.20 m behind a converging lens of focal length 0.15 m. Calculate how far in front the lens the image was formed. [2 marks]

Q2 The length of a seed is 12.5 mm. A lens is placed behind the seed, so that the axis of the lens is parallel to the seed. An image of the seed is projected onto a screen. The image has a length of 47.2 mm.

a) Calculate the linear magnification of the lens. [1 mark]

b) If the seed is 4.0 mm from the lens, calculate how far the screen is from the lens. [2 marks]

c) Calculate the power of the lens in dioptres. [2 marks]

Ray diagrams — not as sunny as they sound...

This is all fairly straightforward — just a few formulas, a handful of diagrams and the odd practical to learn. But don't drop your guard — this topic is a great one for experiment-based questions in exams, so make sure you know how to deal with uncertainties and margins of error. See page 10 for what you need to know on error analysis.

Information in Images

Don't panic if lenses are getting a bit too much for you — it's time for something completely different.

A Single Binary Digit is Called a Bit

1) The **binary number system**, like the **decimal** system, is a way of writing numbers.
2) The difference is that the **decimal** system uses **ten digits** (0-9) while the **binary** system only uses **two** (**0 and 1**). The table shows the first few values in each system.
3) The **zeros** and **ones** that make up binary numbers are called **binary digits** — a **single binary digit** is called a **bit**. A group of **eight bits** is called a **byte**.
4) So to get from **bits to bytes**, **divide** by **8**. To go from **bytes to bits**, multiply by **8**.

Decimal	Binary
0	0
1	1
2	10
3	11
4	100

In binary, 1 kilobyte is 1024 bytes, 1 megabyte is 1024 KB, and 1 gigabyte is 1024 MB.

$$\textbf{bits} = 8 \times \textbf{bytes} \quad \text{and} \quad \textbf{bytes} = \frac{\textbf{bits}}{8}$$

one byte

1 | 0 | 0 | 1 | 0 | 1 | 1 | 0

one bit

10010110 = 150

The Binary System is used to Store Data in Computer Memory

1) When you **save** a file on your computer, the computer stores the data as a **string of bits**.
2) The **number of bits** in a string (***b***) determines how many **alternatives** that string can represent. For example, a **single** bit has only **two** alternatives (0 and 1), while one **byte** (eight bits) has **256** alternatives. The number of alternatives **doubles** with each additional bit, which means:

$$\text{Number of alternatives} = 2^{\text{Number of bits}} \quad \text{or} \quad N = 2^b$$

3) The **number of bits** you need depends on how many **alternatives** you want:

$$\text{Number of bits} = \log_2(\text{Number of alternatives}) \quad \text{or} \quad b = \log_2 N$$

If your calculator doesn't let you calculate $\log_2 N$ directly, you can do ($\log_{10} N \div \log_{10} 2$), which gives you the same thing.

4) For example, if you wanted to represent any letter of the **alphabet**, you'd need a string with **26 alternatives** — one for every letter. Substituting 26 into the **equation** gives $b = \log_2 26 \approx 4.7$ — so you'd need **five bits**.

Images Are Stored as Arrays of Binary Numbers

1) If you **zoom** in on part of a **digital photograph**, you'll see the individual **pixels** (single points that make up the image) — check out the example on the right.
2) When an **image** is stored in a digital camera (e.g. on a **memory card**) or on a **computer**, each pixel is represented by a **binary number**.
3) The **binary numbers** are stored in an **array**. This is a grid of numbers arranged so that the **location of** each **number** in the grid **matches** the location of the **pixel** in the photo.
4) The **value** of the binary number maps to (it gives) the **colour** of the corresponding **pixel**.
5) In **coloured images**, each pixel can be described by **three** binary numbers — one for each of the **primary colours** of light (**red**, **green** and **blue**). The **length** of the binary numbers used depends on **how many** colours are needed. On a typical PC display, each of the numbers for red, green and blue are 8-bits long, giving 256^3 = 16.8 million possible colours.

Image Resolution can mean Different Things

1) Usually **image resolution** refers to the **length** represented by **each pixel**. If an object of width 1.0 m is represented by 200 pixels in an image, then the resolution is 1.0 ÷ 200 = 0.005 metres per pixel.
2) The resolution of an image can also mean the **number of pixels** in the format **width × height** — for example the resolution of a **full HD telly** screen is 1920 × 1080.
3) Sometimes you might see resolution given in terms of **megapixels** — the **total number of pixels** in an image. For example, a **digital camera** that produces images of width 3790 pixels and height 2130 pixels has a resolution of 3790 × 2130 = 8 072 700 pixels or 8 megapixels in total.

Information in Images

The Amount of Information in an Image Depends on the Bits per Pixel

```
0    200  255
100  0    100
255  200  0
```

an array

The **image** shown above can be made up of **256 shades** of grey, from 0 (black) to 255 (white).

The shade of each pixel (in the array on the left) is represented by a number between 0 and 255.

Each of these shade numbers is stored in memory as an **8-bit binary number**.

red (R) green (G) blue (B)

255, 255, 255

```
0,0,0        255,0,0        0,255,0
0,0,255      255,255,0      0,255,255
255,0,255    255,255,255    0,0,0
```

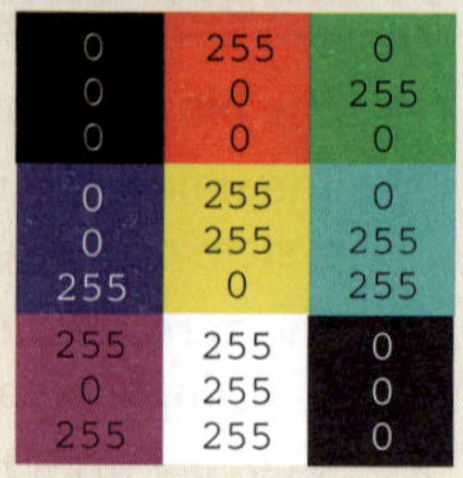

The colours in the image shown are made up of 3 numbers, each providing **256 possible shades** of red, green or blue.

This means there are 256 × 256 × 256 = **16.8 million** possible different RGB combinations in the image.

Each pixel needs **three different 8-bit binary numbers** in the array on the left — so each pixel requires 3 × 8 = **24 bits**.

The more **bits per pixel**, the **more information** is held by each pixel — and the **more pixels** there are in an image, the **more information** is held by the image. So the **total** amount of information in an image is given by:

total amount of information = number of pixels × bits per pixel

Multiplying by a Fixed Value Improves Contrast

The **values** of the binary numbers that make up an **image** determine how it looks — if you **change** the **values**, you **change** the **image**. Take a look at the example below to see what happens.

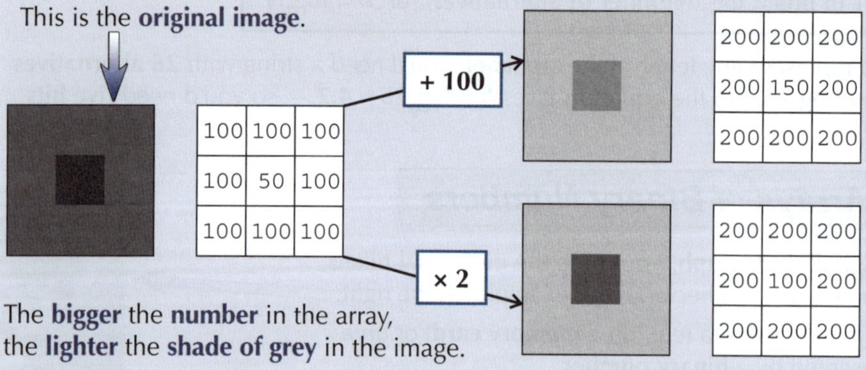

The **bigger** the **number** in the array, the **lighter** the **shade of grey** in the image.

Adding a fixed positive value makes the image **lighter**, but it **doesn't change** the **difference** between the **dark** and **light** regions (**contrast**).

Multiplying by a fixed value (greater than 1) makes the image **lighter**, and **increases** the **contrast** (there's a bigger difference between colour values).

Adding False Colour can Highlight Features

1) In the example above, a value of '**50**' in the array mapped to a **dark shade of grey** in the image, while a value of '**100**' mapped to a slightly **brighter shade**. But you could map '**2**' to a **dark shade of pink** and '**4**' to a **brighter shade** — or '**2**' to **orange** and '**4**' to **green**.
2) This process is called adding **false colour**. You can use **any** colours you like, but they're usually picked to **highlight certain features** — e.g. the **important features** could be made a really **bright** colour.
3) False colour images are often used in **remote sensing** of data. For example, a **thermal imaging camera** shows **heat** in colour instead of visible light. Satellites can be used to map **different parts** of the **electromagnetic spectrum** to red, green and blue to show features that **can't be seen** in **true colour** images.

You can transform an image by changing the colours that the values in its array map to — not the array itself.

A false colour satellite image of a river delta.

Information in Images

Replacing Pixels With the Median of their Neighbours Reduces Noise

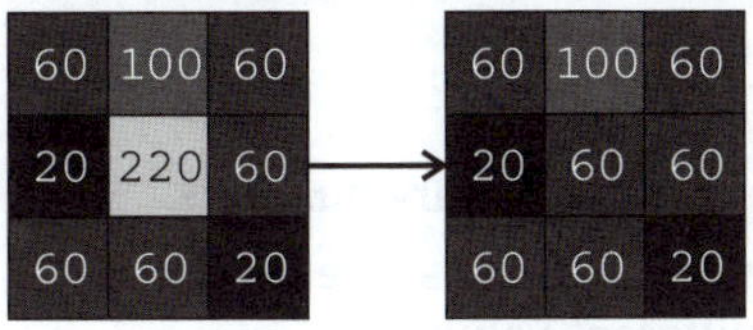

This array shows a bright spot in a uniform darker region — the spot is probably noise. Replacing the central value with the median of its value and the surrounding values reduces the brightness of the region.

1) **Noise** is **unwanted interference** affecting a signal. In images this is usually **bright** or **dark** spots on the picture.
2) One way you can get rid of **noise** is to **replace** each pixel with the **median** of itself and the eight pixels surrounding it.
3) The median is a kind of average, like the mean (p.8). To find the median, write all your values out in **numerical order**. The value in the **middle** of your list is the median.
4) The result is that any **'odd'** (i.e. very **high** or very **low**) values are **removed** and the image is made **smoother**.

Edge Detection Tells You if there is Something in your Image

1) If you're trying to work out if there is **something** in your **image** (rather than just a load of **noise**), using **edge detection** to find any **edges** can be a really **useful** first step.
2) The **Laplace rule** is a method of **finding edges**. To apply the rule, you **multiply** the value of a pixel by **four**, then **subtract** the value of the pixels immediately **above**, **below**, to the **left** and to the **right** of it. If the answer is negative, the pixel is treated as if its value is 0 (ie. it maps to the colour black).
3) Edge-detected images are usually then **inverted** so that **white becomes black** and vice versa.
4) The result is that any pixel **not** on an **edge** goes **white** — so you're left with **just** the **edges**.

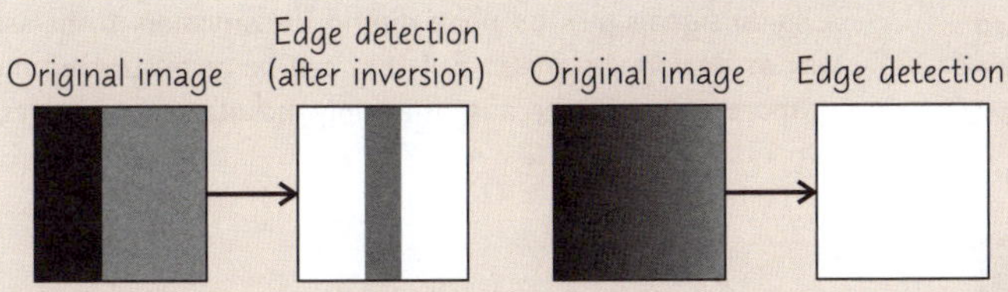

Edge detection will only highlight an edge if there is a steep change in brightness (i.e. an edge). If the change in brightness is gradual (i.e. not an edge), it will not be detected.

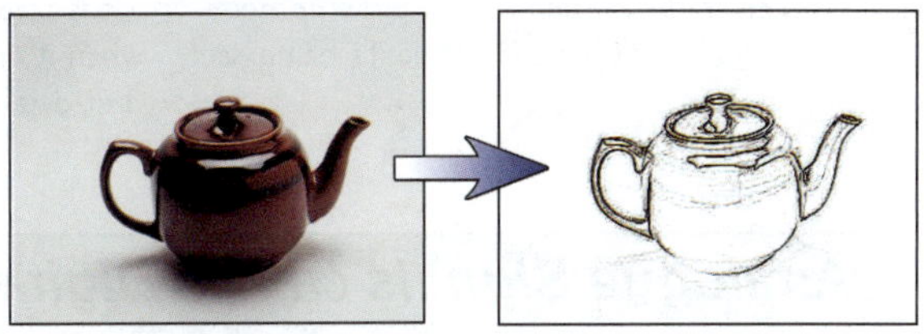

original image — image after edge detection

Practice Questions

Q1 What's the difference between a bit and a byte? Find the number of bits in 1024 bytes.

Q2 A string contains 3 500 000 different alternatives. What is the minimum number of bits required?

Q3 How can the brightness and contrast of an image be changed?

Q4 What is the Laplace rule used for? How do you apply it?

Exam Questions

Q1 A TV can display 65 536 different colours. It displays a still image of width 1920 pixels and height 1080 pixels.

a) Calculate the number of bytes required to store the colour of each pixel. [2 marks]

b) Calculate the total amount of information contained in the image in bits. [1 mark]

c) An image contains a 1.5 m^2 square that spans exactly a quarter of the available width. What is the resolution of the screen? [2 marks]

Q2 The diagram shows part of an array that describes an image. The image is made up of 256 shades of grey — 0 represents black and 255 represents white.

100	99	100
97	185	98
101	101	98

a) Sketch the image that this part of the array describes. [1 mark]

b) Describe how noise can be removed from digital images by finding a median value. [1 mark]

c) Apply this technique to the central value of the array shown. [1 mark]

All this talk of bytes is making me a megabit hungry — mmm, tasty bites...

It doesn't matter how many attempts you take, or how many filters you apply — all your selfies are really just a long list of ones and zeroes (unless you actually print them out on paper). Now — everybody say cheeeese...

Digital and Analogue Signals

You've just seen how information can be stored digitally, but what if you want to send that information? And what if the information isn't digital to start with? So many questions — read on to find out the answers.

Analogue Signals Vary Continuously

1) **Digital signals**, like the images on the previous three pages, are represented by **binary numbers.**
2) The **values** that a **digital signal** can take depend on the **number of bits** used — e.g. a **one bit** signal can only take the values **0 and 1**, but a **one byte** signal can take **256 different values.**
3) **Analogue signals** are **not limited** in the values they can take — they **vary continuously**. For example, **speech** is an **analogue signal** — the **sound waves** produced **vary continuously** over a range of **loudness** and **frequency**.

Digital Signals Are Resistant to the Effects of Noise

When you **transmit** an electronic signal it will pick up **noise** (interference) from **electrical disturbances** or other **signals**. The receiver needs to be able to **reconstruct** the **original signal** from the **noisy signal** if they're to get an **accurate representation** of what was sent. This is **much easier** with **digital** than analogue signals because the **number of values** a digital signal can take is **limited.**

Analogue signal

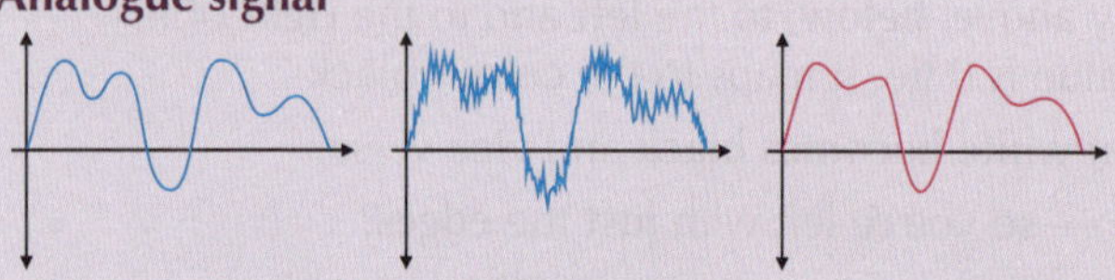

This is the original signal that's sent. | It picks up noise while it's being sent. | It is reconstructed when it's received — but details are lost.

Digital signal

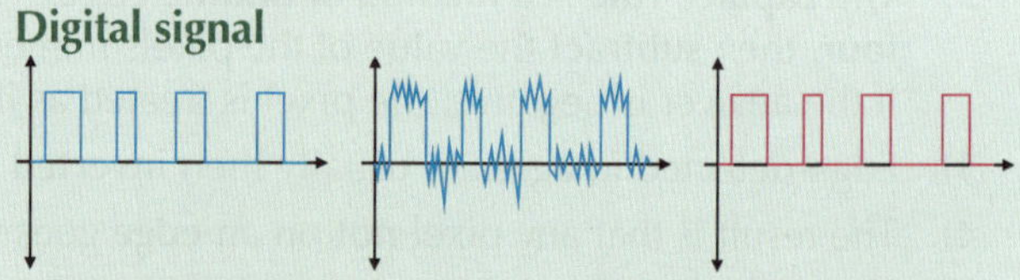

Digital signals pick up noise during transmission in the same way as analogue signals. But they can be reconstructed much more accurately because they only include certain values.

Analogue Signals can be Digitised

1) It's possible to turn an **analogue signal** into a **digital signal** — this is called **digitising** the signal.
2) To digitise a signal, you take the **value** of the signal at **regular time intervals**, then find the **nearest digital value**.
3) Each **digital value** is represented by a **binary number**, so you can **convert** the **analogue** values to **binary** numbers.
4) The **digital signal** you end up with won't be **exactly** the same as the **analogue signal**, but it's usually quite **close**.

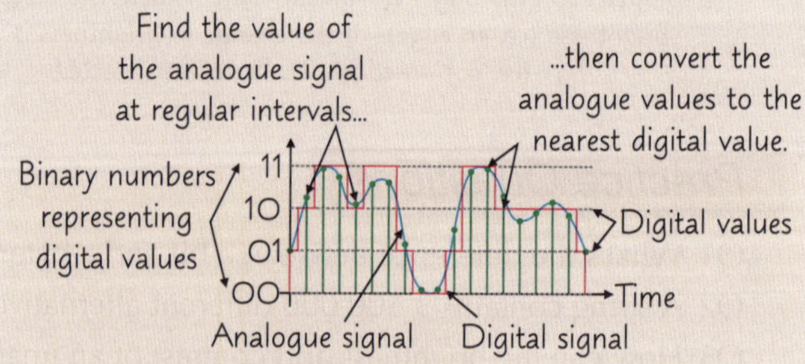

The Quality of a Digitised Signal Depends on its Resolution

1) How well a **digitised** signal matches the original depends on **two** factors — the **difference** between the possible **digital values** (**resolution**) and the **time** from one **sample** to the next (**sampling rate**, see p.108).
2) If a signal is digitised using only a **few**, **widely spaced** digital values, it's likely that a lot of the analogue values sampled will be **far** from the **nearest digital value**. But, if a **large** number of **closely spaced** digital values are used, most of the analogue values will be **very close** to a digital value, so will only change **slightly**.
3) This means that the **higher the resolution** (i.e. the **more possible digital values** there are), the **more closely** the digitised signal will **match** the original.
4) **Resolution** is determined by the **number of bits** in the **binary numbers** representing the digital values — the **greater** the number of **bits**, the **greater** the **resolution**.
5) When **music** is digitised to make **CDs**, a resolution of **16 bits** is used. This gives a total of **65 536 digital values** and means the recorded music is **very similar** to the **original**.

6) **Low resolution** digital signals are often used in **telephone lines**, and other systems where top-quality audio isn't essential. Telephone conversations have to be **audible**, but you **don't need** an **accurate reproduction** of the callers' voices — even if this does mean the music sounds **distorted** when you're put on hold. Using a lower resolution and sampling rate means a **lower rate of transmission** (p.109) can be used to send information.

Digital and Analogue Signals

Digital Signals Have Several Advantages Over Analogue Signals

1) Digital signals can often be **sent**, **received** and **reproduced** more easily than analogue signals because they can only take a limited number of values.
2) Digital files can be **compressed** to reduce their size, and **manipulated** easily for artistic effect.
3) **Noise** is more of a problem for analogue signals than digital signals (see previous page).
4) A digital signal can be used to represent **different** kinds of **information** in the **same way** — for example, **images** and **sounds** can both be represented as a string of bits.
5) **Computers** can be used to **easily process** digital signals, since computers are **digital devices** too.

But they have some Disadvantages too

1) Digital signals can **never** reproduce analogue signals **exactly** — some **information** will always be **lost**.
2) Because digital signals can be copied more easily, digital information like films and music can be **reproduced illegally unlimited times**.
3) **Confidential information**, such as **personal data** and **photographs**, may be **stolen** and **copied** without the owners' knowledge or consent more easily, for example by **hackers**, **infected networks** or **malicious websites**.

The UK Government wants to switch all radio broadcasts over to **digital (DAB) signals** as these can (in theory) have better audio quality and more stations can fit in less space. But **analogue** radio devices remain **very popular** among the public, and the **bit rate** used for DAB radio broadcasts is often **low**, producing a **lower quality sound** than FM radio.

Signals are Made Up of Lots of Different Frequencies

1) The wave on page 78 is the **simplest** kind of **signal** because it contains just **one frequency**. In practice, most **signals** are made up of **several** waves, all with **different frequencies**, added together.
2) When these waves are added together, the **amplitude** of the final signal is the **sum** of the **amplitudes** of the individual waves at each point in time.
3) For example, if you play a **musical note**, the sound you hear contains the **frequency** (pitch) of the **main note** and a load of **other frequencies**. It's these '**other frequencies**' that make instruments **sound different**, even though they're playing the **same note**.
4) The **fundamental frequency** is the **lowest frequency** wave that makes up a sound wave. You can spot it by finding the **shortest repeating part** of the sound wave and calculating the inverse of its period.

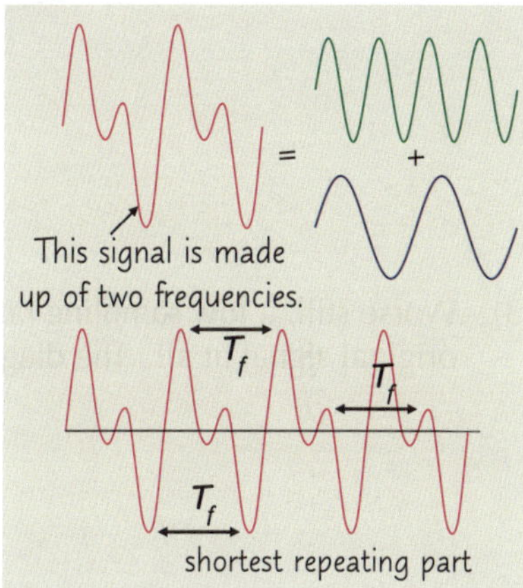

Practice Questions

Q1 Explain how a digital signal is different from an analogue one.

Q2 What is meant by resolution in the context of digitising analogue signals?

Q3 Give three advantages of digital signals over analogue signals.

Exam Question

Q1 The diagram shows the waveform of a musical note played on an instrument.

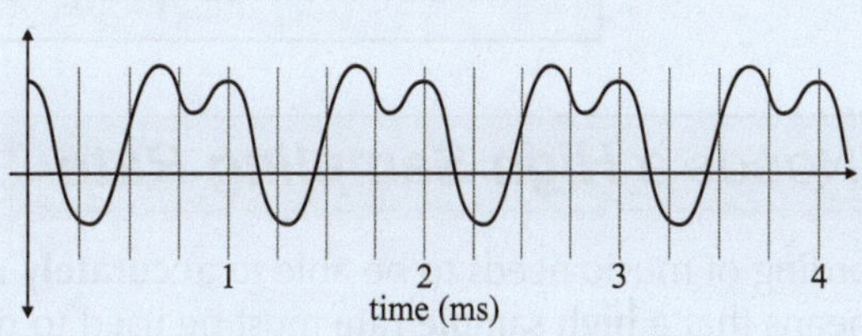

a) Calculate the fundamental frequency of the note. [1 mark]

b) The wave is sampled every 0.001 s.
The first sample occurs at 0.0 ms.
Sketch the waveform of the reconstructed signal. [2 marks]

How do finger puppets communicate? With digital signals...

Digital signals are everywhere these days — CDs, MP3s, DVDs, TV, DAB radio — there's just no escape. Analogue signals are still around, though — lots of radio stations broadcast FM or AM analogue signals (for now at least...). Some people prefer analogue signals for listening to music and use vinyl (ask your dad) instead of MP3s or CDs.

Sampling and Transmitting Signals

Next up, even more fun things to know about signals...

Noise Limits** the **Number of Bits** used for **Sampling

Lesley made sure noise didn't affect the number of sweets sampled.

1) You've just seen that the **higher the resolution**, the **better** a digitised signal **matches** the original.
2) But, if the original signal contains **noise** (as most real signals do), then a really **fine resolution** will reproduce all the little wiggles caused by the **noise** — **not useful**.
3) In practice, the **resolution** is **limited** by the **ratio** of the total **variation** in the **signal** to the **variation** caused by **noise**:

$$\textbf{Maximum number of bits} = \log_2\left(\frac{\textbf{total variation}}{\textbf{noise variation}}\right) \quad \text{or} \quad b = \log_2\left(\frac{V_{total}}{V_{noise}}\right)$$

Here the variation is measured in volts, but the ratio has no units because they cancel.

Minimum Sampling Rate** is **Twice** the **Maximum Frequency

1) When you **digitise** a signal, you **record the value** of (**sample**) the original signal at **regular intervals**. The **rate** at which you **sample** the signal is called the **sampling rate** — imaginative name, I know.
2) The **sampling rate** has to be **high** enough to record all the **high frequency** detail of the signal. The diagram below shows how **detail** can be **lost** if the sampling rate is too **low**.

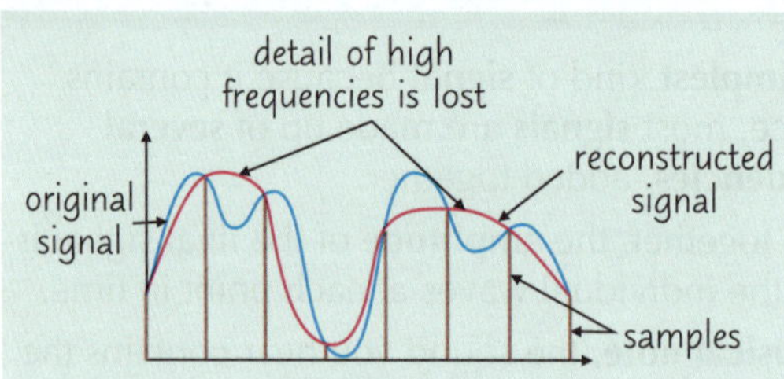

3) Worse still, a **low** sampling rate can **create** low frequency signals — called **aliases** — that **weren't** in the original signal at all. The diagram **below** shows how **aliases** can be **created** by a **low sampling rate**.

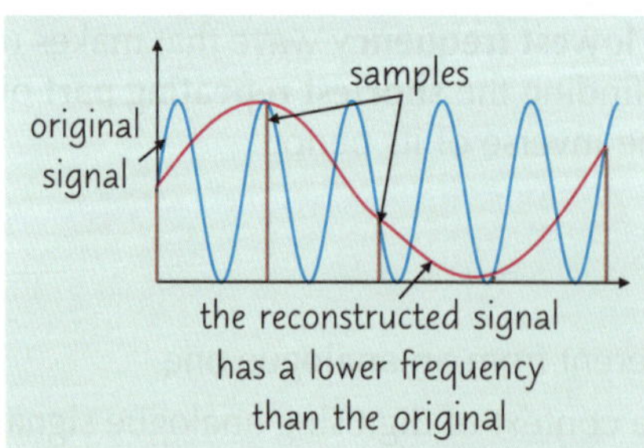

4) To avoid these problems, the **sampling rate** must be at least **twice** the **highest frequency** in the original signal.

Minimum rate of sampling > 2 × maximum frequency of signal

Music** Needs a **High Sampling Rate

1) A recording of music needs to be able to **accurately reproduce** the **original sounds**. This means that a **high sample rate** must be used to make sure **high frequency details** aren't **lost**, and to avoid the creation of **aliases** in the recording.
2) **CD** and **MP3-quality** audio uses a sampling rate of **44 100 Hz** . The maximum frequency sound that can be recorded at this sampling rate is about **20 000 Hz**, which is roughly the highest-frequency sound audible to the human ear.
3) The audio on **video DVDs** and in **digital TV broadcasts** uses a slightly higher sampling rate of **48 000 Hz**, while **Blu-ray Discs**™ can use a sampling rate of up to **192 000 Hz**.

Sampling and Transmitting Signals

Rate of Transmission = Samples per Second × Bits per Sample

The **rate of transmission** of a digital signal depends on **two** factors:

1) The number of **samples per second** — this must be at least **twice** the **highest frequency** in the signal to ensure that all the frequencies within its spectrum are transmitted accurately.
2) The number of **bits per sample** — this must be **high enough** that the transmitted signal **closely** matches the original, but not so high that it is negatively affected by **noise**.

Rate of transmission of a digital information (bits per second) = samples per second × bits per sample

Time Taken to Transmit a Signal = Bits ÷ Rate of Transmission

To find the **time** it will take to **transmit** a signal, you need to know the **number of bits** that need to be transmitted, and the **rate of transmission** (in bits per second):

Time taken to transmit a signal = number of bits to transmit ÷ rate of transmission

The speed of your Internet connection is usually measured in **bits**, or more likely megabits (Mb), **per second**. Note the lower case b — M**b** means mega**bits**, whereas M**B** means mega**bytes**.

Don't get them confused — if you've got an Internet upload connection of 2 megabits per second and want to upload a 2 megabyte image to a social network, it won't take 1 second — it will take at least 8 seconds.

To **speed up** the process, and **use less storage space**, you could **compress** your image to **reduce** the file size.

Practice Questions

Q1 Why is there a maximum number of bits used for sampling?

Q2 Describe two problems that can be caused by an insufficient sampling rate.

Q3 The minimum rate of sampling for a signal is 72 kHz. What is the maximum frequency of the signal?

Q4 Why is music often sampled a rate of 40-50 kHz?

Exam Questions

Q1 A digital signal has a total variation of 160 mV and a noise variation of 10 mV. Find the maximum number of bits that should be used when sampling this signal. [2 marks]

Q2 The telephone system samples your voice 8000 times a second and converts this into an eight bit digital signal.

a) Find the rate of transmission for bits in this telephone system. [1 mark]

b) Calculate how many bytes are sent each second. [1 mark]

Q3 An Internet radio station streams a single channel of digital audio at a rate of 128 kbit s^{-1}.

a) The audio the station broadcasts contains 16 bits per sample. Calculate the sampling rate. [1 mark]

b) A listener wants to stream the radio station using her mobile phone. She tests her mobile data connection and finds that she can download a 2.0 MB file in 110 seconds. State whether her connection is sufficient to stream the radio station. [3 marks]

Music needs a high sampling rate — just don't tell the copyright lawyers...

I don't know about you, but I'm really enjoying this little foray into the physics behind the tech that most of us use every day. Once you've got every last bit learned, you can feel OK about taking a byte out of your revision time to stream a celebratory self array of binary numbers to your favourite social network...

Ultrasound Imaging

I wrote these next two pages for you, because I know what you're thinking — 'This is great, but I want to learn about medical imaging. Also, as a side note, I think CGP books are brilliant.' Oh stop, no really, I'm blushing...

Waves are Reflected and Transmitted at Interfaces

1) The **boundary** between two different media is called an **interface**.
2) When a wave passes from one medium to another, some of its **energy** is **reflected** and some of it is **transmitted** — as shown in the diagram.
3) The proportion of energy reflected or transmitted depends on the two media involved. If the media have very **different densities**, most of the energy is **reflected**. If they are quite **similar**, most of the energy is **transmitted**.

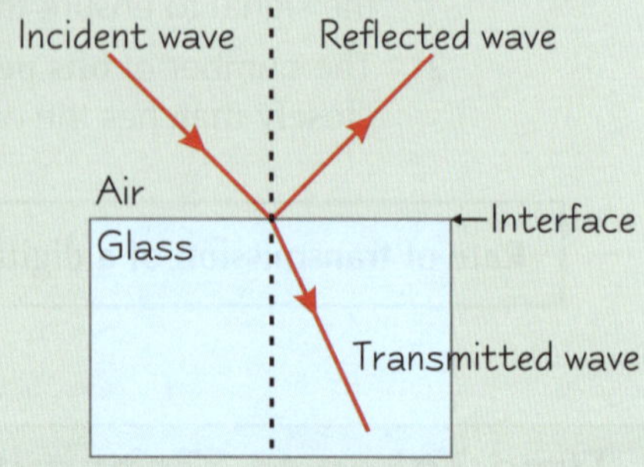

The Reflection of Ultrasound Waves is Used in Ultrasound Scans

1) **Ultrasound waves** are **sound waves** that have too high a **frequency** for humans to **hear**.
2) Ultrasound scans use short pulses of **ultrasound radiation** to form images of the inside of your body.
3) The ultrasound is directed into your body using a **transducer**. If you have air between the transducer and your skin, most of the waves are **reflected** because air has a very **different** density from skin. A **gel** is applied to the skin so there is no air between the transducer and your skin to **increase** the proportion of ultrasound waves that **enter** your body.
4) When the ultrasound waves reach an **interface** inside your body — e.g. between different types of tissue — some of them are **reflected**. A computer attached to the transducer calculates how far from the surface of your skin the interface is by **timing** how long it takes for the reflected waves to **return**.
5) The computer uses the information about the **location** of the boundaries between different tissues to build up an **image** of the inside of your body.

Sending out ultrasound waves and detecting the reflections is known as a pulse-echo technique.

A similar technique is used in **sonar** — e.g. ships send sonar pulses (**ultrasound waves**) down towards the seabed and the pulses are **reflected** back from any **submerged** objects. It can also be used to measure the **speed** of objects.

Shorter Pulses Produce Clearer Images

Ultrasound transducers cannot **transmit** and **receive** pulses at the same time. If **reflected** waves reach the transducer while it is **transmitting**, the information they contain will be **lost** and image **quality** will be reduced. This means:

1) The **pulses** of ultrasound transmitted must be **very short** (a few microseconds long) so that the reflections from nearby interfaces don't reach the transducer before the pulse has ended.
2) The gap between pulses must be **long** (at least 1 millisecond) so that all the reflected waves from one pulse return to the transducer **before** the next pulse is transmitted.

Jack didn't know how to tell Joanna that most of the image quality had been lost.

Ultrasound Imaging

Shorter Wavelengths Produce **Clearer Images**

Ultrasound scanning can be a really useful technique — but only if the images are **clear**.
The **wavelength** of the ultrasound **radiation** used has a big effect on the clarity of the images produced.

1) **Shorter** wavelengths **diffract** much **less** than longer wavelengths.
2) This means that the shorter the **wavelength** of the ultrasound, the less the waves **spread out** as they travel and the more **precisely** the location of the interfaces between tissues can be mapped.
3) So ultrasound scanners use waves with a **high frequency** and **short wavelength**.

For more on diffraction of waves, see p.84-85.

In order for an **object** to be **resolved** (distinguished from other objects), the **wavelength** of the ultrasound wave must be of a **similar size** to the width of the object being resolved.

Example: Ultrasound of frequency 0.95 MHz is used to image a spherical tumour with a diameter of 3.2 mm. If the speed of ultrasound waves in the body is 1540 ms^{-1}, will the tumour be resolved in the ultrasound image?

Find the wavelength of the ultrasound by rearranging $v = f\lambda$:

$$\lambda = \frac{v}{f} = \frac{1540}{0.95 \times 10^6} = 1.6210... \times 10^{-3} \text{ m} = 1.6210... \text{ mm}$$

Take a look back at p.78 for a reminder on the wave equation.

The wavelength used is approximately half the size of the tumour, so the wavelength and tumour are similar in size. This means it is reasonable to assume that **the tumour will be resolved**.

Practice Questions

Q1 What is an interface?

Q2 What happens to a wave when it reaches an interface?

Q3 How is sonar similar to ultrasound scanning?

Q4 State and explain two things you can do to improve the clarity of an ultrasound image.

Q5 How long does the wavelength of ultrasound used to form an image need to be compared to the size of the object in order to be able to resolve it?

Exam Questions

Q1 Explain how pulse-echo techniques are used to image the inside of a patient's body, including what happens when waves meet an interface between media. [4 marks]

Q2 Explain why a transducer used in medical imaging produces pulses of ultrasound waves with the following properties:

a) A short wavelength. [3 marks]

b) A short pulse length and a long time between pulses. [4 marks]

Ultrasound — Mancunian for 'très bien'

Next time you're watching TV and they show a pregnant woman having her unborn baby scanned, turn to the person you're sat next to and explain how the procedure works. You'll also be able to explain why they put a gel on their stomach first. Just make sure you know the person first before you launch into an explanation about transducers.

Atomic Structure

*"So what did you do today, Johnny?" "Particle physics, Mum." "How nice dear — done with times tables then?" Yeah, well, it's not exactly the **easiest** topic in the world, but it's a darn sight more interesting than biology.*

Atoms** are made up of **Protons, Neutrons** and **Electrons

Inside **every atom**, there's a **nucleus** containing **protons** and **neutrons**.
Protons and **neutrons** are both known as **nucleons**. **Orbiting** this core are the **electrons**.
This is the **nuclear model** of the atom.

The diagram shows **neutral oxygen**, with **eight protons** and **eight electrons**.

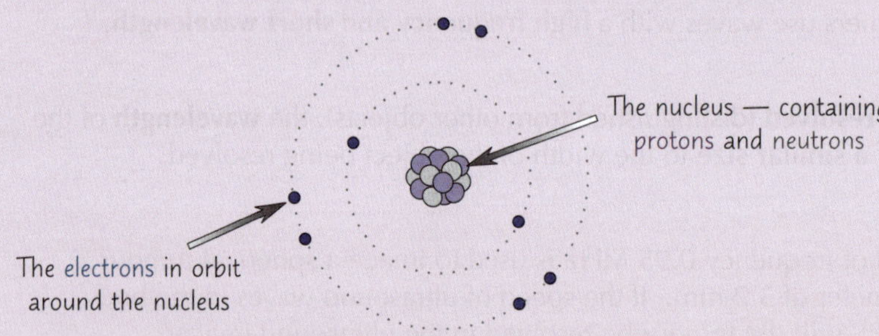

Tom dreamt of becoming a nuclear model when he grew up.

The particles in an atom have different **properties**. Their charges and masses are so **tiny** that it's often easier to talk about their **relative charge** and **relative mass**.

Particle	Charge (coulombs, C)	Mass (kg)	Relative Charge	Relative Mass
Proton	$+1.60 \times 10^{-19}$	$1.67(3) \times 10^{-27}$	+1	1
Neutron	0	$1.67(5) \times 10^{-27}$	0	1
Electron	-1.60×10^{-19}	9.11×10^{-31}	−1	0.0005

*The **Proton Number** is the **Number** of **Protons** in the Nucleus*

No... really.

The **proton number** is sometimes called the **atomic number**, and has the **symbol Z** (I'm sure it makes sense to someone). **Z** is just the **number of protons** in the nucleus.

It's the **proton number** that **defines** the **element** — **no two elements** will have the **same** number of protons.

In a **neutral atom**, the number of **electrons equals** the number of **protons**.
The element's **reactions** and **chemical behaviour** depend on the number of **electrons**.
So the **proton number** tells you a lot about its **chemical properties**.

A particle with a different number of electrons to protons is called an ion.

*The **Nucleon Number** is the **Total Number** of **Protons and Neutrons***

The **nucleon number** is also called the **mass number**, and has the **symbol A** (*shrug*).
It tells you how many **protons** and **neutrons** are in the nucleus. Since each **proton or neutron** has a relative **mass** of (approximately) **1** and the electrons weigh virtually nothing, the **number** of **nucleons** is the same as the **atom's relative mass**.

The nuclide notation of an element summarises information about its atomic structure:

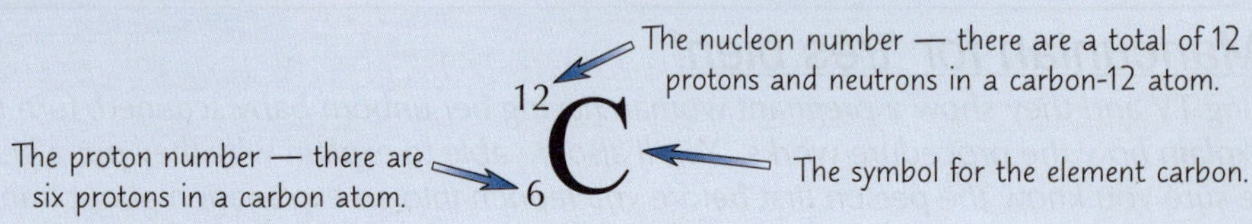

Atomic Structure

Isotopes have the *Same Proton Number*, but *Different Nucleon Numbers*

Atoms with the **same number of protons** but **different numbers of neutrons** are called **isotopes**.

Example: Hydrogen has three natural isotopes — hydrogen, deuterium and tritium.
Hydrogen has 1 proton and 0 neutrons.
Deuterium has 1 proton and 1 neutron.
Tritium has 1 proton and 2 neutrons.

Changing the number of **neutrons doesn't affect** the atom's **chemical** properties.

The **number of neutrons** affects the **stability** of the nucleus though.

Unstable nuclei may be **radioactive** and **decay** over time into different nuclei that are more stable (see p.114 – 115).

Radioactive Isotopes Can be Used to Find Out *How Old* Stuff Is

1) All living things contain the same percentage of radioactive **carbon-14** taken in from the atmosphere.
2) After they die, the amount of carbon-14 inside them **decreases** over time as it **decays** to stable elements.
3) Scientists can calculate the **approximate age** of archaeological finds made from dead **organic matter** (e.g. wood, bone) by using the **isotopic data** (amount of each isotope present) to find the percentage of **radioactive carbon-14** that's **left in** the object.

The *Specific Charge* of a Particle is Equal to its *Charge Over its Mass*

The **specific charge** of a particle is the ratio of its charge to its mass, given in coulombs per kilogram ($C\,kg^{-1}$). To calculate specific charge, you just divide the charge in C by the mass in kg.

$$\textbf{Specific charge} = \frac{\textbf{charge}}{\textbf{mass}}$$

You could be asked to find the specific charge of any particle, from a **fundamental particle** like an electron, to the nucleus of an atom or an ion.

A fundamental particle is one that you can't break up into anything smaller.

Example: Calculate the specific charge of a proton.

A proton has a **charge** of $+1.60 \times 10^{-19}$ C and a **mass** of 1.673×10^{-27} kg (see p.112).

So specific charge = $(+1.60 \times 10^{-19}) \div (1.673 \times 10^{-27}) = 9.563... \times 10^{7}$ = **9.58×10^{7} C kg^{-1} (to 3 s.f.)**

In calculations, always give your answer to the smallest number of significant figures used in the question.

Practice Questions

Q1 List the particles that make up the atom and give their relative charges and relative masses.

Q2 Define the proton number and nucleon number.

Q3 Explain how the amount of carbon-14 in dead organic matter can tell scientists how old it is.

Q4 How could you calculate the specific charge of a particle?

Exam Questions

Q1 Describe the nuclear model of the atom. [2 marks]

Q2 Write down the numbers of protons, neutrons and electrons in a neutral atom of oxygen, ${}^{16}_{8}O$. [2 marks]

Q3 a) State what is meant by an 'isotope'. [1 mark]

b) State the similarities and differences between the properties of two isotopes of the same element. [2 marks]

Q4 An alpha particle is the nucleus of a ${}^{4}_{2}He$ atom. Calculate the specific charge of an alpha particle. [4 marks]

"Proton no. = no. of protons" — hardly nuclear physics is it... oh wait...

Physics is the science of all things great and small — on these pages you saw the small (like, really really small). That's why it's useful to know about things like relative charge and relative mass — 'one point six seven three times ten to the power of negative twenty-seven' is a bit more of a mouthful than 'one'. It all makes perfect sense.

Stable and Unstable Nuclei

Keeping the nucleus stable requires a lot of effort — a bit like Physics then...

*The **Strong Nuclear Force** Binds Nucleons Together*

There are several different **forces** acting on the nucleons in a nucleus. The two you already know about are **electrostatic** forces from the protons' electric charges, and **gravitational** forces due to the masses of the particles.

If you do the calculations (don't worry, *you* don't have to) you find the repulsion from the **electrostatic force** is much, much **bigger** than the **gravitational** attraction. If these were the only forces acting in the nucleus, the nucleons would **fly apart**. So there must be **another attractive force** that **holds the nucleus together** — called the **strong nuclear force**. (The gravitational force is so small, you can just ignore it.)

The **strong nuclear force** is quite **complicated**:

1) To **hold the nucleus together**, it must be an **attractive force** that's **stronger** than the electrostatic force.
2) Experiments have shown that the strong nuclear force has a **very short range**. It can only hold nucleons together when they're separated by up to **a few femtometres** (1 fm = 1×10^{-15} m) — the size of a nucleus.
3) The **strength** of the strong nuclear force **quickly falls** beyond this distance (see the graph below).
4) Experiments also show that the strong nuclear force **works equally between all nucleons**. This means that the size of the force is the same whether it's proton-proton, neutron-neutron or proton-neutron.
5) At **very small separations**, the strong nuclear force must be **repulsive** or it would **crush** the nucleus to a **point**.

*The **Size** of the Strong Nuclear Force **Varies** with **Nucleon Separation***

The **strong nuclear force** can be plotted on a **graph** to show how it changes with the **distance of separation** between **nucleons**. If the **electrostatic force** is also plotted, you can see the **relationship** between these **two forces**.

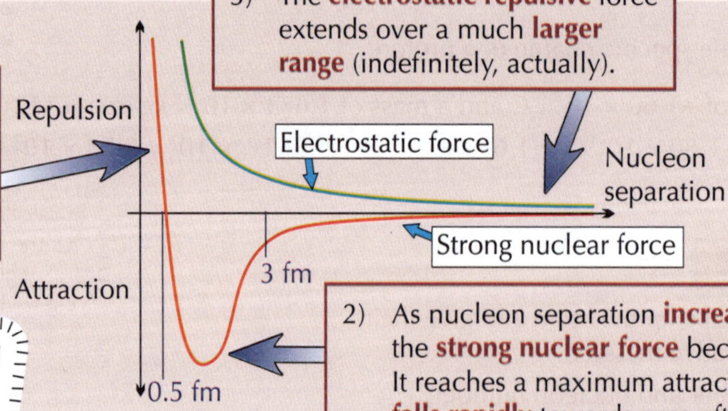

1) The strong nuclear force is **repulsive** for **very small** separations of nucleons.

2) As nucleon separation **increases past about 0.5 fm**, the **strong nuclear force** becomes **attractive**. It reaches a maximum attractive value and then **falls rapidly** towards zero after about **3 fm**.

3) The **electrostatic repulsive** force extends over a much **larger range** (indefinitely, actually).

That's why neutrons are so important in keeping a neutron stable — they increase the strong nuclear force between the nucleons to balance out the electrostatic repulsion between the protons.

You might hear the strong nuclear force called the strong interaction.

*α **Emission** Happens in **Very Big Nuclei***

1) **Alpha emission** only happens in **very big** nuclei, like **uranium** and **radium**.
2) The **nuclei** of these atoms are just **too massive** for the strong nuclear force to keep them stable.
3) When an alpha particle is **emitted**:

nucleon number decreases by 4

$$^{238}_{92}U \rightarrow {}^{234}_{90}Th + {}^{4}_{2}\alpha$$

proton number decreases by 2

The **proton number decreases** by **two**, and the **nucleon number decreases** by **four**.

Alpha particles have a very **short range** — only a few cm in air. This can be seen by observing the tracks left by alpha particles in a **cloud chamber**. You could also use a **Geiger counter** (a device that measures the amount of ionising radiation). Bring it up close to the alpha source, then **move it away** slowly and observe how the **count rate drops**.

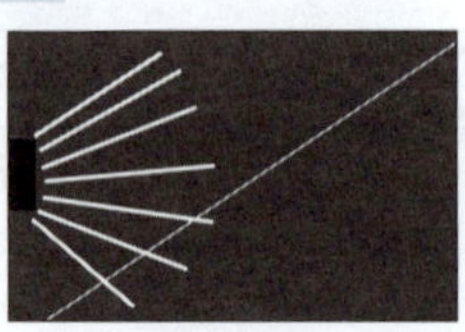

The thin line is a cosmic ray particle

Stable and Unstable Nuclei

β^- Emission Happens in Neutron-Rich Nuclei

1) **Beta-minus** (usually just called beta) decay is the emission of an **electron** from the **nucleus** along with an **antineutrino**.
2) Beta decay happens in isotopes that are unstable due to being **'neutron rich'** (i.e. they have too many more **neutrons** than **protons** in their nucleus).
3) When a nucleus ejects a beta particle, one of the **neutrons** in the nucleus is **changed** into a **proton**.

nucleon number stays the same

$${}^{187}_{75}\text{Re} \longrightarrow {}^{187}_{76}\text{Os} + {}^{0}_{-1}\beta + \bar{\nu}_e$$

proton number increases by 1

The **proton number increases** by **one**, and the **nucleon number stays the same.**

Beta particles have a much greater range than alpha particles.

In beta decay, you get a **tiny neutral particle** called an **antineutrino** released. This antineutrino carries away some **energy** and **momentum**.

Neutrinos Were First *Hypothesised* Due to Observations of *Beta Decay*

1) Scientists originally thought that the **only** particle emitted from the nucleus during beta decay was an **electron**.
2) However, observations showed that the **energy** of the particles **after** the **beta decay** was **less** than it was **before**, which didn't fit with the principle of **conservation of energy** (p. 50).
3) In 1930 Wolfgang Pauli suggested **another particle** was being emitted too, and it carried away the **missing energy**. This particle had to be **neutral** (or charge wouldn't be **conserved** in beta decay) and had to have **zero** or **almost zero** mass (as it had never been **detected**).
4) Other discoveries led to Pauli's theory becoming accepted and the particle was named the **neutrino**. (We now know this particle was an antineutrino — p. 116).
5) The neutrino was eventually observed 25 years later, providing evidence for Pauli's hypothesis.

Practice Questions

Q1 What causes an electrostatic force inside the nucleus?

Q2 What evidence suggests the existence of a strong nuclear force?

Q3 Is the strong interaction attractive or repulsive at a nucleon separation of 2 fm?

Q4 Describe the changes that happen in the nucleus during alpha and beta-minus decay.

Q5 What observations led to the hypothesis of the existence of the neutrino?

Exam Questions

Q1 The strong nuclear force binds the nucleus together.

a) Explain why the force must be repulsive at very short distances. [1 mark]

b) Explain why a nucleus containing two protons in unstable, but one containing two protons and two neutrons is stable. [2 marks]

Q2 Radium-226 and potassium-40 are both unstable isotopes.

a) Radium-226 undergoes alpha decay to radon. Complete the balanced nuclear equation for this reaction:

$${}^{226}_{88}\text{Ra} \longrightarrow \quad \text{Rn} +$$

[3 marks]

b) Potassium-40 (Z = 19, A = 40) undergoes beta decay to calcium. Write a balanced nuclear equation for this reaction (the symbol for potassium is K and for calcium is Ca). [4 marks]

The strong interaction's like nuclear glue...

Energy, momentum, charge and nucleon number (and several other things that you'll find out about in this section) are conserved in every nuclear reaction. That's why the antineutrino in beta decay has to be there.

Particles and Antiparticles

"I cannae do it Cap'n — their electron-antineutrino ray gun's interfering with my antineutron positron reading..."

Photons are *Packets* of *Electromagnetic Radiation*

Visible light is just one type of electromagnetic radiation. The **electromagnetic spectrum** is a continuous spectrum of **all** the possible frequencies of electromagnetic radiation.

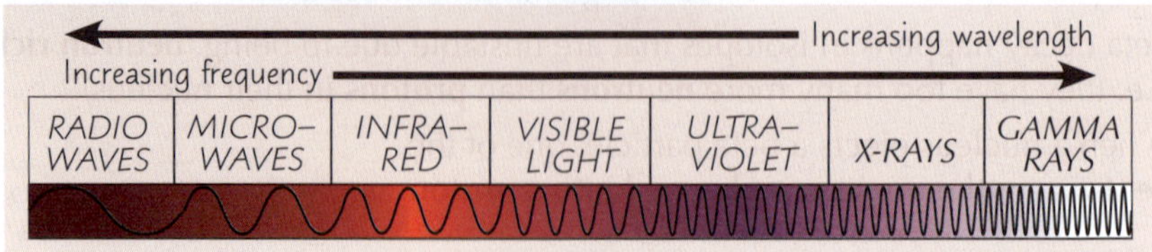

The **frequency** f and **wavelength** λ are linked by $f = \frac{c}{\lambda}$, where $c = 3.00 \times 10^8$ ms^{-1} is the **speed of light** in a vacuum (sometimes called the speed of light **in vacuo**).

Electromagnetic radiation exists as **photons** of energy (page 126).
The **energy** of a photon depends on the frequency of the radiation: $E = hf = \frac{hc}{\lambda}$

h is the Planck constant, equal to 6.63×10^{-34} Js.

Every Particle has an *Antiparticle*

1) Each particle has a **matching antiparticle** with the **same mass** and **rest energy** (more later), but with **opposite charge** (amongst other things).
2) For instance, an **antiproton** is a **negatively-charged** particle with the same mass as the **proton**, and the **antineutrino** is the antiparticle of the **neutrino** — it doesn't do much either.

Particle/Antiparticle	Symbol	Relative Charge	Mass (kg)	Rest Energy (MeV)
proton	p	+1	$1.67(3) \times 10^{-27}$	938(.3)
antiproton	$\bar{p}$	−1		
neutron	n	0	$1.67(5) \times 10^{-27}$	939(.6)
antineutron	$\bar{n}$			
electron	e^-	−1	9.11×10^{-31}	0.51(1)
positron	e^+	+1		
neutrino	ν_e	0	0	0
antineutrino	$\bar{\nu}_e$			

These are actually an electron-neutrino and an electron-antineutrino (p. 122).

Luckily, if you need to use them in the exam, you'll be given all the **masses** in kg and **rest energies** in MeV of each of these particles and their antiparticles. You just need to remember that the **mass** and **rest energy** are the **same** for a particle and its antiparticle. Neutrinos and antineutrinos are incredibly tiny — you can assume they have zero mass and zero rest energy.

1 MeV = 1×10^6 eV. There's more on eV (electron volts) on p. 128.

You can Create Matter and Antimatter from Energy

You've probably heard about the **equivalence** of energy and mass. It all comes out of Einstein's Special Theory of Relativity. **Energy** can turn into **mass** and **mass** can turn into **energy** if you know how. The **rest energy** of a particle is just the 'energy equivalent' of the particle's **mass**, measured in MeV. You can work it all out using the formula $E = mc^2$, but you won't be expected to do the calculations for AS.

When **energy** is converted into **mass** you get **equal amounts** of **matter** and **antimatter**.

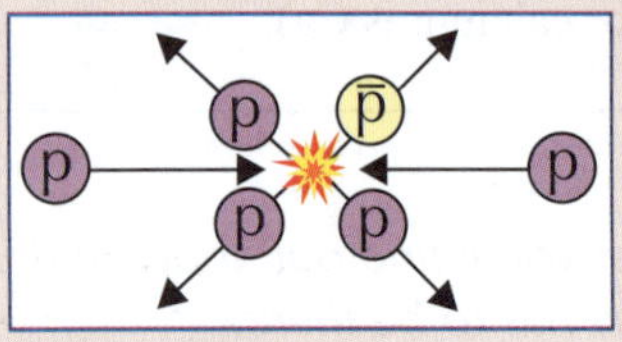

Fire **two protons** at each other at high speed and you'll end up with a lot of **energy** at the point of impact. This energy might be converted into **more particles**.

If an extra **proton** is formed then there will always be an **antiproton** to go with it. It's called **pair production**.

Particles and Antiparticles

Each Particle-Antiparticle Pair is Produced from a Single Photon

Energy that gets **converted** into **matter** and **antimatter** is in the form of a **photon** (p.116). Pair production only happens if **one photon** has enough energy to produce that much mass — only **gamma ray** photons have enough energy. It also tends to happen near a **nucleus**, which helps conserve momentum.

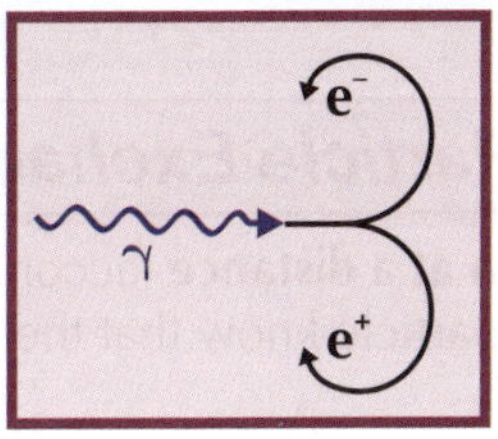

The particle tracks are curved because there's usually a magnetic field present in particle physics experiments. They curve in opposite directions because of the opposite charges on the electron and positron.

You usually get **electron-positron** pairs produced (rather than any other pair) — because they have a relatively **low mass**.

The **minimum energy** for a photon to undergo **pair production** is the **total rest energy** of the particles produced. The particle and antiparticle each have a rest energy of E_0, so:

$$E_{min} = hf_{min} = 2E_0$$

The Opposite of Pair-Production is Annihilation

When a **particle** meets its **antiparticle** the result is **annihilation**. All the **mass** of the particle and antiparticle gets converted back to **energy**. Antiparticles can usually only exist for a fraction of a second before this happens, so you don't get them in ordinary matter.

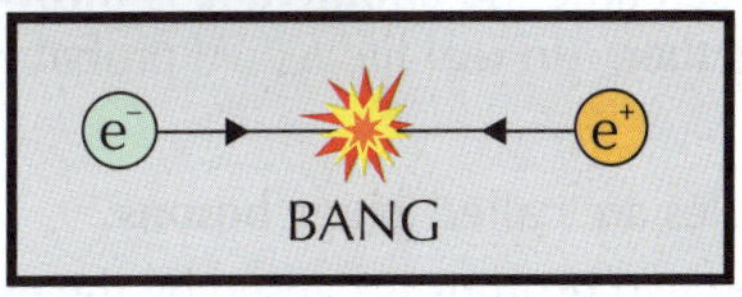

OR

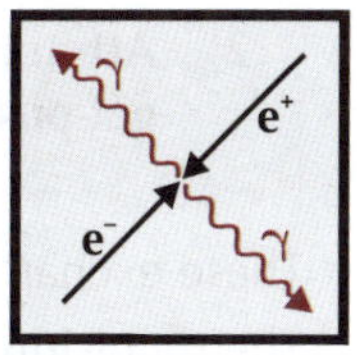

The electron and positron annihilate and their mass is converted into the energy of a pair of gamma ray photons to conserve momentum.

An annihilation is between a particle-antiparticle pair, which both have a rest energy E_0. **Both** photons need to have a **minimum energy**, E_{min}, which when added together equals at least $2E_0$ for **energy** to be **conserved** in this interaction. So $2E_{min} = 2E_0$ and:

$$E_{min} = hf_{min} = E_0$$

Example: Calculate the maximum wavelength of one of the photons produced when an electron and positron annihilate each other.

For annihilation, minimum photon energy $E_{min} = hf_{min} = E_0$. Remember $f = \frac{c}{\lambda}$, so $\frac{hc}{\lambda_{max}} = E_0$.

So $\lambda_{max} = \frac{hc}{E_0} = \frac{(6.63\times10^{-34})\times(3.00\times10^{8})}{(0.511\times10^{6})\times(1.60\times10^{-19})} = 2.432... \times 10^{-12} = \mathbf{2.43 \times 10^{-12}}$ **m (to 3 s.f.)**

The Planck constant is in J, so you need to convert E_0 from MeV to J.

Practice Questions

Q1 Describe the properties of an electron-antineutrino.

Q2 Give one similarity and one difference between a proton and an antiproton.

Q3 What is pair production?

Q4 What happens when a proton collides with an antiproton?

Exam Questions

Q1 Write down an equation for the reaction between a positron and an electron and state the name for this type of reaction. [2 marks]

Q2 Explain what causes extra particles to be created when two particles collide. [2 marks]

Q3 Give a reason why the reaction: $p + p \rightarrow p + p + n$ is not possible. [1 mark]

Q4 A photon produces an electron-positron pair, each with 9.84×10^{-14} J of energy. Calculate the frequency of the photon. [2 marks]

This really is Physics at its ~~hardest~~ grooviest...

Inertial dampers are off-line Captain.........oops, no — it's just these false ears making me feel dizzy. Anyway — you'd need to carry an awful lot of antimatter to provide enough energy to run a spaceship. Plus, it's not the easiest to store...

Forces and Exchange Particles

*Having learnt about all those lovely particles and antiparticles, you now have the esteemed privilege of learning about yet another weirdy thing called a **gauge boson**. To the casual observer this might not seem **entirely fair**. And I have to say, I'd be with them.*

Forces are Caused by Particle Exchange

You can't have **instantaneous action at a distance** (according to Einstein, anyway). So, when two particles **interact**, something must **happen** to let one particle know that the other one's there. That's the idea behind **exchange particles**.

1) **Repulsion — Each time the ball is thrown or caught** the people get **pushed apart**. It happens because the ball carries **momentum**.

Particle exchange also explains **attraction**, but you need a bit more imagination.

2) **Attraction — Each time the boomerang is thrown or caught** the people **get pushed together**. (In real life, you'd probably fall in first.)

These exchange particles are called **gauge bosons**.

The **repulsion** between two **protons** is caused by the **exchange** of **virtual photons**, which are the gauge bosons of the **electromagnetic** force. Gauge bosons are **virtual** particles — they only exist for a **very short time**.

There are Four Fundamental Forces

All forces in nature are caused by four **fundamental** forces — the strong nuclear force, the weak nuclear force, the electromagnetic force and gravity. Each one has its **own gauge boson**:

Type of Interaction	Gauge Boson	Particles Affected
electromagnetic	virtual photon (symbol, γ)	charged particles only
weak	W^+, W^-	all types
strong	pions (π^+, π^-, π^0)	hadrons only (see p.120)

Particle physicists never **bother** about **gravity** because it's so incredibly **feeble** compared with the other types of interaction. Gravity only really **matters** when you've got **big masses** like **stars and planets**.

In the **strong nuclear force**, pions are described as being exchanged between **nucleons**. You might also see it described as **gluons** being exchanged between **quarks** (p. 123-125).

*The **Larger** the **Mass** of the **Gauge Boson**, the **Shorter** the **Range** of the **Force***

1) The **W bosons** have a **mass** of about **100 times that of a proton**, which gives the weak force a **very short range**. Creating a **virtual W particle** uses **so much energy** that it can only exist for a **very short time** and it **can't travel far**.
2) On the other hand, the **photon** has **zero mass**, which gives you a force with **infinite range**.

*You can use **Diagrams** to Show What's **Going In** and What's **Coming Out***

Particle interactions can be hard to get your head around. A **neat way** of **solving problems** is by **drawing simple diagrams** of particle interactions rather than doing **calculations**.

1) **Gauge bosons** are represented by **wiggly lines** (technical term).
2) Other **particles** are represented by **straight lines**.

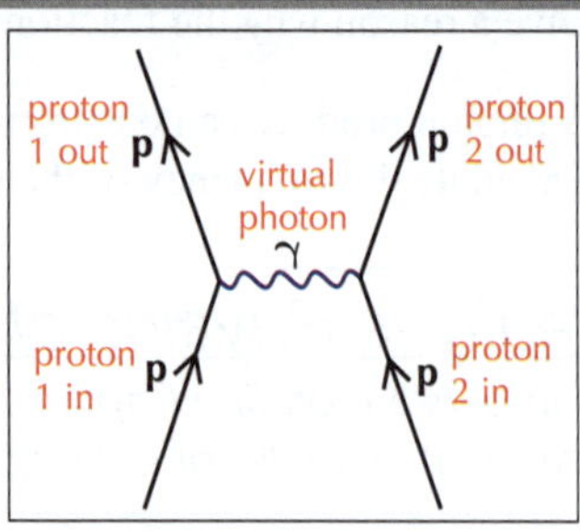

Forces and Exchange Particles

*You can draw simple diagrams of **loads** of interactions...*

You Can **Draw Diagrams** of these **Interactions**

Electromagnetic Repulsion

This is the easiest of the lot. When two particles with **equal charges** get close to each other, they **repel**.

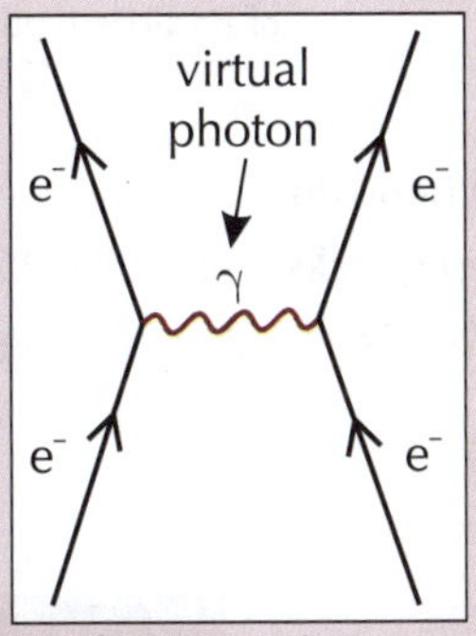

two electrons repelling each other

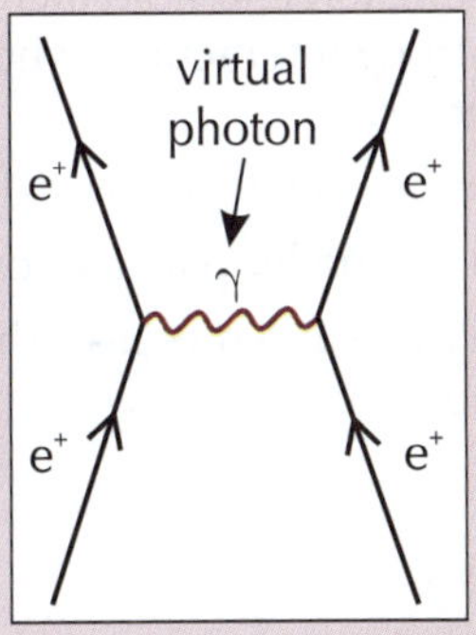

two positrons repelling each other

RULES FOR DRAWING PARTICLE INTERACTION DIAGRAMS:

1) **Incoming** particles start at the bottom of the diagram and move upwards.
2) The **baryons** (p.120) and **leptons** (p.122) can't cross from one side to the other.
3) Make sure the charges on both sides balance. The **W** bosons carry **charge** from one side of the diagram to the other.
4) A **W^-** particle going to the **left** has the same effect as a **W^+** particle going to the **right**.

Electron Capture and Electron-proton Collisions

Electrons and protons are of course attracted by the **electromagnetic interaction** between them, but if a proton **captures** an electron, the **weak interaction** can make this interaction happen.

$$\mathbf{p} + \mathbf{e}^- \rightarrow \mathbf{n} + \nu_e$$

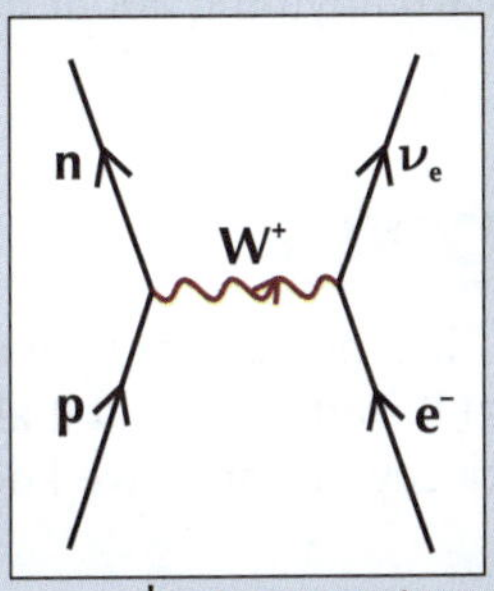

electron capture

You can also have **electron-proton collisions**, where an electron **collides** with a proton. The equation is just the same as electron capture but in the diagram a **W^- boson** goes from the **electron** to the **proton** instead of a W^+ travelling the other way.

Beta-plus and Beta-minus Decay

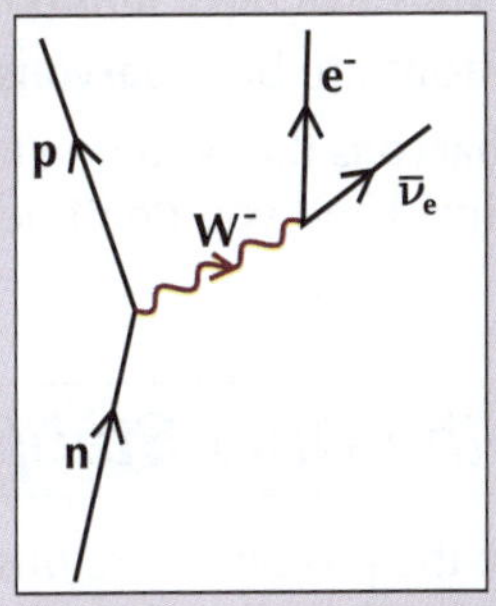

β^- decay

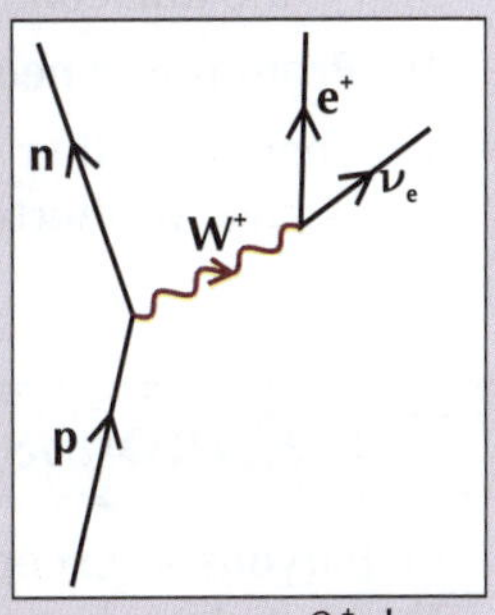

β^+ decay

$$\mathbf{n} \rightarrow \mathbf{p} + \mathbf{e}^- + \bar{\nu}_e$$

$$\mathbf{p} \rightarrow \mathbf{n} + \mathbf{e}^+ + \nu_e$$

You get an **antineutrino** in β^- decay and a **neutrino** in β^+ decay so that **lepton number** (p.122) is conserved.

You'll see on p.124 that when a proton changes to a neutron or vice versa, it has to be the weak interaction.

Practice Questions

Q1 List the four fundamental forces in nature.

Q2 Explain what a virtual particle is.

Q3 Draw a simple diagram to show the particles involved in a beta-minus decay interaction.

Q4 Which gauge bosons are exchanged in weak interactions?

Exam Questions

Q1 Describe how the force of electromagnetic repulsion between two protons is explained by particle exchange. [2 marks]

Q2 Draw a diagram to show the particle interaction when an electron and a proton collide. Label all the particles involved and state clearly which type of interaction is involved. [3 marks]

I need a drink...

Urrrgghhhh... eyes... glazed... brain... melting... ears... bleeding... help me... help me...

help me...

Classification of Particles

There are loads of different types of particle apart from the ones you get in normal matter (protons, neutrons, etc.). They only appear in cosmic rays and in particle accelerators, and they often decay very quickly so they're difficult to get a handle on. The next three pages go over a load of them and their properties. Stick with it — you'll get there.

Hadrons are Particles that Feel the Strong Nuclear Force (e.g. Protons and Neutrons)

1) The **nucleus** of an atom is made up of **protons** and **neutrons** (déjà vu).
2) Since the **protons** are **positively charged** they need a strong force to hold them together. This is called the **strong nuclear force** or the **strong interaction** (who said physicists lack imagination...). See page 114 for details.
3) **Not all particles** can **feel** the **strong nuclear force** — the ones that **can** are called **hadrons**.
4) Hadrons aren't **fundamental** particles. They're made up of **smaller particles** called **quarks** (see pages 123–125).
5) There are **two** types of **hadrons** — **baryons** (and anti-baryons) and **mesons**. They're classified according to the number of **quarks** that make them up, but don't worry about that for now.

(Leptons are an example of particles that can't. See page 122.)

Protons and Neutrons are Baryons

1) It's helpful to think of **protons** and **neutrons** as **two versions** of the **same particle** — the **nucleon**. They just have **different electric charges**.
2) **Protons** and **neutrons** are both **baryons**.
3) There are **other baryons** that you don't get in normal matter — like **sigmas** (Σ) — they're **short-lived** and you **don't** need to **learn** anything about them (woohoo!).

Baryon and Meson felt the strong interaction.

The Proton is the Only Stable Baryon

All **baryons** — except the proton — are **unstable**. This means that they **decay** to become other **particles**. The **particles** a baryon ends up as depends on what it started as, but it **always** includes a **proton**. **Protons** are the only **stable baryons** — they don't decay (as far as we know).

All baryons except protons decay to a **proton**.

Some theories predict that protons should decay with a very long half-life, but there's no evidence for it at the moment.

Antiprotons and Antineutrons are Antibaryons

The **antiparticles** of protons and neutrons — **antiprotons** and **antineutrons** — are **antibaryons**. But, if you remember from page 117, **antiparticles** are **annihilated** when they meet the corresponding **particle** — which means that you **don't** find **antibaryons** in ordinary matter.

The Number of Baryons in an Interaction is called the Baryon Number

The **baryon number** is the number of baryons. (A bit like **nucleon number** but including unusual baryons like Σ too.) The **proton** and the **neutron** each have a baryon number **B = +1**. **Antibaryons** have a baryon number **B = –1**. **Other particles** (i.e. things that aren't baryons) are given a baryon number **B = 0**.

Baryon number is a **quantum number** that must be **conserved** in any interaction — that means it can only take on a **certain set of values** (so you can't have 2.7981 baryons, or 1.991112 baryons... you get the idea).

When an **interaction** happens, the **baryon number** on either side of the interaction has to be the **same**. You can use this fact to **predict** whether an **interaction** will **happen** — if the numbers don't match, it can't happen.

The **total baryon number** in **any** particle interaction **never changes**.

Classification of Particles

Neutrons are Baryons that Decay into Protons

You saw on pages 115 and 119 that **beta decay** involves a **neutron** changing into a **proton**. This happens when there are many **more neutrons** than **protons** in a nucleus or when a **neutron** is **by itself**, **outside** of a nucleus. **Beta decay** is caused by the **weak interaction** (see page 124).

When a neutron decays, it forms a **proton**, an **electron** and an **antineutrino**:

$$n \rightarrow p + e^- + \bar{\nu}_e$$

Electrons and **antineutrinos** aren't baryons (they're **leptons**, as you'll see on the next page), so they have a baryon number **B = 0**. **Neutrons** and **protons** are baryons, so have a baryon number **B = 1**. This means that the **baryon numbers** on both sides are **equal** (to 1), so the interaction **can** happen.

Pions and Kaons are Types of Mesons

There is a second type of hadron called the **meson**.

1) **All mesons** are **unstable** and have **baryon number B = 0** (because they're not baryons).
2) **Pions** (π-mesons) are the **lightest mesons**. You get **three versions** with different **electric charges** — π^+, π^0 and π^-. You get **loads** of pions in **high-energy particle collisions** like those studied at the **CERN** particle accelerator.
3) **Kaons** (**K**-mesons) are **heavier** and more **unstable** than **pions**. You get different ones like K^+ and K^0. Kaons have a very **short lifetime** and **decay** into **pions**.
4) Pions and kaons were **discovered** in **cosmic rays** — cosmic ray showers are a source of both particles. You can observe the tracks of these particles with a **cloud chamber** (see p.114).
5) Mesons **interact** with **baryons** via the **strong force**.

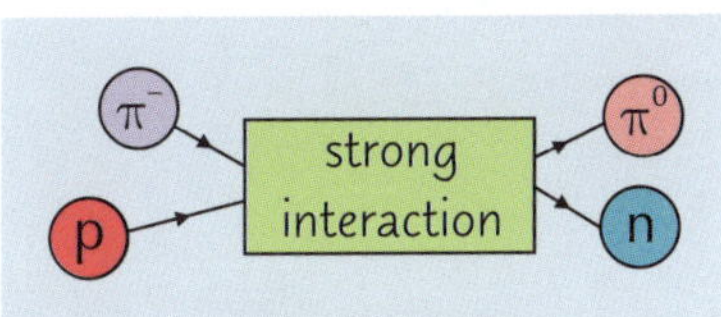

Pion interactions swap protons with neutrons and neutrons with protons, but leave the overall baryon number unchanged.

Summary of Hadron Properties

DON'T PANIC if you don't understand all this yet. For now, just **learn** these properties. You'll need to work through to the end of page 125 to see how it **all fits together**.

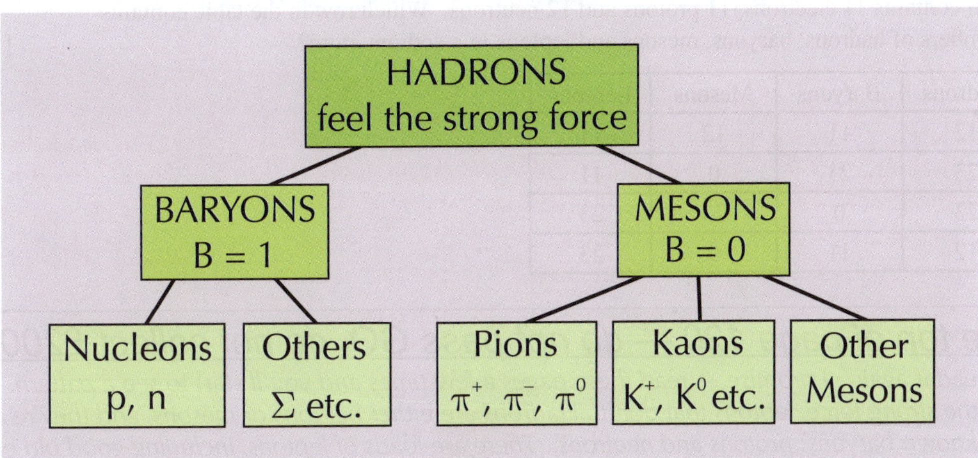

Classification of Particles

Leptons (e.g. Electrons and Neutrinos) Don't Feel the Strong Nuclear Force

1) **Leptons** are **fundamental particles** and they **don't** feel the **strong nuclear force.** They only really **interact** with other particles via the **weak interaction** (along with a bit of gravitational force and the electromagnetic force as well if they're charged).
2) **Electrons** (e^-) are **stable** and very **familiar**, but — you guessed it — there are also **other leptons**, such as the **muon** (μ^-), that are just like **heavy electrons**.
3) **Muons** are **unstable**, and **decay** eventually into **ordinary electrons**.
4) The **electron** and **muon** leptons each come with their **own neutrino**, ν_e and ν_μ.
5) **Neutrinos** have **zero** or **almost zero mass**, and **zero electric charge** — so they don't do much. **Neutrinos** only take part in **weak interactions** (see p.118). In fact, a neutrino can **pass right through the Earth** without **anything** happening to it.

You Have to Count the Types of Lepton Separately

Like the baryon number, the **lepton number** is just the number of **leptons. Each lepton** is given a **lepton number** of **+1**, but the **electron** and **muon** types of lepton have to be **counted separately**. You get **different** lepton numbers, L_e and L_μ.

All the leptons and lepton-neutrinos have their own **antiparticle** too — no surprises there. They have the **opposite charge** and **lepton numbers** to their matching particles. For example, the antimuon μ^+ has charge = +1, L_e = 0 and L_μ = –1.

Name	Symbol	Charge	L_e	L_μ
electron	e^-	–1	+1	0
electron-neutrino	ν_e	0	+1	0
muon	μ^-	–1	0	+1
muon-neutrino	ν_μ	0	0	+1

Practice Questions

Q1 List the differences between a hadron and a lepton.

Q2 Which is the only stable baryon (probably)?

Q3 A particle collision at CERN produces 2 protons, 3 pions and 1 neutron. What is the total baryon number of these particles?

Q4 Which two particles have lepton number L_μ = +1?

Exam Questions

Q1 List all the decay products of the neutron.
Explain why this decay cannot be due to the strong interaction. [3 marks]

Q2 Initially, the muon was incorrectly identified as a meson.
Explain why the muon is not a meson. [3 marks]

Q3 A sodium atom contains 11 electrons, 11 protons and 12 neutrons. Which row in the table contains the correct numbers of hadrons, baryons, mesons and leptons in a sodium atom? [1 mark]

	Hadrons	Baryons	Mesons	Leptons
A	12	11	12	11
B	23	23	0	11
C	23	0	23	23
D	12	11	0	23

Go back to the top of page 120 — do not pass GO, do not collect £200...

Do it. Go back and read it again. I promise — read these pages a few times and you'll start to see a pattern. There are hadrons that feel the strong force, leptons that don't. Hadrons are either baryons or mesons, and they're all weird except for those well-known baryons: protons and neutrons. There are loads of leptons, including good old electrons.

Quarks

Quarks may sound like a bizarre concept, but they weren't just made up willy-nilly. Large teams of scientists and engineers all over the world worked for years to come up with the info on these pages.

Quarks are Fundamental Particles

Quarks are the **building blocks** for **hadrons** (baryons and mesons). Antiparticles of hadrons are made from **antiquarks**.

1) To make **protons** and **neutrons** you only need two types of quark — the **up** quark (**u**) and the **down** quark (**d**).
2) An extra one called the **strange** quark (**s**) lets you make more particles with a property called **strangeness**.

Strangeness is Only Conserved Some of the Time

1) **Strangeness**, like baryon number, is a **quantum number** (see p.120) — it can only take a certain set of values.
2) Strange particles, such as kaons, are **created** via the **strong** interaction but **decay** via the **weak** interaction.
3) Here's the catch — strangeness is **conserved** in the **strong interaction**, but **not** in the **weak interaction** (p.124).
4) That means strange particles are **always produced in pairs** (e.g. K^+ and K^-).
   One has a strangeness of +1, and the other has a strangeness of –1, so the overall strangeness of 0 is **conserved**.

Quarks and Antiquarks have Opposite Properties

The **antiquarks** have **opposite properties** to the quarks — as you'd expect.

QUARKS

Name	Symbol	Charge	Baryon number	Strangeness
up	u	$+\frac{2}{3}$	$+\frac{1}{3}$	0
down	d	$-\frac{1}{3}$	$+\frac{1}{3}$	0
strange	s	$-\frac{1}{3}$	$+\frac{1}{3}$	–1

ANTIQUARKS

Name	Symbol	Charge	Baryon number	Strangeness
anti-up	$\bar{u}$	$-\frac{2}{3}$	$-\frac{1}{3}$	0
anti-down	$\bar{d}$	$+\frac{1}{3}$	$-\frac{1}{3}$	0
anti-strange	$\bar{s}$	$+\frac{1}{3}$	$-\frac{1}{3}$	+1

Baryons are Made from Three Quarks

Evidence for quarks came from **hitting protons** with **high-energy electrons**.
The way the **electrons scattered** showed that there were **three concentrations of charge** (quarks) **inside** the proton.

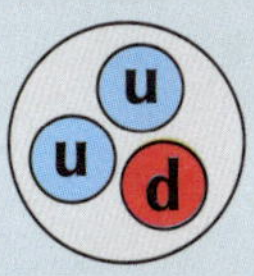

Proton = **uud**

Total charge
$= \frac{2}{3} + \frac{2}{3} - \frac{1}{3} = 1$
Baryon number
$= \frac{1}{3} + \frac{1}{3} + \frac{1}{3} = 1$

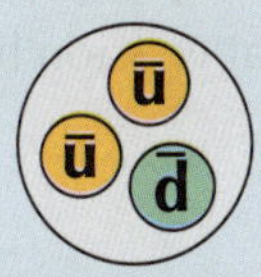

Antiproton = $\mathbf{\bar{u}\bar{u}\bar{d}}$

Total charge
$= -\frac{2}{3} - \frac{2}{3} + \frac{1}{3} = -1$
Baryon number
$= -\frac{1}{3} - \frac{1}{3} - \frac{1}{3} = -1$

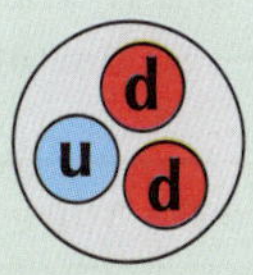

Neutron = **udd**

Total charge
$= \frac{2}{3} - \frac{1}{3} - \frac{1}{3} = 0$
Baryon number
$= \frac{1}{3} + \frac{1}{3} + \frac{1}{3} = 1$

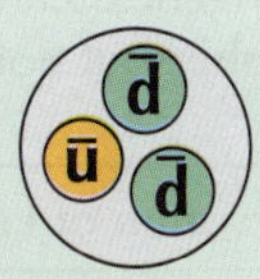

Antineutron = $\mathbf{\bar{u}\bar{d}\bar{d}}$

Total charge
$= -\frac{2}{3} + \frac{1}{3} + \frac{1}{3} = 0$
Baryon number
$= -\frac{1}{3} - \frac{1}{3} - \frac{1}{3} = -1$

Quarks

Mesons are a Quark and an Antiquark

Pions are just made from combinations of **up**, **down**, **anti-up** and **anti-down** quarks.
Kaons have **strangeness** so you need to put in **s** quarks as well (remember, the **s** quark has a strangeness of S = –1).

Before we move on, it's worth mentioning that the π^- meson is just the **antiparticle** of the π^+ meson, the K^- meson is the antiparticle of the K^+ meson, and the **antiparticle** of a π^0 meson is **itself**. It all makes sense when you look at the quark compositions to the right...

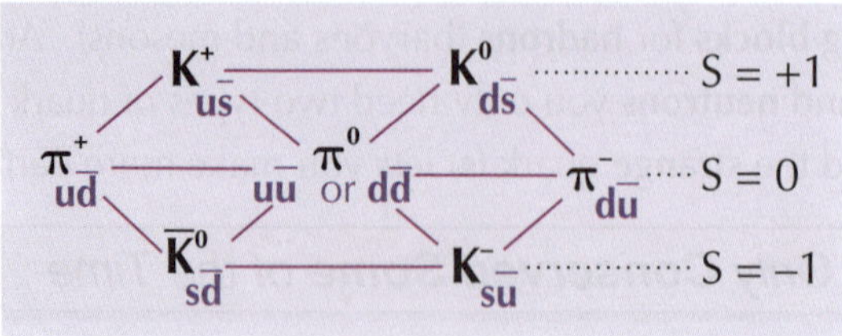

Physicists love patterns. Gaps in patterns like this predicted the existence of particles that were actually found later in experiments. Great stuff.

The Weak Interaction is something that Changes the Quark Type

In β^- decay a **neutron** is changed into a **proton** — in other words **udd** changes into **uud**. It means turning a **d** quark into a **u** quark. Only the weak interaction can do this.

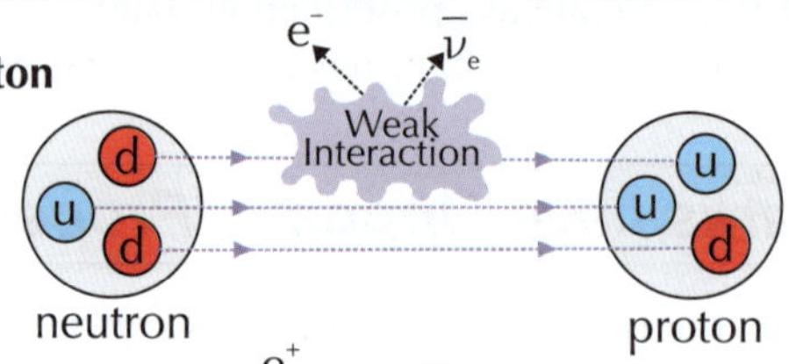

Some unstable isotopes like **carbon-11** decay by β^+ emission. In this case a **proton** changes to a **neutron**, so a **u** quark changes to a **d** quark and we get:

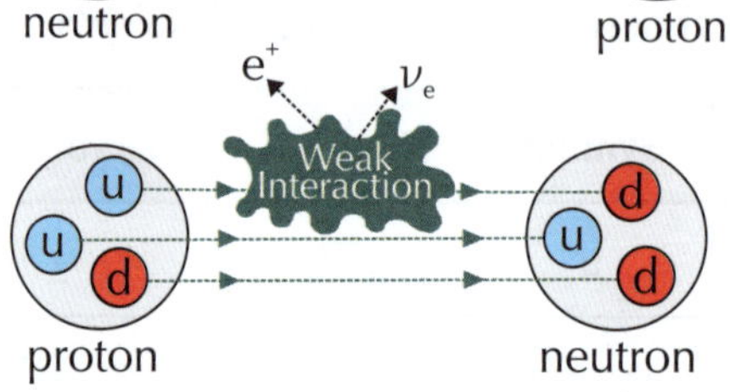

Four Properties are Conserved in Particle Interactions

Charge is Always Conserved

In **any** particle interaction, the **total charge** after the interaction must equal the total charge before the interaction.

Energy and momentum are also always conserved in particle interactions.

Baryon Number is Always Conserved

Just like with charge, in **any** particle interaction, the **baryon number** after the interaction must equal the baryon number before the interaction.

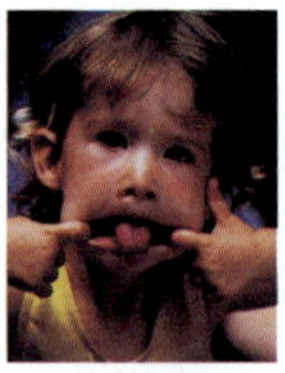

Dylan was committed to conserving strangeness.

Strangeness is Conserved in Strong Interactions

The **only** way to change the **type** of quark is with the **weak interaction**, so in strong interactions there has to be the same number of strange quarks at the beginning as at the end. In weak interactions, strangeness can change by –1, 0 or +1. The interaction $K^- + p \rightarrow n + \pi^0$ is fine for **charge** and **baryon number** but not for **strangeness** — so it won't happen. The negative kaon has an **s** quark in it.

Conservation of Lepton Number is a Bit More Complicated

The **different types** of lepton number have to be conserved **separately**.

1) For example, the interaction
$\pi^- \rightarrow \mu^- + \bar{\nu}_\mu$ has $L_\mu = 0$ at the start and $L_\mu = 1 - 1 = 0$ at the end, so it's OK.
Similarly, $n \rightarrow p + e^- + \bar{\nu}_e$ is fine. $L_e = 0$ at the start and $L_e = 1 - 1 = 0$ at the end.
2) On the other hand, the interaction $\nu_\mu + \mu^- \rightarrow e^- + \nu_e$ can't happen.
At the start $L_\mu = 2$ and $L_e = 0$, but at the end $L_\mu = 0$ and $L_e = 2$.

Lepton number is a quantum number, just like baryon number and strangeness.

Quarks

There's No Such Thing as a Free Quark

What if you **blasted** a **proton** with **enough energy** — could you **separate out** the quarks? Nope. Your energy just gets changed into more **quarks and antiquarks** — it's **pair production** again and you just make **mesons**. It's not possible to get a quark by itself — this is called **quark confinement**.

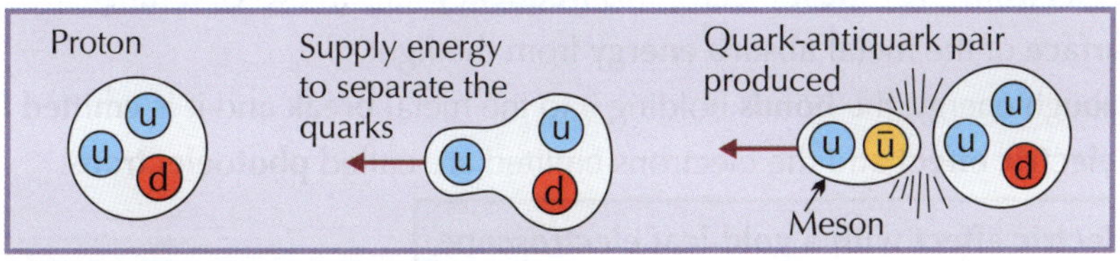

Simon was practising quack confinement.

We're Still Searching for Particles

As time goes on, our knowledge and understanding of particle physics **changes**.

1) **New theories** are created to try to explain observations from experiments. Sometimes, physicists hypothesise a new **particle** and the **properties** they expect it to have. E.g. the **neutrino** was hypothesised due to observations of beta decay.
2) **Experiments** to try to find the existence of this new particle are then carried out. Results from different experiments are **combined** to try to **confirm** the new particle. If it exists, the theory is **more likely** to be correct and the scientific community start to accept it — it's **validated**.
3) It's not quite that simple though. Experiments in particle physics often need particles travelling at incredibly **high speeds** (close to the speed of light). This can only be achieved using **particle accelerators**. These huge pieces of equipment are very **expensive** to build and run. This means that **large groups** of scientists and engineers from all over the **world** have to **collaborate** to be able to fund these experiments.

Example: Paul Dirac predicted the existence of **antimatter** in 1928. His theory was **validated** with the observation of the **positron** and, over the years, more and more observations of antiparticles. Nowadays, it's **accepted** that antimatter exists, but there are still **questions**. For example, there should have been **equal amounts** of matter and antimatter created when the universe was formed, but **almost everything** we observe is made of **matter**.

Scientists are trying to figure out what happened to all the antimatter by studying the differences in behaviour of matter and antimatter particles using the **Large Hadron Collider (LHC)** at **CERN**. The LHC is a **17 mile long** particle accelerator costing around **£3 billion** to build and **£15 million** per year to run. Some **10,000** scientists from **100** countries are involved.

ATLAS, just one of many experiments the LHC at CERN is used for, involves around 3000 scientists from 38 different countries.

Practice Questions

Q1 What is a quark?

Q2 Kaons are produced by the strong interaction. Why must they be produced in pairs?

Q3 Describe how a neutron is made up from quarks.

Q4 Which type of particle is made from a quark and an antiquark?

Q5 By how much can the strangeness change in a weak interaction?

Q6 List six quantities that are conserved in strong particle interactions.

Exam Questions

Q1 a) Write down the quark composition of the π^-. [1 mark]

b) Explain how the charges of the quarks give rise to its charge. [1 mark]

Q2 Explain how the quark composition is changed in the β^- decay of the neutron. [2 marks]

Q3 Give two reasons why the reaction $p + p \rightarrow p + K^+$ does not happen. [2 marks]

A physical property called strangeness — how cool is that...

True, there's a lot of information here, but this page really does tie up a lot of the stuff on the last few pages. Learn as much as you can from this three-page topic, then go back to page 123, and work back through to here. Don't expect to understand it all — but you'll definitely find it much easier to learn when you can see how all the bits fit together.

The Photoelectric Effect

I think they should rename 'the photoelectric effect' as 'the piece-of-cake effect' — it's not easy, I just like cake.

Shining Light on a Metal can Release Electrons

If you shine **electromagnetic waves** (e.g. light) of a **high enough frequency** onto the **surface of a metal**, it will **instantaneously eject electrons**. For **most** metals, this **frequency** falls in the **U.V.** range.

1) **Free electrons** on the **surface** of the metal **absorb energy** from the light.
2) If an electron absorbs **enough** energy, the **bonds** holding it to the metal **break** and it is **emitted from the surface**.
3) This is called the **photoelectric effect** and the electrons emitted are called **photoelectrons**.

Demonstrating the photoelectric effect with a gold-leaf electroscope

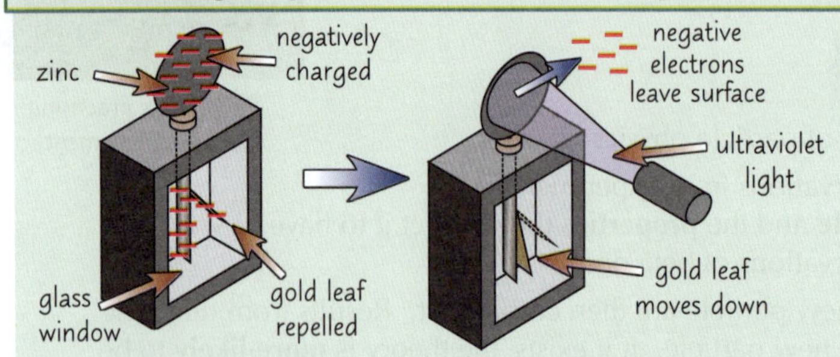

1) The electroscope plate is initially **negatively** charged, so the gold leaf is **repelled**.
2) The zinc plate is then exposed to **ultraviolet light** and the photoelectric effect causes its free electrons to be **ejected**. This causes it to lose its negative charge — the gold leaf is no longer repelled and so drops down.

These are the two that had scientists puzzled. They can't be explained using wave theory.

Conclusion 1: For a given metal, **no photoelectrons are emitted** if the radiation has a frequency **below** a certain value — called the **threshold frequency**.

Conclusion 2: The photoelectrons are emitted with a variety of kinetic energies ranging from zero to some maximum value. This value of **maximum kinetic energy** increases with the **frequency** of the radiation, and is **unaffected** by the **intensity** of the radiation.

Conclusion 3: The **number** of photoelectrons emitted per second is **proportional** to the **intensity** of the radiation.

The Photoelectric Effect Couldn't be Explained by Wave Theory...

According to wave theory:

1) For a particular frequency of light, the **energy** carried is **proportional** to the **intensity** of the beam.
2) The energy carried by the light would be **spread evenly** over the wavefront.
3) **Each** free electron on the surface of the metal would gain a **bit of energy** from each incoming wave.
4) Gradually, each electron would gain **enough energy** to leave the metal.

SO... The **higher the intensity** of the wave, the **more energy** it should transfer to each electron — the kinetic energy should increase with **intensity**. There's **no explanation** for the **kinetic energy** depending only on the **frequency**.

There is also **no explanation** for the **threshold frequency**. According to **wave theory**, the electrons should be emitted **eventually**, no matter what the **frequency** is.

...But it Could be Explained by Einstein's Photon Model of Light

1) **Einstein** suggested that **EM waves** (and the energy they carry) **exist** in discrete packets — called **photons**.
2) The **energy carried** by one of these **photons** is: $E = hf = \frac{hc}{\lambda}$ where h = the Planck constant = 6.63×10^{-34} Js and c = speed of light in a vacuum = 3.00×10^{8} ms^{-1}
3) Einstein saw these photons of light as having a **one-on-one**, **particle-like** interaction with **an electron** in a **metal surface**. A photon would **transfer all** its **energy** to **one**, **specific electron**.

You might have seen this formula before on p.116.

According to the photon model:

1) When light hits its surface, the metal is **bombarded** by photons.
2) If one of these photons **collides** with a free electron, the electron will gain energy equal to ***hf***.

Before an electron can **leave** the surface of the metal, it needs enough energy to **break the bonds holding it there**. This energy is called the **work function** (which has the symbol ϕ (phi)) and its **value** depends on the **metal**.

The Photoelectric Effect

The **Photon Model** Explains the **Threshold Frequency**...

1) If the energy **gained** by an electron (on the surface of the metal) from a photon is **greater** than the **work function**, the electron is **emitted**.
2) If it **isn't**, the metal will heat up, but **no electrons** will be emitted.
3) Since, for **electrons** to be released, $hf \geq \phi$, the **threshold frequency** must be:

$$f = \frac{\phi}{h}$$

...and the **Maximum Kinetic Energy**

1) The **energy transferred** to an electron is hf.
2) The **kinetic energy** the electron will be carrying when it **leaves** the metal is hf **minus** any energy it's **lost** on the way out. Electrons **deeper** down in the metal lose more energy than the electrons on the **surface**, which explains the **range** of energies.
3) The **minimum** amount of energy it can lose is the **work function**, so the **maximum kinetic energy** of a photoelectron, $E_{k\,(max)}$, is given by the photoelectric equation:

$$hf = \phi + E_{k\,(max)} \quad \text{where} \quad E_{k\,(max)} = \frac{1}{2}mv_{max}^2$$

You might sometimes see the maximum kinetic energy given the symbol KE_{max}.

4) The **kinetic energy** of the electrons is **independent of the intensity** (the **number** of photons **per second** on an **area**, p.80), as they can **only absorb one photon** at a time. Increasing the **intensity** just means **more photons per second** on an **area** — each photon has the **same energy** as before.

The **Stopping Potential** Gives the **Maximum Kinetic Energy**

1) The **maximum kinetic energy** can be measured using the idea of **stopping potential**.
2) The **emitted electrons** are made to lose their energy by **doing work** against an applied **potential difference**.
3) The **stopping potential**, V_s, is the p.d. needed to stop the **fastest** moving electrons, with $E_{k\,(max)}$.
4) The **work done** by the p.d. in **stopping** the fastest electrons is equal to the **energy** they were carrying:

work done = p.d. × charge (see p.64)

$$eV_s = E_{k\,(max)}$$

where e = charge on the electron = 1.60×10^{-19} C, V_s = stopping potential in V, and $E_{k\,(max)}$ is measured in J.

Work done = zero.

Practice Questions

Q1 Explain what the photoelectric effect is.

Q2 What three conclusions were drawn from experimentation on the photoelectric effect?

Q3 What is meant by the work function of a metal?

Q4 How is the maximum kinetic energy of a photoelectron related to the work function?

Q5 Explain what is meant by the stopping potential. Write down a formula relating stopping potential and $E_{k\,(max)}$.

Exam Questions

Q1 An isolated zinc plate with neutral charge is exposed to high-frequency ultraviolet light. State and explain the effect of the ultraviolet light on the charge of the plate. [2 marks]

Q2 Explain why photoelectric emission from a metal surface only occurs when the frequency of the incident radiation exceeds a certain threshold value. [2 marks]

I'm so glad we got that all cleared up...

The most important bits here are why wave theory doesn't explain the phenomenon of the photoelectric effect, and why the photon theory does. A good way to learn conceptual stuff like this is to try to explain it to someone else.

Electron Levels and Photon Emission

So you've seen that light can be explained by the photon model. Prepare yourself for lots more photon fun...

Electrons** in **Atoms** Exist in **Discrete Energy Levels

1) **Electrons** in an **atom** can **only exist** in certain **well-defined energy levels**. Each level is given a **number**, with **n = 1** representing the **ground state**.
2) Electrons can **move down** energy levels by **emitting** a **photon**.
3) Since these **transitions** are between **definite energy levels**, the **energy** of **each photon** emitted can **only** take a **certain allowed value**.
4) The diagram on the right shows the **energy levels** for **atomic hydrogen**.
5) The **energies involved** are **so tiny** that it makes sense to use a more **appropriate unit** than the **joule**. The **electronvolt (eV)** is defined as:

The **kinetic energy carried** by an **electron** after it has been **accelerated** through a **potential difference** of **1 volt**.

energy gained by electron (eV) = accelerating voltage (V)

1 eV = 1.60 × 10^{-19} J

The ground state is the lowest energy state of the atom.

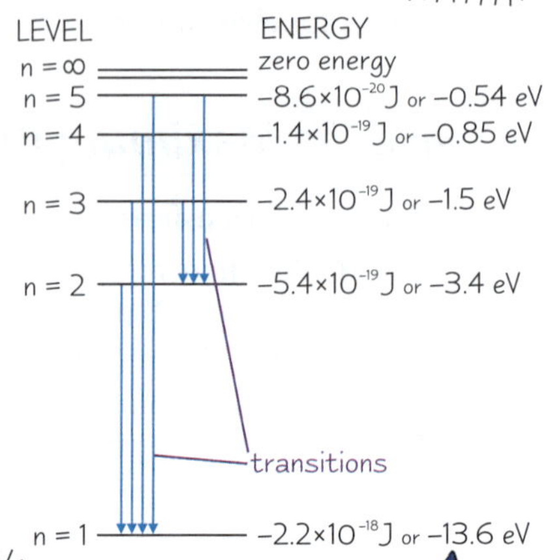

Photon energies are usually given in eV.

The energies are only negative because of how "zero energy" is defined. Just one of those silly convention things — don't worry about it.

6) On the diagram, energies are labelled in **both units** for **comparison's** sake.
7) The **energy** carried by each **photon** is **equal** to the **difference in energies** between the **two levels**. The equation below shows a **transition** between a higher energy level, n = 2, where the electrons have energy E_2, and a lower energy level, n = 1, with electrons of energy E_1: $\Delta E = E_2 - E_1 = hf = \frac{hc}{\lambda}$
8) Electrons can also **move up** energy levels if they **absorb a photon** with the **exact energy difference** between the two levels. The movement of an electron to a higher energy level is called **excitation**.
9) If an electron is **removed** from an atom, we say the atom is **ionised**. The energy of each **energy level** within an atom gives the amount of **energy** needed to **remove an electron** in that level from the atom. The **ionisation energy** of an atom is the amount of energy needed to completely remove an electron from the atom from the **ground state** (**n = 1**).

Fluorescent Tubes** use **Excited Electrons** to Produce **Light

1) **Fluorescent tubes** contain **mercury vapour**, across which an initial **high voltage** is applied. This **high voltage** accelerates **fast-moving free electrons** that **ionise** some of the **mercury atoms**, producing **more** free electrons.
2) When this flow of free electrons **collides** with electrons in **other mercury atoms**, the electrons in the mercury atoms are **excited** to **higher energy levels**.
3) When these **excited electrons** return to their **ground states**, they emit **photons** in the **UV** range.
4) A **phosphorus coating** on the **inside** of the tube **absorbs** these **photons**, exciting its **electrons** to **much higher levels**. These electrons then **cascade** down the **energy levels**, **emitting** many **lower energy photons** in the form of **visible light**.

Fluorescent Tubes** Produce **Line Emission Spectra

1) If you **split** the light from a **fluorescent tube** with a **prism** or a **diffraction grating** (see pages 94-95), you get a **line spectrum**.
2) A line spectrum is seen as a **series** of **bright lines** against a **black background**.
3) Each **line** corresponds to a **particular wavelength** of light **emitted** by the source.
4) Since only certain photon energies are allowed, you only see the **wavelengths** corresponding to these energies.

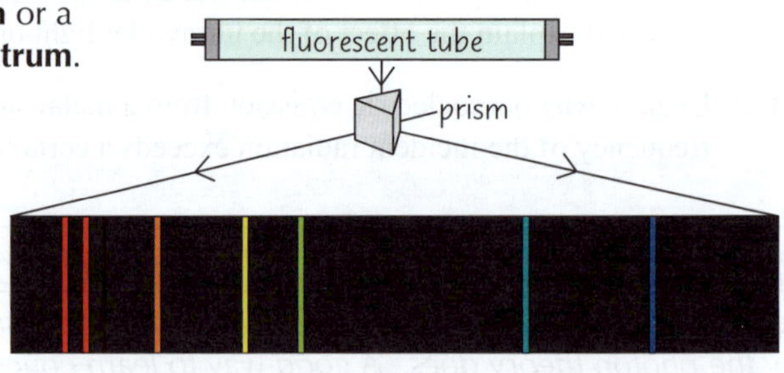

Electron Levels and Photon Emission

Continuous Spectra Contain All Possible Wavelengths

1) The **spectrum** of **white light** is **continuous**.
2) If you **split** the **light** up with a **prism**, the **colours** all **merge** into each other — there **aren't** any **gaps** in the spectrum.

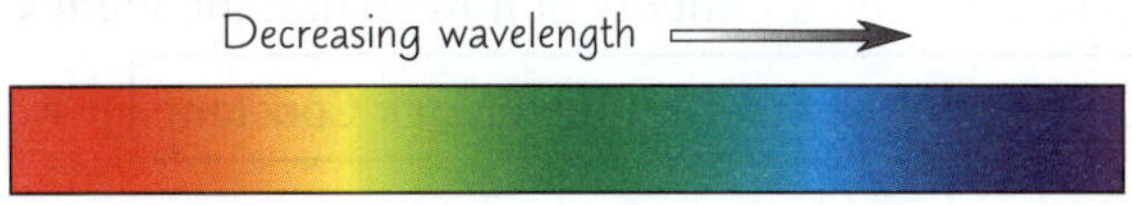

Hot Gases Produce Line Emission Spectra

1) If you heat a gas to a **high temperature**, the **atoms** become excited (one or more of their **electrons** is moved to a **higher** energy level).
2) As electrons within the atoms **fall** back to the **lower energy levels**, they **emit energy** as **photons** (see previous page).
3) If you **split** the light from a **hot gas** with a **prism** or a **diffraction grating** (see pages 94-95), you get a **line emission spectrum** (see previous page).

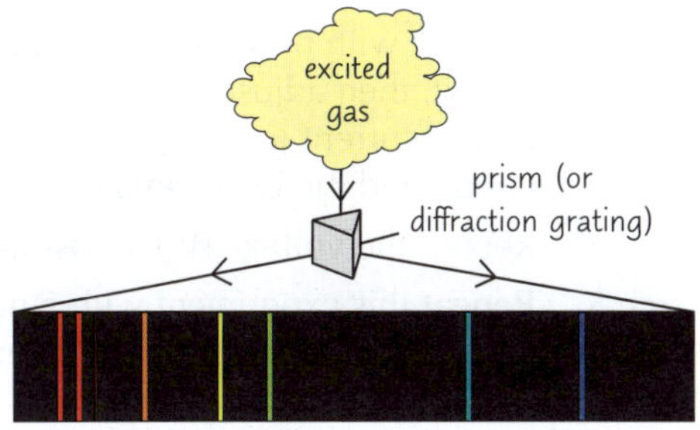

Cool Gases Produce Line Absorption Spectra

1) You get a **line absorption spectrum** when **white light** passes through a cool gas and then a **diffraction grating** or **prism**.
2) At **low temperatures**, **most** of the **electrons** in the **gas atoms** will be in their **ground states**.
3) **Photons** of the **correct wavelength** are **absorbed** by the **electrons** to **excite** them to **higher energy levels**.
4) These **wavelengths** are then **missing** from the **continuous spectrum** when it **comes out** the other side of the gas.
5) You see a **continuous spectrum** with **black lines** in it corresponding to the **absorbed wavelengths**.
6) If you **compare** the **absorption** and **emission spectra** of a **particular gas**, the **black lines** in the **absorption spectrum match up** to the **bright lines** in the **emission spectrum**.

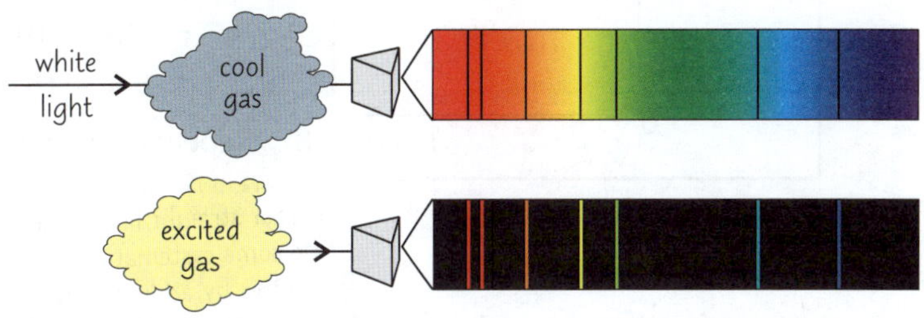

The Threshold Voltage is the Minimum Voltage Needed for an LED to Work

1) The energy of light is calculated using the **Planck constant** (see page 126) — but it's not just some random number plucked out of the air. You can find its value by doing a simple experiment with **light-emitting diodes** (**LEDs**) (see next page).
2) Current will only pass through an LED after a **minimum voltage** is placed across it — **the threshold voltage V_0**.
3) This is the voltage needed to give the electrons the **same energy** as a photon emitted by the LED. **All** of the electron's **kinetic energy** after it is accelerated over this potential difference is **transferred** into a **photon**.

$$E = hf = eV_0 \Rightarrow \frac{V_0}{f} = \frac{h}{e}$$

LEDs — long-eared dogs.

4) So by finding the **threshold voltage** for LEDs with **varying frequencies**, you can **estimate** the Planck constant.

Electron Levels and Photon Emission

You can Use LEDs to Estimate the Planck Constant

You've just seen the **theory** of how to find the **Planck constant** — now it's time for the **practicalities**.

Experiment to Measure the Planck Constant

1) Connect an LED of **known wavelength** in the electrical circuit shown.
2) Close any **blackout blinds** and place a **shaded tube** over the LED to look through. The room should be as dark as possible so you can see when the LED **first** begins to emit light.
3) Start off with no **current** flowing through the circuit, then adjust the **variable power source** until a current **just** begins to flow through the circuit and the LED **lights up**.
4) Record the **voltage** (V_0) across the LED.
5) **Repeat** this experiment with a number of LEDs of different colours that emit light at different wavelengths.

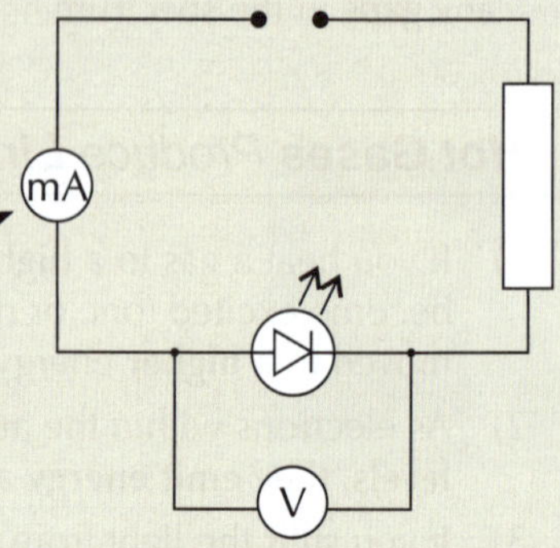

This is a milliammeter — used for measuring small currents.

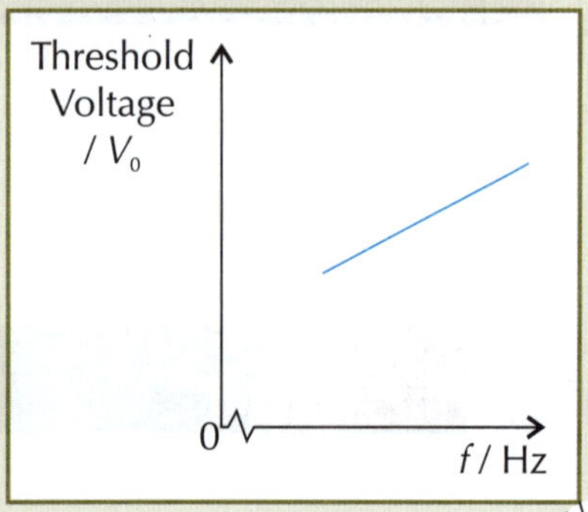

6) Plot a **graph** of **threshold voltages** (V_0) against frequency ($f = c/\lambda$) (where λ is the wavelength of light emitted by the LED in metres).
7) You should get a **straight line graph** with a **gradient** of h/e — which you can then use to find the value of h:

$$\text{E.g. gradient} = \frac{h}{e} = 4.1 \times 10^{-15} \text{ so } h = 4.1 \times 10^{-15}\text{e}$$
$$= (4.1 \times 10^{-15}) \times (1.60 \times 10^{-19}) = 6.56 \times 10^{-34}$$
$$= \mathbf{6.6 \times 10^{-34}\ Js\ (to\ 2\ s.f.)}$$

8) Repeat the experiment to find an **average** value of h.

If your straight line doesn't go through the origin, there could be some systematic errors you need to account for. You can do this by adding or taking away the difference between the origin and the vertical intercept from all of your data.

After careful measurements, Fluffles determined that her plank was indeed constant.

Practice Questions

Q1 What is an electronvolt? What is 1 eV in joules?

Q2 Write down the two formulas you can use to find the energy of a photon. Define all the symbols you use.

Q3 Explain how the coating on a fluorescent tube converts UV into visible light.

Q4 Describe line absorption and line emission spectra. How are these two types of spectra produced?

Q5 Describe an experiment to determine the Planck constant using different coloured LEDs.

Exam Question

Q1 An electron is accelerated through a potential difference of 12.1 V.

a) How much kinetic energy has it gained in i) eV and ii) joules? [2 marks]

b) This electron hits a hydrogen atom in its ground state and excites it.

i) Explain what is meant by excitation. [1 mark]

ii) Using the energy values on the right, calculate which energy level the electron from the hydrogen atom is excited to. [1 mark]

iii) Calculate the energies and frequencies of all the photons that might be emitted as the electron returns to its ground state. [6 marks]

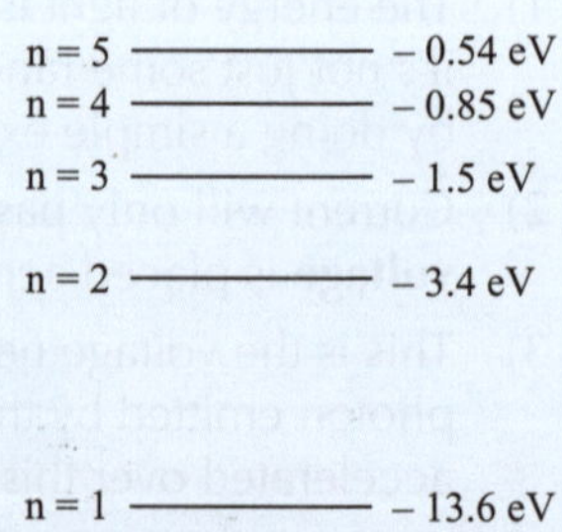

I can honestly say I've never got so excited that I've produced light...

This is heavy stuff. Quite interesting though, as I was just saying to Dom a moment ago. He's doing a psychology book. Psychology's probably quite interesting too — and easier. But it won't help you become an astrophysicist.

The "Sum Over Paths" Theory

You met phasors back on p.99 ... now here's where the really weird stuff kicks in. Buckle your seatbelts... and prepare to be amazed as the magical world of quantum reveals why light travels in straight lines, and why probability is a bit more useful than guessing what coloured ball you're likely to pick out of a bag...

Photons try Every Possible Path

1) A rather clever bloke called Richard Feynman came up with a completely different idea of how photons (or any subatomic particles, **quanta**) get from a source to a detector.
2) Feynman reckoned that instead of just taking one route to the detector, a photon will take **all** of the **possible paths** to the detector in one go. You can keep track of this photon whizzing along every possible route using **phasors** (see p.99).

A phasor shows the amplitude (size) and phase (direction) of a point on a wave.

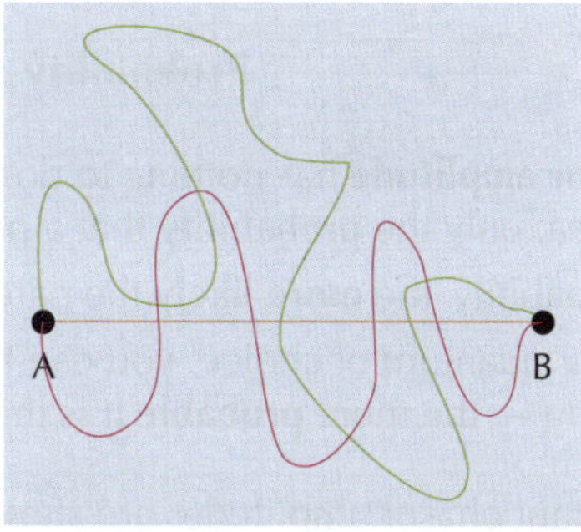

And all means <u>all</u> — the paths between A and B can be as squiggly as you like.

Andrew wasn't lost, he was just trying every path.

You can use Phasors to Track Quanta

1) Take **Young's double-slit experiment** (see p.92). You can use **phasors** to show how light or dark a certain spot on a screen will be. In quantum mechanics, you can use phasors to tell you how **probable** it is that a quantum (in this case a photon) will arrive there.
2) Take a photon travelling down **one particular path**.
3) As it travels, its phasor will rotate (anticlockwise) until it reaches the detector. By knowing the energy of the photon, you can work out the **frequency** of the phasor's rotation, *f*, by rearranging the equation on p.126.

$$f = \frac{E}{h}$$

Remember — *E* is the photon's energy and *h* is the Planck constant.

4) You want to **record** the position of the phasor at the **end** of every path — you could then **sum** these phasors to find the **resultant phasor** for the photon making the journey from a source to a detector.
5) Of course you can't find the final phasor for every path as there's an **infinite** number of them. When you do the maths, nearly all the phasors cancel each other out — so you only need to consider the straightest/quickest possible paths (see next page).

Young's Double-Slit Experiment (again...)

1) Imagine that a photon is emitted by the source and hits point X on the screen. Take two of its possible paths and say it follows **both** of them, as shown.
2) The **phasor** of the photon along each path rotates at the **same rate** (because it's the **same photon** so the phasors will have the same frequency).
3) Because the photon has to travel slightly **further** on the green path, it takes slightly **longer** to reach point X. This means the final phasor for the green path will have **rotated** slightly **further** than that for the blue path.
4) You can find the **resultant** phasor arrow for the photon reaching point X by **adding** the final phasor position for each path, **tip-to-tail** (just like a normal vector sum (see p.14)).

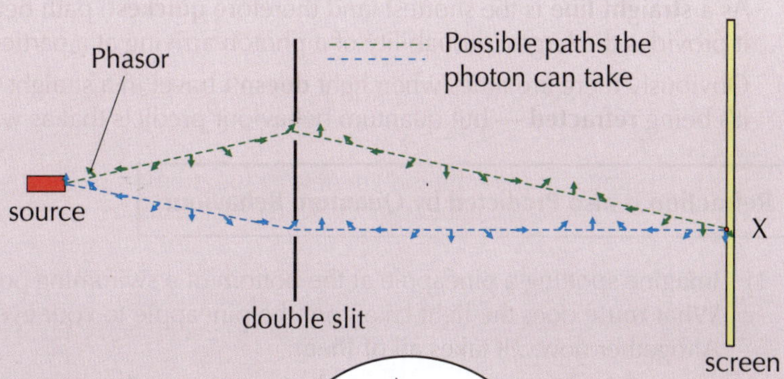

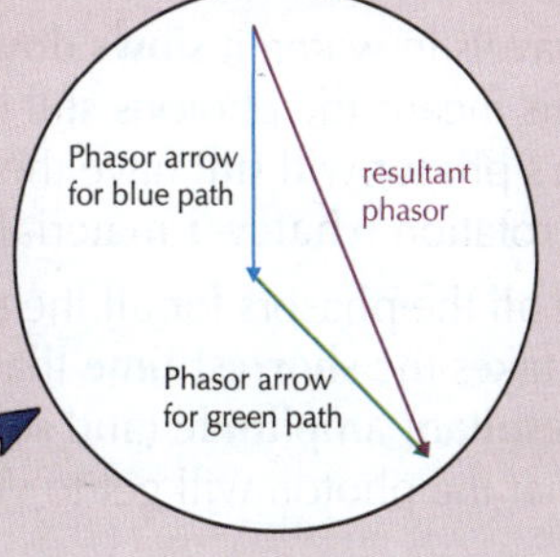

The "Sum Over Paths" Theory

You can Calculate *Probability* from the *Resultant Phasor*

1) You can find the **probability** that a quantum will arrive at a point from **squaring** the **resultant phasor amplitude**.

Probability ∝ (Resultant phasor)^2

2) The **resultant phasor amplitude** has nothing to do with the **amplitude** of the light wave hitting the area, only the **probability** that a particle will arrive there.
3) The **higher** the probability, the **more likely** the particle will arrive there (well duh....).
4) If the **photon** is your quantum of choice, you can think of the **probability** and the **brightness** of the area as pretty much the same thing — the more **probable** it is that a photon will arrive at a point, the **brighter** it will appear.

Example: The resultant phasor amplitudes are shown for the paths a photon could take to points X and Y. How many times brighter does point X appear than point Y? Explain your answer.

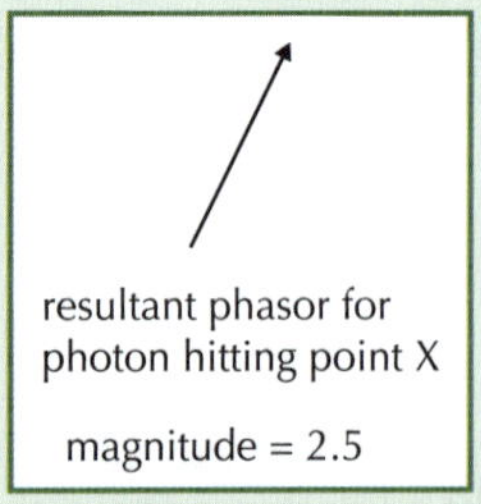

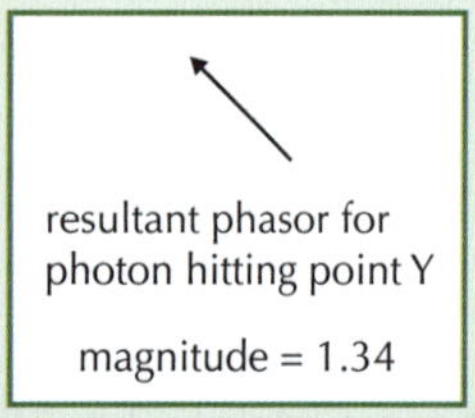

Square the magnitude of each phasor to find a number proportional to the probability of the photon arriving at each point.

Probability of photon hitting point X $\propto (2.5)^2 = 6.25$

Probability of photon hitting point Y $\propto (1.34)^2 = 1.7956$

The more probable a photon will arrive at a point, the brighter it will be. So the relative probability of a photon arriving at the two points will be the relative brightness between the points.
So, point X appears 6.25 ÷ 1.7956 = 3.48... = **3.5 times (to 2 s.f.)** brighter than point Y.

The *Path* that gives the *Highest Probability* is the *Quickest Route*

1) The sum over paths rule (finding the path with the **highest probability**) predicts all sorts of physics laws we take for granted. And each time it seems to be down to the **same reason**:

The final phasor of the **quickest path** will contribute the **most** to the **resultant amplitude** and the **probability** of a quantum arriving at a point.

2) It even predicts one of the most fundamental light behaviours — that **light** travels in a **straight line**. As a **straight line** is the shortest (and therefore **quickest**) path between two points — it provides the largest probability of a photon arriving at a particular point.
3) Obviously there are times when light **doesn't** travel in a straight line, like when it's being **refracted** — but quantum behaviour predicts that as well...

Refraction is also Predicted by Quantum Behaviour

1) Imagine spotting a pineapple at the bottom of a swimming pool. What **route** does the light take from the pineapple to your eye? Altogether now... it takes **all of them**.
2) When light travels in water, it **slows down**, but its **frequency stays the same**. This means the photons still have the **same energy**, and a photon's phasor will still have the **same amplitude** and **frequency** of rotation **whatever** material it's travelling through.
3) If you **add** up all the phasors for all the possible paths, it's the path that takes the **shortest time** that contributes the most to the **resultant amplitude** (and so gives the **highest probability** that the photon will get to your eye).

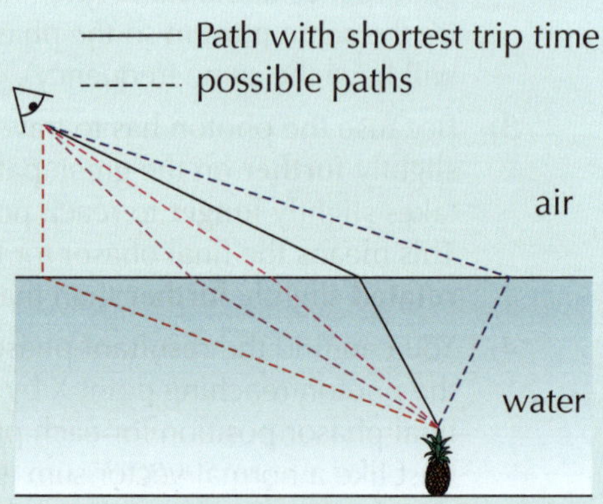

The "Sum Over Paths" Theory

*You Need **All Paths** to Take the **Same Time** to Focus **Quanta***

To **focus** photons (or any other quanta), you need to make sure all straight line paths (that follow the reflection or refraction rule) from the source to the focus point take the **same amount of time** — so the final phasors for every path will be in the same direction.

You met lenses on p.100.

A Convex Lens

The paths towards the edges of the lens are **longer** than those that go through the middle. You make the time taken for each path the same by **increasing** the amount of **glass** in the **middle part** of the lens to increase the time it takes to travel along the shorter paths between the source and detector.

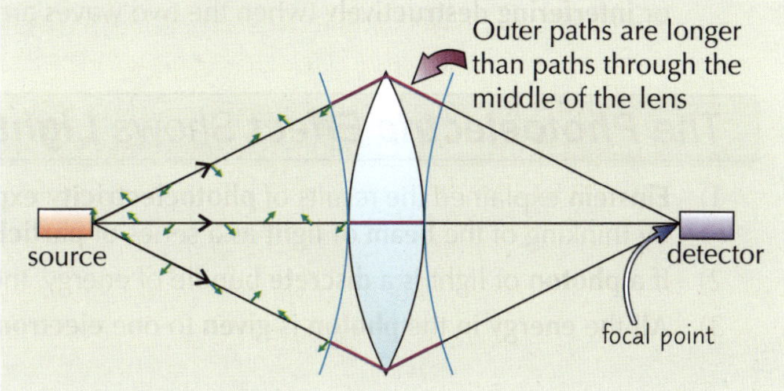

Practice Questions

Q1 What equation would you use to find the frequency of rotation of a photon phasor?

Q2 How are two phasor arrows combined to give a resultant?

Q3 How will a point appear if the probability for a light photon is zero there?

Q4 What path gives the highest probability?

Q5 Explain why light travels in straight lines.

Q6 Describe in terms of phasors how light is focused through a convex lens.

Exam Questions

Q1 The resultant phasors for an electron reaching points A and B have magnitudes of 6.3 and 4.5 respectively. How many times more likely is it that an electron will arrive at point A than point B? [3 marks]

Q2 A light photon has a frequency of 6.0×10^{14} Hz. How many times does the photon's phasor arrow rotate as it moves along a path 120 mm long from a source to a detector? ($c = 3.00 \times 10^8$ ms^{-1}.) [2 marks]

Q3 Using the idea of photons trying every possible path and phasors, explain why very little (if any) light would be detected around the corner from the source in the diagram shown. [3 marks]

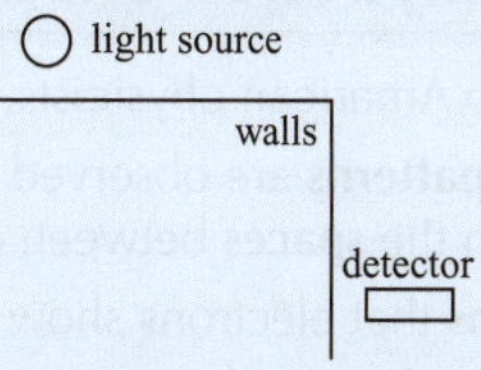

Q4 A photon takes all possible paths through a convex lens. By considering three possible straight line paths (that can reflect and refract), choose the option which correctly describes the phasors of each path at the focal point of the lens.

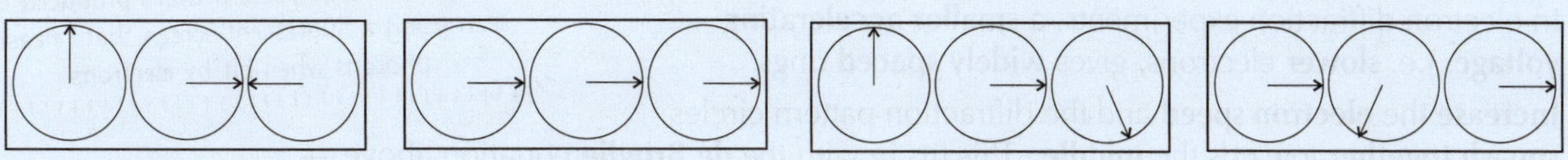

Point A Point B Point C Point D

[1 mark]

I ate all the pies — some would call it greed, I say it's photon thinking...

OK, so that's some wacky sounding physics... but it does seem to work and agree with the standard physics laws we know and love... well, maybe not love. It's tricky, but you just need to follow the same method each time: sum the final phasors for all possible paths to get the resultant, then use the resultant to find the probability the quanta will get there.

Wave-Particle Duality

Is it a wave? Is it a particle? No, it's a wave. No, it's a particle. No it's not, it's a wave. No don't be daft, it's a particle. (etc.)

Interference and ***Diffraction*** show ***Light*** as a ***Wave***

1) Light produces **interference** and **diffraction** patterns — **alternating bands** of **dark** and **light**.
2) These can **only** be explained using **waves interfering constructively** (when two waves overlap in phase) or **interfering destructively** (when the two waves are out of phase). (See p.90.)

The ***Photoelectric Effect*** Shows ***Light*** Behaving as a ***Particle***

1) **Einstein** explained the results of **photoelectricity experiments** (see p.126-127) by thinking of the **beam of light** as a series of **particle-like photons**.
2) If a **photon** of light is a **discrete** bundle of energy, then it can **interact** with an **electron** in a **one-to-one way**.
3) **All** the **energy** in the **photon** is **given** to one **electron**.

De Broglie Came up With the ***Wave-Particle Duality Theory***

1) Louis de Broglie made a **bold suggestion** in his **PhD thesis**:

> If **'wave-like' light** showed **particle properties** (photons), **'particles'** like **electrons** should be expected to show **wave-like properties**.

This means you can model electrons as phasors too. The frequency of a free electron phasor is given by $f = E_k \div h$, where E_k is the kinetic energy of the electron and h is the Planck constant.

2) The **de Broglie equation** relates a **wave property** (**wavelength, λ**) to a **moving particle property** (momentum, p, see p.44). h = Planck constant = 6.63×10^{-34} Js.

$$\lambda = \frac{h}{p} = \frac{h}{mv}$$

where m = mass of the particle in kg and v = velocity of the particle in ms^{-1}.

3) The de Broglie wave of a particle can be interpreted as a **'probability wave'** — the **likelihood** of finding a particle at a point is **directly proportional** to the **square** of the **amplitude** of the wave at that point.
4) Many physicists at the time **weren't very impressed** — his ideas were just **speculation**. But later experiments **confirmed** the wave nature of electrons.

I'm not impressed — this is just speculation. What do you think Dad?

Electron Diffraction shows the ***Wave Nature*** of ***Electrons***

1) In **1927**, two American physicists, **Clinton Davisson** and **Lester Germer**, succeeded in diffracting **electrons**.
2) **Diffraction patterns** are observed when **accelerated electrons** in a vacuum tube **interact** with the **spaces** between **carbon atoms** in **polycrystalline graphite**.
3) This **confirms** that electrons show **wave-like** properties.
4) According to wave theory, the **spread** of the **lines** in the diffraction pattern **increases** if the **wavelength** of the wave is **greater**.
5) In electron diffraction experiments, a **smaller accelerating voltage**, i.e. **slower** electrons, gives **widely spaced** rings.
6) **Increase** the **electron speed** and the diffraction pattern circles **squash together** towards the **middle**. This fits in with the **de Broglie** equation above — if the **momentum** is **higher**, the **wavelength** is **shorter** and the **spread** of lines is **smaller**.
7) Electron diffraction was a **huge** discovery — this was the first **direct evidence** for de Broglie's theory.

You can repeat Young's double-slit experiment (see p.92) using electrons instead of photons. The interference pattern that's produced is seen using a fluorescent screen that releases photons when hit by electrons.

> In general, λ for **electrons** accelerated in a **vacuum tube** is about the **same size** as **electromagnetic waves** in the **X-ray** part of the spectrum.

8) If particles with a **greater mass** (e.g. **neutrons**) were travelling at the **same speed** as the electrons, they would show a more **tightly-packed diffraction pattern**. That's because a neutron's **mass** (and therefore its **momentum**) is **much greater** than an electron's, and so a neutron has a **shorter de Broglie wavelength**.

Wave-Particle Duality

Particles Don't show Wave-Like Properties All the Time

1) You **only** get **diffraction** if a particle interacts with an object of about the **same size** as its **de Broglie wavelength**.
2) A **tennis ball**, for example, with **mass 0.058 kg** and **speed 100 ms^{-1}** has a **de Broglie wavelength** of **10^{-34} m**. That's **10^{19} times smaller** than the **nucleus** of an **atom**! There's nothing that small for it to interact with.

Example: An electron of mass 9.11×10^{-31} kg is fired from an electron gun at 7.00×10^{6} ms^{-1}. What size object will the electron need to interact with in order to diffract?

$$\lambda = \frac{h}{mv} = \frac{6.63 \times 10^{-34}}{(9.11 \times 10^{-31}) \times (7.00 \times 10^{6})} = \mathbf{1.04 \times 10^{-10}\ m\ (to\ 3\ s.f.)}$$

Electrons with a wavelength of around **1×10^{-10} m** are **likely** to be diffracted by the atoms in **polycrystalline structures**.

3) A **shorter wavelength** gives **less diffraction effects**. This fact is used in the **electron microscope**.
4) **Diffraction** effects **blur detail** on an image. If you want to **resolve tiny detail** in an **image**, you need a **shorter wavelength**. **Light** blurs out detail more than **'electron-waves'** do, so an **electron microscope** can resolve **finer detail** than a **light microscope**. They can let you look at things as tiny as a single strand of DNA... which is nice.

Wave-Particle Duality Wasn't Accepted Straight Away

De Broglie first **hypothesised** wave-particle duality to explain **observations** of light acting as both a particle and a wave. But his theory **wasn't accepted** straight away. **Other scientists** had to **evaluate** de Broglie's theory (by a process known as **peer review**) before he **published** it, and then it was **tested with experiments**. Once enough evidence was found to back it up, the theory was accepted as **validated** by the scientific community.

Scientists' understanding of the nature of matter has changed over time through this process of **hypothesis and validation**. De Broglie's theory is **accepted** to be true — that is, until any new conflicting evidence comes along.

Practice Questions

Q1 Which observations show light to have a 'wave-like' character?

Q2 Which observations show light to have a 'particle' character?

Q3 What happens to the de Broglie wavelength of a particle if its momentum increases? How does this affect the particle's diffraction pattern?

Q4 Which observations show electrons to have a 'wave-like' character?

Exam Questions

Q1 a) State what is meant by the wave-particle nature of electromagnetic radiation. [1 mark]

b) Calculate the momentum of an electron with a de Broglie wavelength of 590 nm. [2 marks]

Q2 Electrons travelling at a speed of 3.50×10^{6} ms^{-1} exhibit wave properties.

a) Calculate the wavelength of these electrons. (Mass of an electron = 9.11×10^{-31} kg) [2 marks]

b) Calculate the speed of protons with the same wavelength as these electrons. (Mass of a proton = 1.67×10^{-27} kg) [2 marks]

c) Some electrons and protons were accelerated from rest by the same potential difference, giving them the same kinetic energy. Explain why they will have different wavelengths. [3 marks]

Don't hide your wave-particles under a bushel...

Right — I think we'll all agree that quantum physics is a wee bit strange when you come to think about it. What it's saying is that electrons and photons aren't really waves, and they aren't really particles — they're both... at the same time. It's what quantum physicists like to call a 'juxtaposition of states'. Well they would, wouldn't they...

Answers

How Science Works

Page 3 — The Scientific Process

1 ***A maximum of two marks can be awarded for structure and reasoning of the written response:***
2 marks:
The answer is coherent and constructed logically, and displays clear reasoning and links between points throughout.
1 mark:
The answer is partially logical, with some reasoning and links between points.
0 marks:
The answer has no structure and no links between points.
Here are some points your answer may include:

- The proposed power station would produce electricity without producing greenhouse gases, and so replacing a nearby coal-fired power station could benefit the environment by helping to limit human contribution to climate change.
- As the proposed power station generates energy renewably, it won't use up resources, which will also benefit the environment.
- However, the power station could have a negative effect on the natural environment in the local area (for example by interfering with local habitats or through pollution during its construction).
- As well as damaging the environment, if the power station spoils the scenic coastline it may be unpopular with local people.
- As the coastline is important for tourism in the area, this could also damage the local economy.
- Building the new power station (and closing down the coal-fired power station) may be expensive.
- If the proposed power station causes a coal-fired power station to close, people there could lose their jobs.
- On the other hand, building a new power station could create other new jobs in the area.

[4 marks — 4 marks if 6 points mentioned including social, environmental, and economic factors, 3 marks if 4-5 points mentioned, covering at least two aspects of the question, 2 marks if 2-3 points mentioned, 1 mark if 1 point mentioned.]

Development of Practical Skills

Page 5 — Quantities and Units

1 a) S.I. unit of mass = kg. S.I. unit of volume = m^3.
$\rho = m / V$ and volume is length cubed, so in S.I. base units
$\rho = kg / m^3 =$ **$kg\,m^{-3}$** ***[1 mark]***

b) $m = 9.8$ g $= 9.8 \times 10^{-3}$ kg
$V = (11 \text{ mm})^3 = (11 \times 10^{-3} \text{ m})^3 = 1.331 \times 10^{-6}$ m^3 ***[1 mark]***
$\rho = m / V = (9.8 \times 10^{-3}) \div (1.331 \times 10^{-6})$
$= 7362.88... =$ **7400 $kg\,m^{-3}$ (to 2 s.f.)** ***[1 mark]***
You could also have worked this out in $g\,mm^{-3}$ then converted your answer, but this way is much easier.

Page 7 — Planning and Implementing

1 a) Independent variable: light level / distance from the light source, dependent variable: resistance of the LDR ***[1 mark]***.

b) Any two of: e.g. the light source used / the background lighting in the room / the temperature of the room/LDR/wires / the potential difference / the power supply the LDR is connected to / the length of wires in the circuit / the type of wires in the circuit / the multimeter used to measure the resistance.
[2 marks available — 1 mark for each correct answer.]

Page 9 — Analysing Results

1 a)

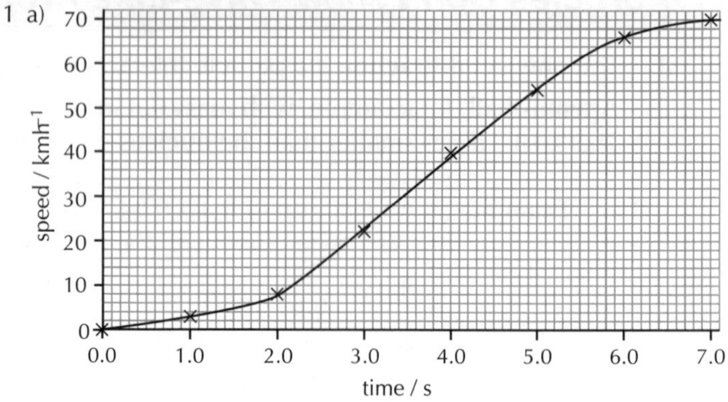

[1 mark for both axes drawn to a sensible scale, 1 mark for labelling both axes correctly, 1 mark for all the points drawn correctly and 1 mark for a sensible line of best fit.]

b) The graph is linear between 2 and 5 seconds ***[1 mark]***.
Accept an answer of 6 s as the upper limit if the graph in part a) agrees.

c) The maximum acceleration is the value of the steepest gradient, which is the linear portion of the graph ***[1 mark]***:

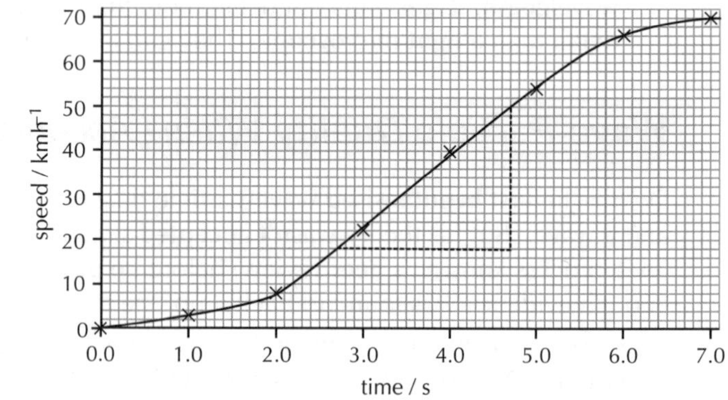

change in speed $= 50 - 18 = 32$ $km\,hr^{-1}$ $= 32 \div (60 \times 60)$
$= 0.0088...$ $km\,s^{-1}$ ***[1 mark]***
change in time $= 4.7 - 2.7 = 2.0$ s ***[1 mark]***
acceleration $= 0.0088... \div 2.0 = 0.0044...$ $km\,s^{-2}$
$=$ **0.0044 $km\,s^{-2}$** or **57 000 $km\,hr^{-2}$ (to 2 s.f.)** ***[1 mark]***
Accept an answer in the range 0.0040-0.0046 $km\,s^{-2}$ or 54 000-60 000 $km\,hr^{-2}$.

Page 11 — Measurements and Uncertainties

1 a) $(0.02 \div 0.52) \times 100 = 3.846... =$ **4% (to 1 s.f.)** ***[1 mark]***

b) $(0.02 \div 0.94) \times 100 = 2.127... =$ **2% (to 1 s.f.)** ***[1 mark]***

c) acceleration = change in velocity / time $= (0.94 - 0.52) \div 2.5$
$= 0.168$ ms^{-2} ***[1 mark]***
Absolute error in change of velocity $= 0.02 + 0.02 = 0.04$ ms^{-1}
Percentage uncertainty in
change of velocity: $(0.04 \div (0.94 - 0.52)) \times 100 = 9.523...\%$
[1 mark]
Percentage uncertainty in time taken $= (0.5 \div 2.5) \times 100 = 20\%$
Percentage uncertainty in acceleration $= 9.523...\% + 20\%$
$= 29.523...\%$ ***[1 mark]***
Absolute error in acceleration $= 0.168 \times (29.523... \div 100)$
$= 0.0496$ ms^{-2}
So the acceleration = **0.17 ± 0.05 ms^{-2} (to 2 s.f.)** ***[1 mark]***

Page 13 — Evaluating and Concluding

1 a) $t = 0.32$ seconds, $v = 2.0$ ms^{-1} ***[1 mark]***

b) E.g. The results do not support this conclusion ***[1 mark]***, because the student has only collected data for a small range of times so he cannot draw conclusions about times longer than those he measured ***[1 mark]*** / because the student has only investigated one object so he cannot draw conclusions about other objects ***[1 mark]***.

Answers

Topic 1 — Motion

Page 15 — Scalars and Vectors

1

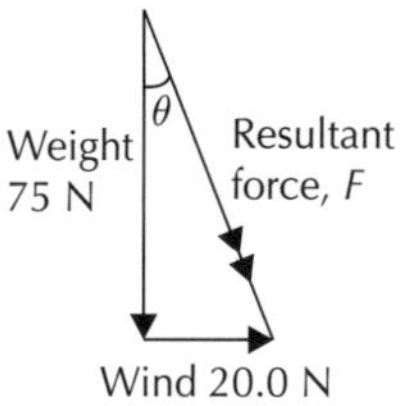

$F^2 = 20.0^2 + 75^2 = 6025$
So $F = 78$ N (to 2 s.f.)
$\tan\theta = 20.0 / 75 = 0.266...$
So $\theta = \tan^{-1} 0.266... = 15°$ (to 2 s.f.)
The resultant force on the rock is **78 N (to 2 s.f.)** ***[1 mark]*** at an angle of **15° (to 2 s.f.)** ***[1 mark]*** to the vertical.
Make sure you know which angle you're finding — and label it on your diagram.

2

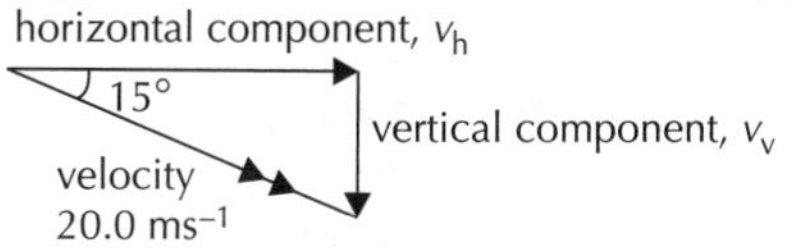

horizontal component $v_h = 20.0 \times \cos 15.0°$
$=$ **19.3 ms^{-1} (to 3 s.f.)** ***[1 mark]***
vertical component $v_v = 20 \times \sin 15.0°$
$=$ **5.18 ms^{-1} (to 3 s.f.)** ***[1 mark]***

3 E.g. current, $v_{river} = 0.20$ ms^{-1}

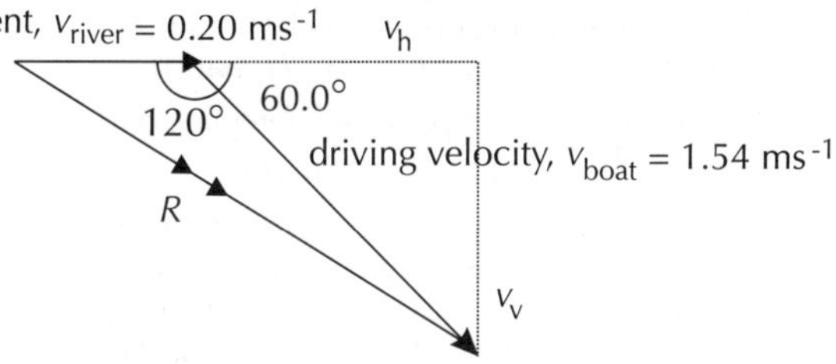

horizontal component $v_{river} = 0.20$ ms^{-1}
vertical component $v_{river} = 0$ ms^{-1}
horizontal component $v_{boat} = 1.54 \times \cos 60.0 = 0.77$ ms^{-1}
vertical component $v_{boat} = 1.54 \times \sin 60.0 = 1.333...$ ms^{-1}
So, horizontal $v_{resultant} = 0.20 + 0.77 = 0.97$ ms^{-1} ***[1 mark]***
vertical $v_{resultant} = 0 + 1.333... = 1.333...$ ms^{-1} ***[1 mark]***
Combine the vertical and horizontal components of R.

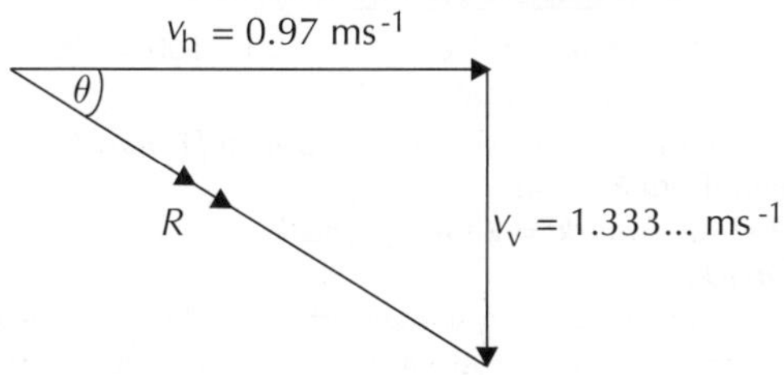

$v_{resultant} = \sqrt{0.97^2 + 1.333...^2} = 1.649...$ ms^{-1}
$=$ **1.6 ms^{-1} (to 2 s.f.)** ***[1 mark]***

$\tan\theta = 1.333... \div 0.97 = \tan^{-1} 1.374...$ so $\theta =$ **54° (to 2 s.f.)**
So the resultant velocity of the boat is 1.6 ms^{-1} at an angle of 54° to the current ***[1 mark]***.

Page 17 — Displacement-Time Graphs

1 Split graph into four sections:
A: Acceleration ***[1 mark]***
B: Constant velocity ***[1 mark]***
C: Stationary ***[1 mark]***
D: Constant velocity in opposite direction to A and B ***[1 mark]***

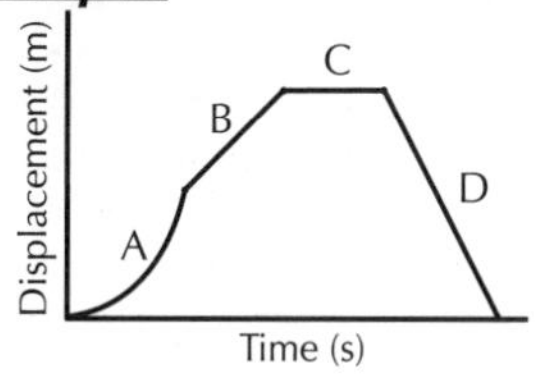

2 a)

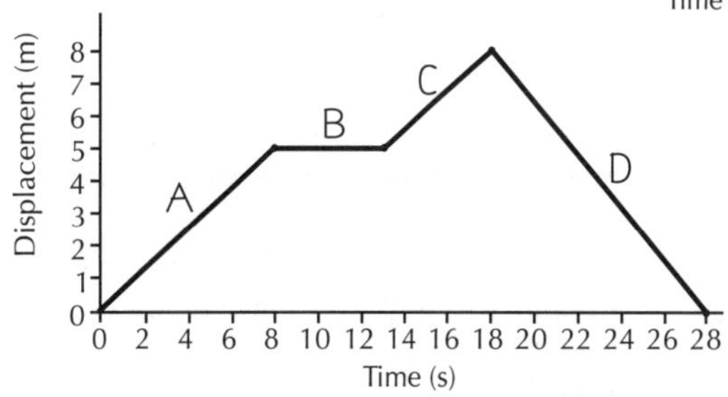

[4 marks available — 1 mark for each section correctly drawn.]

b) At A: $v = \frac{\text{displacement}}{\text{time}} = \frac{5}{8} = 0.625 =$ **0.6 ms^{-1} (to 1 s.f.)**

At B: $v = 0$

At C: $v = \frac{\text{displacement}}{\text{time}} = \frac{3}{5} =$ **0.6 ms^{-1}**

At D: $v = \frac{\text{displacement}}{\text{time}} = \frac{-8}{10} =$ **−0.8 ms^{-1}**

[2 marks for all correct or just 1 mark for 2 or 3 correct.]

Page 19 — Velocity-Time and Acceleration-Time Graphs

1 a)

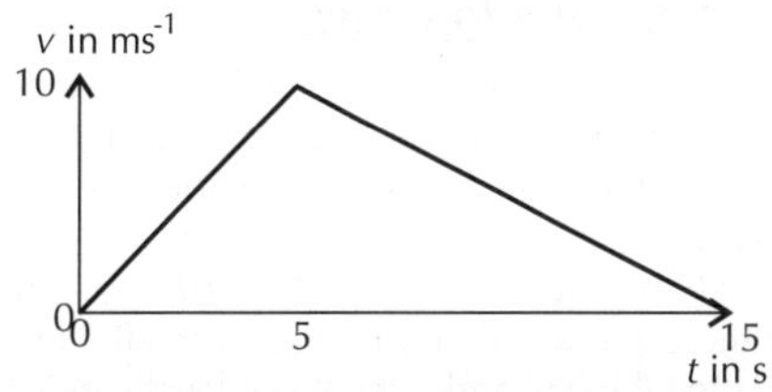

[1 mark for drawing a correctly labelled pair of axes, with a straight line between v = 0 ms^{-1}, t = 0 s and v = 10 ms^{-1}, t = 5 s. 1 mark for correctly drawing a straight line between v = 10 ms^{-1}, t = 5 s and v = 0 ms^{-1}, t = 15 s.]

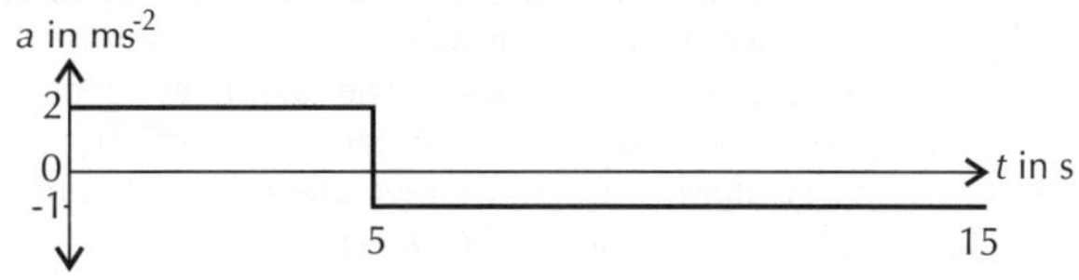

[1 mark for drawing a correctly labelled pair of axes, with a straight line between a = 2 ms^{-2}, t = 0 s and a = 2 ms^{-2}, t = 5 s. 1 mark for correctly drawing a straight line from a = 2 ms^{-2}, t = 5 s and a = −1 ms^{-2}, t = 5 s, and a straight line between a = −1 ms^{-2}, t = 5 s and a = −1 ms^{-2}, t = 15 s.]

b) Distance travelled is equal to the area under v-t graph between $t = 0$ and $t = 5$ ***[1 mark]*** $= 0.5 \times 5 \times 10 =$ **25 m** ***[1 mark]***

Answers

Page 21 — Motion With Uniform Acceleration

1 a) $a = -9.81$ ms^{-2}, $t = 5$ s, $u = 0$ ms^{-1}, $v = ?$
Use: $v = u + at$, $v = 0 + 5 \times -9.81$ ***[1 mark]***
$v =$ **-49 ms^{-1} (to 2 s.f.)** ***[1 mark]***
NB: It's negative because she's falling downwards and we took upwards as the positive direction.

b) Use: $s = \left(\frac{u+v}{2}\right)t$ or $s = ut + \frac{1}{2}at^2$
$s = \frac{-49}{2} \times 5 = -120$ m (to 2 s.f.)
or $s = 0 + \frac{1}{2} \times -9.81 \times 5^2 = -120$ m (to 2 s.f.)
So she fell **120 m (to 2 s.f.)** ***[1 mark for working, 1 mark for answer]***

2 a) $v = 0$ ms^{-1}, $t = 3.2$ s, $s = 40$ m, $u = ?$ Use: $s = \left(\frac{u+v}{2}\right)t$
$40 = 3.2u \div 2$ ***[1 mark]***, so $u = \frac{80}{3.2} =$ **25 ms^{-1}** ***[1 mark]***

b) E.g. use: $v^2 = u^2 + 2as$
$0 = 25^2 + 80a$ ***[1 mark]***
$-80a = 625$, so $a =$ **-7.8 ms^{-2} (to 2 s.f.)** ***[1 mark]***

3 a) Take upstream as negative:
$v = 5$ ms^{-1}, $a = 6$ ms^{-1}, $s = 1.2$ m, $u = ?$ Use: $v^2 = u^2 + 2as$
$5^2 = u^2 + 2 \times 6 \times 1.2$ ***[1 mark]***
$u^2 = 25 - 14.4 = 10.6$
$u = -3.255... =$ **-3 ms^{-1} (to 1 s.f.)** ***[1 mark]***
Take the negative root as we've defined downstream to be positive and initially the boat was travelling upstream.

b) From furthest point: $u = 0$ ms^{-1}, $a = 6$ ms^{-2}, $v = 5$ ms^{-1}, $s = ?$
Use: $v^2 = u^2 + 2as$
$5^2 = 0 + 2 \times 6 \times s$ ***[1 mark]***
$s = 25 \div 12 = 2.083... =$ **2 m (to 1 s.f.)** ***[1 mark]***

Page 24 — Acceleration Due to Gravity

1 a) The air resistance on a falling small steel ball will be less than that on a beach ball. The air resistance on the ball used in this experiment needs to be negligible in order to be able to calculate the value of g ***[1 mark]***.

b) E.g. the ball's fall might be affected by wind ***[1 mark]***. To remove this error, conduct the experiment indoors and close all windows ***[1 mark]***. / Not aligning the ball and ruler at eye level can lead to a measuring error ***[1 mark]***. This can be reduced by making sure your eye is perpendicular to the measuring scale being used when taking measurements ***[1 mark]***.

c) E.g. there may be a delay on the stopwatch/light gates ***[1 mark]***. To remove this, ensure they are properly calibrated before conducting the experiment ***[1 mark]***. / The ruler may not be aligned properly so would give slightly incorrect vertical height measurements ***[1 mark]***. To remove this, use a clamp to ensure the rule is straight and unmoving ***[1 mark]***.

d) Use: $s = ut + \frac{1}{2}at^2$ ***[1 mark]***
$u = 0$, so $s = \frac{1}{2}at^2$ or $\frac{1}{2}a = \frac{s}{t^2}$ ***[1 mark]***
So the gradient of a graph of s against t^2, $\frac{\Delta s}{\Delta t^2}$, is equal to half the acceleration, i.e. $\frac{1}{2}g$ ***[1 mark]***.

2 a) You know $s = 5.0$ m, $a = -g$, $v = 0$
You need to find u, so use $v^2 = u^2 + 2as$
$0 = u^2 - 2 \times 9.81 \times 5.0$ ***[1 mark]***
$u^2 = 98.1$, so $u = 9.90... =$ **9.9 ms^{-1} (to 2 s.f.)** ***[1 mark]***

b) You know $a = -g$, $v = 0$ at highest point, $u = 9.90...$ ms^{-1}
You need to find t, so use $v = u + at$
$0 = 9.90... - 9.81t$ ***[1 mark]***
$t = 9.90.../9.81 =$ **1.0 s (to 2 s.f.)** ***[1 mark]***

c) Her velocity as she lands back on the trampoline will be **-9.9 ms^{-1} (to 2 s.f.)**. ***[1 mark]***
She has a negative velocity because she's moving downwards and the same magnitude as her initial speed because no energy has been lost due to air resistance.

Page 27 — Projectile Motion

1 a) You only need to worry about the vertical motion of the stone.
$u = 0$ ms^{-1}, $s = -560$ m, $a = -g = -9.81$ ms^{-2}, $t = ?$
You need to find t, so use: $s = ut + \frac{1}{2}at^2$
$-560 = 0 + \frac{1}{2} \times -9.81 \times t^2$
$t = \sqrt{\frac{2 \times (-560)}{-9.81}}$ ***[1 mark]***
$t = 10.68... =$ **11 s (to 2 s.f.)** ***[1 mark]***

b) You know that in the horizontal direction:
$v = 20.0$ m/s, $t = 10.68...$ s, $a = 0$, $s = ?$
$s = v \times t = 20.0 \times 10.68...$ ***[1 mark]***
$s = 213.69... =$ **210 m (to 2 s.f.)** ***[1 mark]***

2 C ***[1 mark]***
Use $v^2 = u^2 + 2as$ to find the vertical displacement when $v = 0$. The arrow was fired from 1 m above the ground so don't forget to include the extra metre in your calculations.

Page 29 — Motion Experiments and Stopping Distances

1 a) Reaction time is 0.50 s, speed is 20.0 ms^{-1}
$s = vt = 20.0 \times 0.50$ ***[1 mark]*** = **10 m** ***[1 mark]***

b) Use $F = ma$ to get a: $a = \frac{F}{m} = (-1.0 \times 10^4) \div 850$
$= -11.76...$ ***[1 mark]***
Use $v^2 = u^2 + 2as$, and rearrange
$s = (0^2 - 20^2) \div (2 \times -11.76...)$ ***[1 mark]***
= **17 m** ***[1 mark]***
Remember that a force against the direction of motion is negative.

c) No. Total stopping distance = 10 + 17 = 27 m
She stops 3 m before the cow. ***[1 mark]***

2 a) If it takes 1 frame to pass the reference point,
$t = 1 \times \frac{1}{26} = 0.03846...$ s ***[1 mark]***
$v = L \div t = 0.15 \div 0.03846...$ ***[1 mark]*** = **3.9 ms^{-1}** ***[1 mark]***

b) E.g. any one of: Use a camera with a higher frame rate ***[1 mark]*** so the time can be measured more precisely, which decreases the overall uncertainty of the velocity ***[1 mark]*** / Use longer trollies marked with distances ***[1 mark]*** to make sure the trolley takes at least one frame to pass the reference point so more accurate time assumptions can be made ***[1 mark]***.

Topic 2 — Forces and Energy

Page 32 — Newton's Laws of Motion

1 a) Force perpendicular to river flow = 500 – 100 = 400 N ***[1 mark]***
Force parallel to river flow = 300 N
Resultant force = $\sqrt{400^2 + 300^2}$ = **500 N** ***[1 mark]***

b) $a = F/m$ (from $F = ma$)
= 500/250 ***[1 mark]*** = **2 ms^{-2}** ***[1 mark]***

2 B ***[1 mark]***
The overall acceleration is a, so ma must be equal to the resultant force, which is the force John is pushing with minus the resistance caused by friction. So $F_{John} - F = ma$, and $F_{John} = ma + F$.

3 The only force acting on each of them is their weight = mg ***[1 mark]***. Since $F = ma$, this gives $ma = mg$, or $a = g$ ***[1 mark]***. Their acceleration doesn't depend on their mass — it's the same for both of them — so they reach the water at the same time ***[1 mark]***.

Page 35 — Drag, Lift and Terminal Speed

1 a) The velocity increases at a steady rate, which means the acceleration is constant ***[1 mark]***.
Constant acceleration means there must be no atmospheric resistance (atmospheric resistance would increase with velocity, leading to a decrease in acceleration). So there must be no atmosphere ***[1 mark]***.

Answers

b) 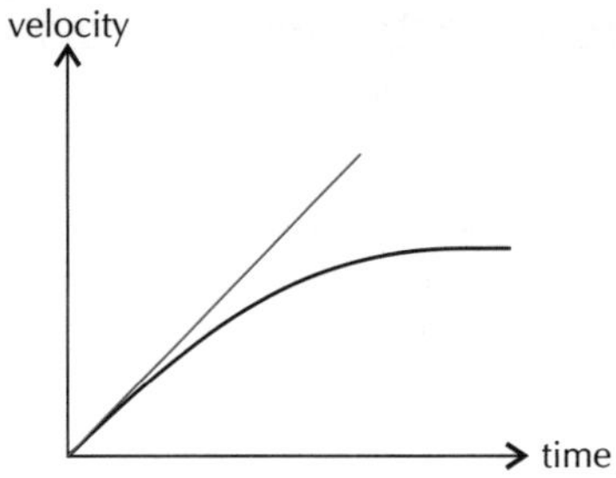

[1 mark for a smooth curve that levels out, 1 mark for correct position relative to existing line]

Your graph must be a smooth curve which levels out. It must NOT go down at the end.

c) (The graph becomes less steep)
because the acceleration is decreasing ***[1 mark]***
because air resistance increases with speed ***[1 mark]***.
(The graph levels out)
because air resistance has become equal to weight ***[1 mark]***.
If the question says 'explain', you won't get marks for just describing what the graph shows — you have to say why it is that shape.

2 a) The 15 cm cone will have the lowest terminal speed ***[1 mark]*** because it has the largest surface area and therefore the largest drag ***[1 mark]***.

b) 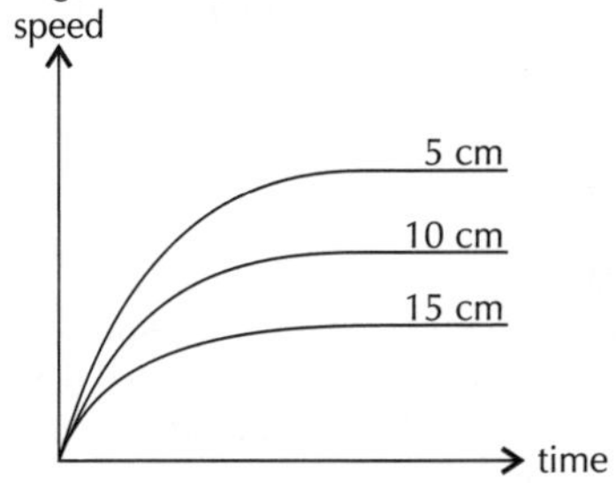

[3 marks, 1 mark for each correct line.]

c) E.g. The shape/slant/height of the cone because it would affect the amount of air resistance ***[1 mark]***.

d) The curve for the largest cone would reach a higher terminal speed ***[1 mark]*** because the shape is more streamlined ***[1 mark]*** so the air resistance would be lower at a given speed ***[1 mark]***.

Page 37 — Resolving Forces

1 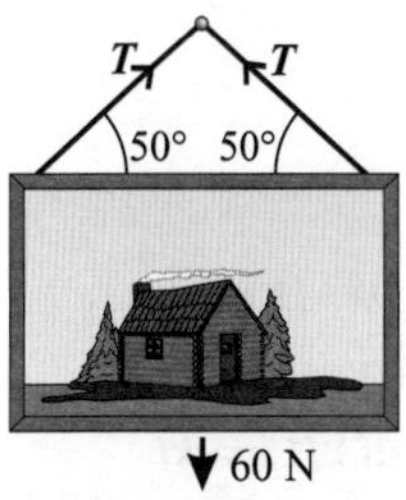

Weight = vertical component of tension × 2
$60 = 2T \sin 50°$ ***[1 mark]***
$60 = 2T \times 0.766...$
$78.3... = 2T$
$T = 39.1... =$ **40 N (to 1 s.f.)** ***[1 mark]***

2 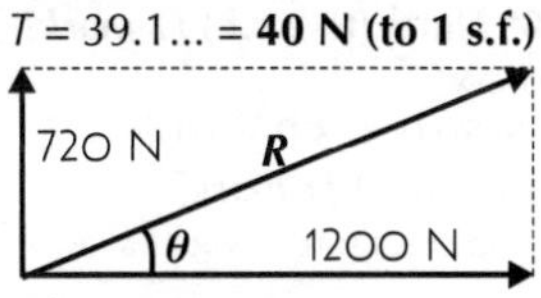

By Pythagoras:
$R = \sqrt{1200^2 + 720^2} = 1399.42... =$ **1400 N (to 2 s.f.)** ***[1 mark]***
$\tan\theta = \frac{720}{1200}$, so $\theta = \tan^{-1} 0.6 = 30.96... = 31°$ (to 2 s.f.)
So the direction of the resultant force is **31° (to 2 s.f.) from the horizontal** ***[1 mark]***

Page 39 — Density, Upthrust and Viscosity

1 a) F is the viscous drag (N), η is the viscosity of the fluid (Nsm^{-2} or Pa s), r is the radius of the object (m) and v is the speed the object is moving at (ms^{-1}) ***[1 mark]***.

b) Any three of: The object moving through the fluid must be small ***[1 mark]*** / spherical ***[1 mark]*** / moving through the fluid with laminar flow ***[1 mark]*** / moving slowly through the fluid ***[1 mark]***.

c) Weight – drag – upthrust = 0 ***[1 mark]***
So you can work out the weight by calculating the drag and upthrust acting on the ball bearing.
Upthrust = weight of fluid displaced = $m_{liquid} \times g$
$m_{liquid} = \rho_{liquid} \times V_{liquid}$
$= 1400 \times \frac{4}{3}\pi(3.0\times10^{-3})^3 = 1.583... \times10^{-4}$ kg ***[1 mark]***
So upthrust $= m_{liquid} \times g = (1.583... \times 10^{-4}) \times 9.81$
$= 1.553... \times 10^{-3}$ N ***[1 mark]***
Drag $= F = 6\pi\eta rv = 6\pi \times 3.0 \times (3.0 \times10^{-3}) \times 0.040$
$= 6.785... \times 10^{-3}...$ N ***[1 mark]***
So weight $= (1.553... \times 10^{-3}) + (6.785... \times 10^{-3}) = 8.339... \times 10^{-3}$
$=$ **8.4 × 10^{-3} N (to 2 s.f.)** ***[1 mark]***

Page 41 — Mass, Weight and Centre of Mass

1 a) Experiment:
Hang the object freely from a point. Hang a plumb bob from the same point, and use it to draw a vertical line down the object ***[1 mark]***. Repeat for a different point and find the point of intersection ***[1 mark]***. The centre of mass is halfway through the thickness of the object (by symmetry) at the point of intersection ***[1 mark]***.
Identifying and reducing error: e.g.
Source: the object and/or plumb line might move slightly while you're drawing the vertical line ***[1 mark]***.
Reduced by: hang the object from a third point to confirm the position of the point of intersection ***[1 mark]***.

b) You can find the centre of mass of any regular shape using symmetry. The centre of mass will be at the centre where the lines of symmetry cross, halfway through its thickness ***[1 mark]***.

Page 43 — Moments

1 Moment = force × distance
$60.0 = 0.40F$ ***[1 mark]***, so $F =$ **150 N** ***[1 mark]***

2 Clockwise moment = anticlockwise moment
$W \times 2.0 = T \times 0.3$
[1 mark for either line of working]
$60 \times 9.81 \times 2.0 = T \times 0.3$
$T = 3924 =$ **4000 N (to 1 s.f.)** ***[1 mark]***

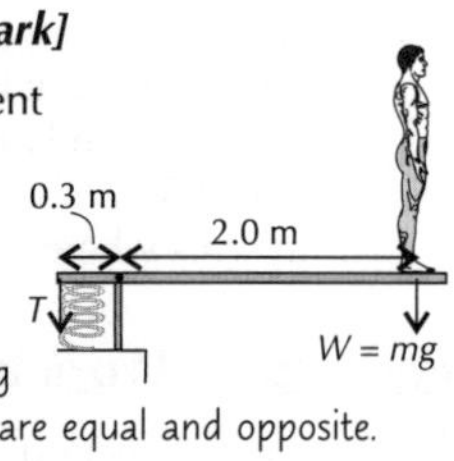

The moments caused by the tension in the spring and the force exerted by the diver on the spring are equal and opposite.

Page 45 — Momentum and Impulse

1 a) total momentum before collision = total momentum after ***[1 mark]***
$(0.60 \times 5.0) + 0 = (0.60 \times -2.4) + 2.0v$
$3 + 1.44 = 2.0v$ ***[1 mark]*** $v = 2.22 =$ **2.2 ms^{-1} (to 2 s.f.)** ***[1 mark]***

b) Kinetic energy before collision
$= ½ \times 0.6 \times 5^2 + ½ \times 2 \times 0^2 =$ **7.5 J** ***[1 mark]***
Kinetic energy after the collision
$= ½ \times 0.6 \times 2.4^2 + ½ \times 2 \times 2.22^2 = 1.728 + 4.9284$
$=$ **6.7 J (to 2 s.f.)** ***[1 mark]***
The kinetic energy of the two balls is greater before the collision than after (i.e. it's not conserved), so the collision must be inelastic ***[1 mark]***.

Answers

Page 47 — Car Safety

1 ***5-6 marks:***
The answer clearly explains more than one way that air bags protect passengers, with reference to Newton's second law of motion to explain the forces involved where appropriate. The answer clearly explains the risks associated with air bags when used incorrectly, and suggests how to reduce the risks.
The answer is structured in a logical way, with relevant information supporting it throughout.
3-4 marks:
The answer describes ways in which air bags protect passengers, and explains them with reference to forces involved. The explanation is partly incomplete. The answer gives the risks associated with air bags when used incorrectly, and includes some explanation of these risks or ways to reduce the risks.
The answer has some logical structure, with mostly relevant information supporting it.
1-2 marks:
The answer includes some description of how air bags protect passengers, but no relevant explanation involving forces. The answer has some limited description of the risks associated with air bags when used incorrectly.
The answer is basic, poorly structured and unsupported by relevant information.
0 marks:
No relevant information is given.
Here are some points your answer may include:
- Air bags protect passengers by stopping them from hitting the dashboard/steering wheel/hard surfaces inside the car in a crash.
- Air bags increase the time taken for passengers to come to a complete stop.
- Newton's 2nd Law, $F = \Delta p \div \Delta t$, so by increasing the time that passengers take to come to a stop, air bags reduce the forces acting on passengers in a crash (for a given momentum change).
- Air bags can be dangerous if a passenger isn't properly restrained / isn't wearing a seat belt.
- This is because passengers who aren't properly restrained keep moving forwards in a crash and hit the air bag whilst it is still inflating / with a lot of force.
- This risk can be reduced by wearing a seat belt at all times.
- Air bags are also dangerous if used with rear-facing child seats.
- This is because the air bag inflates behind the child seat, so can throw the child into the car seat with a lot of force.
- This risk can be reduced by not placing rear-facing child seats in seats where an air bag is fitted.

Page 49 — Work and Power

1 a) Force in direction of travel
$= 100 \cos 40° = 76.6...$ N ***[1 mark]***
$W = Fs = 76.6... \times 1500 =$
110 000 J (to 2 s.f.) ***[1 mark]***

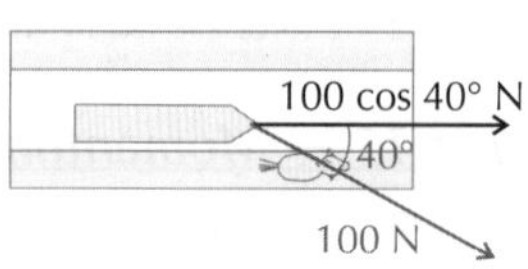

b) Use $P = Fv$
$= 100 \cos 40° \times 0.80$ ***[1 mark]*** = **61 W (to 2 s.f.)** ***[1 mark]***

2 a) Use $W = Fs$
$= (20 \times 9.81) \times 3.0$ ***[1 mark]*** = **590 J (to 2 s.f.)** ***[1 mark]***
Remember that 20 kg is not the force — it's the mass. So you need to multiply it by 9.81 Nkg^{-1} to get the weight.

b) Use $P = Fv$
$= (20 \times 9.81) \times 0.25$ ***[1 mark]*** = **49 W (to 2 s.f.)** ***[1 mark]***

Page 51 — Conservation of Energy and Efficiency

1 a) Use $E_k = \frac{1}{2}mv^2$ and $\Delta E_p = mg\Delta h$ ***[1 mark]***
$\frac{1}{2}mv^2 = mg\Delta h$
$\frac{1}{2}v^2 = g\Delta h$
$v^2 = 2g\Delta h = 2 \times 9.81 \times 2.0 = 39.24$ ***[1 mark]***
$v = \sqrt{39.24}$ = **6.3 ms^{-1} (to 2 s.f.)** ***[1 mark]***
'No friction' allows you to say that the changes in kinetic and potential energy will be the same.

b) 2.0 m — no friction means the kinetic energy will all change back into potential energy, so he will rise back up to the same height as he started ***[1 mark]***.

c) Put in some more energy by actively 'skating' / 'pumping the board' ***[1 mark]***.

2 a) If there's no air resistance, $E_k = \Delta E_p = mg\Delta h$ ***[1 mark]***
$E_k = 0.0200 \times 9.81 \times 8.0 = 1.5696$ = **1.6 J (to 2 s.f.)** ***[1 mark]***

b) If the ball rebounds to 6.5 m, it has gravitational potential energy:
$\Delta E_p = mg\Delta h = 0.0200 \times 9.81 \times 6.5 = 1.2753$ ***[1 mark]***
So $1.5696 - 1.2753 = 0.2943$ = **0.29 J (to 2 s.f.)** is converted to other forms ***[1 mark]***.
You could also work out the loss of E_p from the difference in height — $\Delta E_p = 0.0200 \times 9.81 \times (8.0 - 6.5) = 0.29$ J (to 2 s.f.).

3 Use efficiency $= \dfrac{\text{useful energy output}}{\text{total energy input}}$
$= (140 - 65) \div 140 = 0.535...$ = **0.54 (to 2 s.f.)** ***[1 mark]***

Topic 3 — Materials

Page 53 — Hooke's Law

1 a) Force is proportional to extension. The force is 1.5 times as great, so the new extension will also be 1.5 times the original extension.
new extension = 1.5×4.0 mm = **6.0 mm** ***[1 mark]***

b) $\Delta F = k\Delta x$ and so $k = \Delta F/\Delta x$
$k = 10.0 \div (4.0 \times 10^{-3})$ = **2500 Nm^{-1}** or **2.5 Nmm^{-1}** ***[1 mark]***

c) Any one from e.g. the string now stretches much further for small increases in force ***[1 mark]***. / When the string is loosened it is longer than at the start ***[1 mark]***.

2 The rubber band does not obey Hooke's law ***[1 mark]*** because when the force is doubled from 2.5 N to 5.0 N, the extension increases by a factor of 2.3. ***[1 mark]***.
Or you could show that k is different for 2.5 N and 5.0 N.

Page 55 — Stress and Strain

1 a) Area $= \pi r^2$ or $\pi\left(\frac{d}{2}\right)^2$
So area $= \pi \times \dfrac{(1.0 \times 10^{-3})^2}{4} = 7.853... \times 10^{-7}$ ***[1 mark]***
Stress, $\sigma = \dfrac{F}{A} = 300 \div (7.853... \times 10^{-7}) = 3.819... \times 10^8$
= **3.8 × 10^8 Nm^{-2} (or Pa) (to 2 s.f.)** ***[1 mark]***

b) Strain, $\varepsilon = \dfrac{\Delta x}{x}$
$= (4.0 \times 10^{-3}) \div 2.00$ = **2.0 × 10^{-3}** ***[1 mark]***

2 a) $\Delta F = k\Delta x$ and so rearranging $k = \Delta F \div \Delta x$ ***[1 mark]***
$k = 50.0 \div (3.0 \times 10^{-3}) = 1.666... \times 10^4$
= **1.7 × 10^4 Nm^{-1} (to 2 s.f.)** ***[1 mark]***

b) Elastic strain energy, $\Delta E_{elastic} = \frac{1}{2}F\Delta x$
$= \frac{1}{2} \times 50.0 \times (3.0 \times 10^{-3})$
= **7.5 × 10^{-2} J** ***[1 mark]***
You could also use $\Delta E_{elastic} = \frac{1}{2}k\Delta x^2$ and substitute in your value of k.

Answers

3 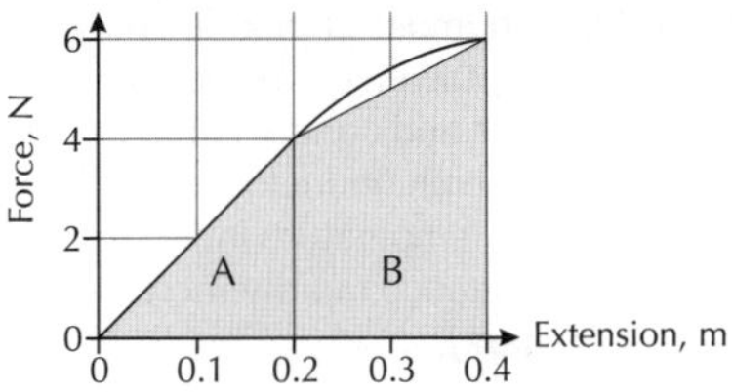

Divide the graph into shapes, then find the area of each:
Area A = 0.5 × base × height = 0.5 × 0.2 × 4 = 0.4 J
Area B = 0.5 × (*a* + *b*) × *h* = 0.5 × (4 + 6) × 0.2 = 1 J
[1 mark for sensibly splitting the area under the graph up and correctly calculating at least one of these areas]
Total elastic strain energy stored = 0.4 + 1
= **1.4 J** ***[1 mark]***

Shape B is a trapezium. The area of a trapezium is given by $A = \frac{1}{2}(a + b) \times h$, where *a* is the length of the first side, *b* is the length of the second side and *h* is the width of the strip. You won't be given this formula in your exam, so if you get stuck you can always break the trapeziums up into rectangles and triangles.
You could also calculate this area by estimating the number of grid squares the area under the graph covers. The area of each grid square is equal to 2 × 0.1 = 0.2 J. There are approximately 7 grid squares under the graph, so the total energy stored is approximately 7 × 0.2 = 1.4 J.

4 The force needed to compress the spring is:
$\Delta F = k\Delta x = 40.8 \times 0.0500 = 2.04$ N ***[1 mark]***
The elastic strain energy in the spring is then:
$E_{elastic} = \frac{1}{2} F\Delta x = \frac{1}{2} \times 2.04 \times 0.05 = 0.051$ J ***[1 mark]***
Assume all this energy is converted to kinetic energy in the ball.
$E_{elastic} = E_{kinetic} =$ **0.051 J** ***[1 mark]***.
You could also begin by using Hooke's law to replace *F* in the formula for elastic strain energy, to give $\Delta E_{elastic} = \frac{1}{2}k\Delta x^2$, and then substituting into this.

Page 57 — The Young Modulus

1 Cross-sectional area = πr^2 or $\pi\left(\frac{d}{2}\right)^2$

So area = $\pi \times \frac{(0.60 \times 10^{-3})^2}{4} = 2.827... \times 10^{-7}$ m² ***[1 mark]***

Stress, $\sigma = \frac{F}{A} = 80.0 \div (2.827... \times 10^{-7})$
$= 2.829... \times 10^{8}$ Nm⁻² ***[1 mark]***

Strain, $\varepsilon = \frac{\Delta x}{x} = (3.6 \times 10^{-3}) \div 2.50$
$= 1.44 \times 10^{-3}$ ***[1 mark]***

Young modulus, $E = \frac{\sigma}{\varepsilon} = (2.829... \times 10^{8}) \div (1.44 \times 10^{-3})$
$= 1.964... \times 10^{11}$
$=$ **2.0×10^{11} Nm⁻² (to 2 s.f.)** ***[1 mark]***

2 $E = \frac{\sigma}{\varepsilon} = \frac{F}{A} \div \frac{\Delta x}{x} = \frac{Fx}{\Delta x A}$. Force, original length and extension are the same for both wires, so $E \propto \frac{1}{A}$.
The wire B has half the cross-sectional area of the wire A. So the Young modulus of wire B (E_B) must be twice that of the wire A ***[1 mark]***.
$E_B = 2 \times (7.0 \times 10^{10}) =$ **1.4×10^{11} Nm⁻²** ***[1 mark]***

3 a) Young modulus, $E = \frac{\sigma}{\varepsilon}$
$\varepsilon = \frac{\sigma}{E} = (2.6 \times 10^{8}) \div (1.3 \times 10^{11})$ ***[1 mark]***
$=$ **2.0×10^{-3}** ***[1 mark]***

b) $\sigma = \frac{F}{A}$ and so $A = \frac{F}{\varepsilon} = 100 \div (2.6 \times 10^{8})$ ***[1 mark]***
$= 3.846... \times 10^{-7}$
$=$ **3.8×10^{-7} m² (to 2 s.f.)** ***[1 mark]***

c) Strain energy per unit volume
$= \frac{1}{2} \times \sigma \times \varepsilon = \frac{1}{2} \times (2.6 \times 10^{8}) \times (2.0 \times 10^{-3})$ ***[1 mark]***
$=$ **2.6×10^{5} Jm⁻³** ***[1 mark]***
Give the mark if answer is consistent with the value calculated for strain in part a).

Page 60 — Structures and Properties of Materials

1 a) Instead of increasing the space between the ions in the metal, the applied force causes planes of ions to slip across each other. This slipping is what causes the plastic deformation ***[1 mark]***. Dislocations or imperfections lower the stress needed to cause slipping, meaning that a lower stress is needed in order to cause plastic deformation ***[1 mark]***.

b) Alloying a metal is the process of adding ions from a second material into the crystal lattice ***[1 mark]***. These ions are placed inside dislocations to 'pin' them down ***[1 mark]***. This means a greater stress is needed to cause slipping, and so the chance of slipping is reduced ***[1 mark]***. Alloying the sample will make it harder and less ductile ***[1 mark]***.

2 C ***[1 mark]***

3 ***5-6 marks:***
The answer thoroughly describes two methods and how they measure the size and/or spacings of atoms. References may be made to other additional methods. The time frame has been referenced, showing an appreciation that increased knowledge has lead to the creation of new methods which can directly observe atomic size.
The answer is structured in a logical way, with relevant information supporting it throughout.
3-4 marks:
The answer correctly identifies two methods for calculating atomic size and/or spacing. Brief explanations of these methods are given. No reference to how changing methods over time have lead to direct measurements of atomic sizes/spacing. The answer has some logical structure, with mostly relevant information supporting it.
1-2 marks:
The answer attempts to identify experiments to measure atomic size and spacing, but doesn't describe them.
The answer is basic, poorly structured and unsupported by relevant information.
0 marks:
No relevant information is given.
Here are some points your answer may include:

- Rayleigh's oil drop method was used to estimate atomic sizes.
- This involved comparing the volume of an oil drop to the area and thickness of a thin sheet of oil floating on water.
- From this, the height of an oil molecule could be found.
- The size of an atom is roughly equal to the size of an oil molecule divided by the number of atoms in an oil molecule, so Rayleigh's measurements could be used to find an upper limit for the size of individual atoms.
- Modern day methods include X-ray crystallography, where X-rays are diffracted by a sample.
- The diffraction patterns are then analysed, in order to investigate atomic spacing.
- Scanning Tunnelling Microscopes are used to probe a surface of a sample.
- A voltage is applied to the microscope's fine tip. Electrons then tunnel from the sample to the tip, creating a current.
- The height of the tip is adjusted to keep this current constant, so the surface of the sample can be mapped to find atomic size and spacing.

Answers

Page 63 — Stress-Strain and Force-Extension Graphs

1 E.g.

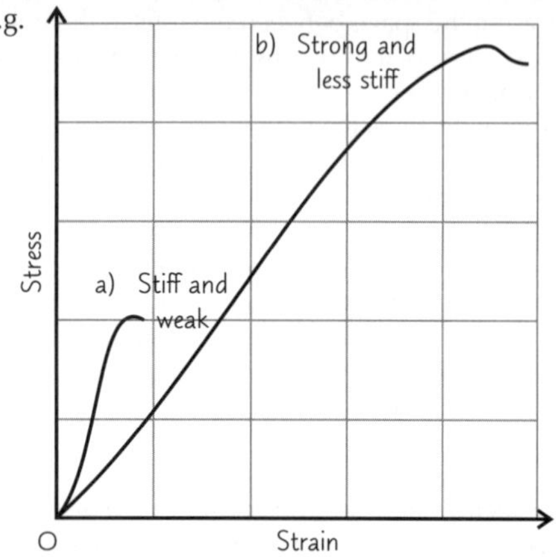

[1 mark for a stiff and weak curve that stops at a low stress and a low strain. 1 mark for a strong and less stiff curve that stops at a high stress and a high strain]

2 a) A ***[1 mark]***

b) B ***[1 mark]***

Topic 4 — Electricity

Page 65 — Charge, Current and Potential Difference

1 Time in seconds = 10 × 60 = 600 s
$I = \Delta Q / \Delta t$ = 4500 / 600 = 7.5 = **8 A (to 1 s.f.)** ***[1 mark]***

2 Energy transferred to water = 0.88 × electrical energy input so the energy input will be 308 / 0.88 = 350 J ***[1 mark]***
$W = VQ$ so $Q = W / V$ = 350 / 230 = **1.5 C (to 2 s.f.)** ***[1 mark]***
The heat energy that the kettle transfers to the water is less than the electrical energy input because the kettle is less than 100% efficient.

3 $I = nqvA$ so $v = I / Anq$
so $v = 13 \div ((5.0 \times 10^{-6}) \times (1.0 \times 10^{29}) \times (-1.60 \times 10^{-19}))$ ***[1 mark]***
$= \mathbf{-1.6 \times 10^{-4}\ ms^{-1}}$ **(to 2 s.f.)** ***[1 mark]***

Page 67 — Resistance and Conductance

1 a) $R = \frac{V}{I} = \frac{1.5}{2.8}$ = 0.5357... = **0.54 Ω (to 2 s.f.)** ***[1 mark]***

b) $G = \frac{I}{V} = \frac{0.15}{1.5}$ = **0.10 S** (or 0.10 Ω^{-1}) ***[1 mark]***

c) The wires have a resistance that causes power dissipation ***[1 mark]***. The power that is dissipated is transferred to heat, making the wires warm ***[1 mark]***.

Page 69 — Electrical Properties of Solids

1 a) E.g.

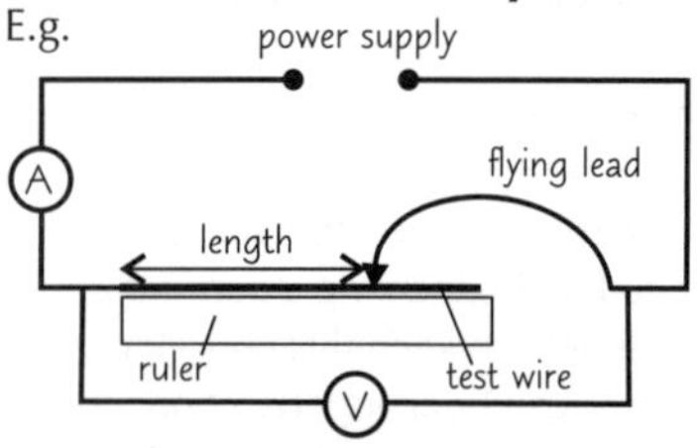

[1 mark for sensible circuit diagram]
Measure the diameter of the wire using a micrometer (to calculate the cross-sectional area) ***[1 mark]***. Connect the wire in a circuit as shown above. Use the flying lead to vary the length of wire connected to the circuit ***[1 mark]***. Take a measurement from the voltmeter and the ammeter for each length of wire used, and use these to calculate the wire's resistance at each length using the formula $R = \frac{V}{I}$ ***[1 mark]***. Plot a graph of resistance against length, then calculate the gradient of the line of best fit ***[1 mark]***. Multiply the gradient of the graph by the cross-sectional area to get ρ (as $\rho = \frac{RA}{L}$). ***[1 mark]***

b) Copper is a metal and has a high number of mobile charge carriers. As temperature increases, the electrons become less free to move so resistance increases and conductance decreases ***[1 mark]***. An NTC thermistor is a semiconductor so has fewer free electrons to begin with, but as temperature increases more electrons are freed so conductance rapidly increases and resistance rapidly decreases ***[1 mark]***.

Page 71 — Power, Work and Electrical Energy

1 a) Energy transferred = $E = VIt$ = 12 × 48 × 2.0
= 1152 = **1200 J (to 2 s.f.)** ***[1 mark]***

b) Energy wasted in wires = $E = I^2Rt = 48^2 \times 0.010 \times 2.0$ = 46.08
= **46 J (to 2 s.f.)** ***[1 mark]***

2 a) $E = Pt$
Model A: E = 0.47 × (135 ÷ 60)
= 1.0575 = **1.1 kWh (to 2 s.f.)** ***[1 mark]***
Model B: E = 0.41 × (125 ÷ 60)
= 0.85416... = **0.85 kWh (to 2 s.f.)** ***[1 mark]***

b) Weeks in a year = 52, so the customer will do 52 × 2 = 104 loads.
Model A:
Energy used per year = 1.0575 × 104 = 109.98 kWh ***[1 mark]***
Cost per year = 109.98 × 16.2 = **1800p (to 2 s.f.)** (or £18.00) ***[1 mark]***
Model B:
Energy used per year = 0.854... × 104 = 88.833... kWh ***[1 mark]***
Cost per year = 88.833... × 16.2 = **1400p (to 2 s.f.)** (or £14.00) ***[1 mark]***

Page 73 — Internal Resistance and E.m.f.

1 a) $\varepsilon = I(R + r)$ so $I = \varepsilon / (R + r)$
= 24 / (4.0 + 0.8) = **5 A** ***[1 mark]***

b) $V = IR$ = 5 × 4.0 = **20 V** ***[1 mark]***
You could also have used $\varepsilon = V + Ir$.

2 C ***[1 mark]***
$\varepsilon = I(R + r)$, but since there are two cells in series replace r with $2r$, and ε with 2ε, then rearrange to find I.

Page 75 — Conservation of Energy & Charge in Circuits

1 a) Resistance of parallel resistors:
$1/R_{parallel}$ = 1/6.0 + 1/3.0 = 1/2 ⇒ $R_{parallel}$ = 2.0 Ω ***[1 mark]***
Total resistance:
R_{total} = 4.0 + $R_{parallel}$ = 4.0 + 2.0 = **6.0 Ω** ***[1 mark]***

b) $V = I_3R_{total} \Rightarrow I_3 = V / R_{total}$ = 12 / 6.0 = **2.0 A** ***[1 mark]***

c) $V = IR$ = 2.0 × 4.0 = **8.0 V** ***[1 mark]***

d) E.m.f. = sum of p.d.s in circuit, so 12 = 8.0 + $V_{parallel}$
$V_{parallel}$ = 12 – 8.0 = **4.0 V** ***[1 mark]***

e) $I = V/R$, so I_1 = 4.0 / 3.0 = **1.3 A (to 2 s.f.)** ***[1 mark]***
I_2 = 4.0 / 6.0 = **0.67 A (to 2 s.f.)** ***[1 mark]***

You can check your answers by making sure that $I_3 = I_2 + I_1$.

Page 77 — The Potential Divider

1 Parallel circuit, so p.d. across both sets of resistors is 12 V.

a) There are two equal resistors in the top branch of the circuit. The p.d. between points A and B is equal to the potential difference across one of these resistors:
$V_{AB} = \frac{1}{2} \times 12$ = **6.0 V** ***[1 mark]***

b) There are three equal resistors in the bottom branch of the circuit. The p.d. between points A and C is equal to the potential difference across two of them:
$V_{AC} = \frac{2}{3} \times 12$ = **8.0 V** ***[1 mark]***

c) $V_{BC} = V_{AC} - V_{AB}$ = 8.0 – 6.0 = **2.0 V** ***[1 mark]***

Answers

2 a) $V_{AB} = 50/80 \times 12 =$ **7.5 V** ***[1 mark]***
(ignore the 10 Ω — no current flows that way)
b) Total resistance of the parallel circuit:
$1/R_T = 1/50 + 1/(10 + 40.0) = 1/25 \Rightarrow R_T = 25\ \Omega$ ***[1 mark]***
p.d. over the whole parallel arrangement = 25/(30 + 25) × 12
= 5.45... V ***[1 mark]***
p.d. across AB = 40.0/(40.0 + 10) × 5.45...
= 4.36... V = **4.4 V (to 2 s.f.)** ***[1 mark]***
current through 40.0 Ω resistor = V/R
= 4.36.../40.0 = **0.11 A (to 2 s.f.)** ***[1 mark]***

Topic 5 — Waves

Page 79 — Progressive Waves

1 a) The gain is set to 2.0 volts/div and the trace has a maximum amplitude of 2 divisions.
So, the maximum voltage = 2 × 2.0 = **4.0 V** ***[1 mark]***
b) One cycle spans 3 divisions, so
$T = 3 \times 3.0\text{ ms} = 9.0\text{ ms} = 9.0 \times 10^{-3}\text{ s}$ ***[1 mark]***
$f = 1/T = 1/9.0 \times 10^{-3}$ ***[1 mark]***
$f = 111.1...$ = **110 Hz (to 2 s.f.)** ***[1 mark]***
$v = f\lambda$ so $\lambda = v/f$
$\lambda = 280/111.1...$ ***[1 mark]***
$\lambda = 2.52$ m = **2.5 m (to 2 s.f.)** ***[1 mark]***

Page 81 — Longitudinal and Transverse Waves

1 a) $I = P \div A = 10.0 \div 0.002 =$ **5000 Wm^{-2}** ***[1 mark]***
b) D ***[1 mark]***

Page 83 — Polarisation of Waves

1 a) They are at right angles to one another (90°, 270° etc.) ***[1 mark]***.
b) It would be half of the intensity ***[1 mark]***.
This is because at 45° the vertical and horizontal contributions are equal, so the intensity is halved between them.
c) E.g. Polaroid sunglasses or 3D film glasses ***[1 mark]***.

Page 86 — Reflection and Diffraction

1 When a wavefront meets an obstacle, the waves will diffract round the corners of the obstacle. When the obstacle is much bigger than the wavelength, little diffraction occurs. In this case, the mountain is much bigger than the wavelength of short-wave radio. So the "shadow" where you cannot pick up short wave is very long ***[1 mark]***.

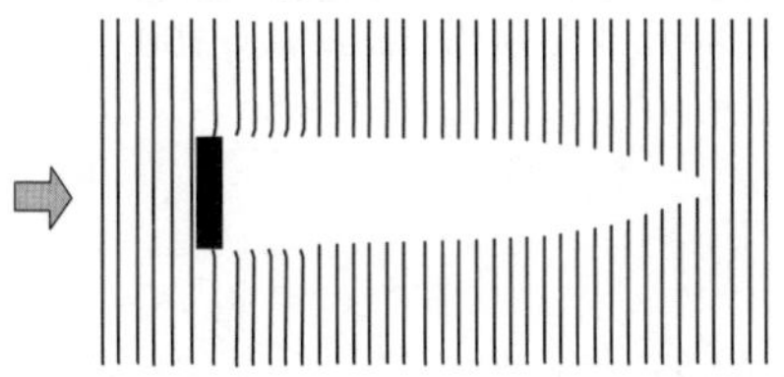

[1 mark]

When the obstacle is comparable in size to the wavelength, as it is for the long-wave radio waves, more diffraction occurs. The wavefront re-forms after a shorter distance, leaving a shorter "shadow" ***[1 mark]***.

2 a) E.g. Laser light is monochromatic/only contains one wavelength/frequency of light ***[1 mark]***. This provides a clearer pattern than non-monochromatic light sources as different wavelengths diffract by different amounts ***[1 mark]***.
b) The central maximum will be wider and less intense ***[1 mark]***. Using a narrower slit means more diffraction ***[1 mark]*** and so there will be fewer photons per unit area, so the intensity will be lower ***[1 mark]***.

Page 89 — Refraction

1 a) $n_{diamond} = c / c_{diamond} = (3.00 \times 10^8) / (1.24 \times 10^8) = 2.419...$
= **2.42 (to 3 s.f.)** ***[1 mark]***
b) $n_{air} \sin\theta_1 = n_{diamond} \sin\theta_2$, $n_{air} = 1$
So, $\sin\theta_1 = n_{diamond} \sin\theta_2$ ***[1 mark]***
$\sin\theta_2 = \sin 50° / 2.419... = 0.316...$
$\theta_2 = \sin^{-1}(0.316...) = 18.459...$ = **18° (to 2 s.f.)** ***[1 mark]***
You can assume the refractive index of air is 1, and don't forget to write the degree sign in your answer.

2 a) When the light is pointing steeply upwards some of it is refracted and some reflected — the beam emerging from the surface is the refracted part ***[1 mark]***.
However, when the beam hits the surface at more than the critical angle (to the normal to the boundary) refraction does not occur. All the beam is totally internally reflected to light the tank, hence its brightness ***[1 mark]***.
b) The critical angle is 90° – 41.25° = 48.75° ***[1 mark]***.
$\sin\theta_c = n_{air} / n_{water} = 1 / n_{water}$,
so $n_{water} = 1 / \sin\theta_c = 1 / \sin 48.75°$
= 1.330... = **1.330 (to 4 s.f.)** ***[1 mark]***
The question talks about the angle between the light beam and the floor of the aquarium. This angle is 90° minus the incident angle — measured from a normal to the surface of the water.

3 a) The cladding has a lower refractive index than the fibre, to allow total internal reflection ***[1 mark]***. It also protects the cable from scratches and damage which may let light escape ***[1 mark]***.
b) How to mark your answer (pick the level description that best matches your answer):
5-6 marks:
The answer fully describes several potential causes of signal degradation and correctly explains how to alter the design or operation to reduce the effects.
3-4 marks:
One potential cause of signal degradation has been fully described with a full explanation of how to reduce the effect. Another potential cause is mentioned with incomplete information or suggestions for design/operation alterations.
1-2 marks:
Potential reason(s) for signal degradation suggested but with incomplete description of the effects or of suggestions of how to reduce the effects with design changes.
0 marks:
No relevant information is given.
Here are some points your answer may include:
- Absorption of light by the material the fibre is made from causes a loss of signal amplitude.
- Dispersion within the fibre causes pulse broadening.
- Modal dispersion is caused by light rays taking different paths of different lengths down the fibre.
- Material dispersion is caused by different wavelengths of light refracting by different amounts in the fibre.
- Designing the fibre so that it only allows light to take one path through it (known as a single-mode fibre) reduces pulse broadening due to modal dispersion.
- Sending signals using monochromatic light reduces material dispersion because monochromatic light only has one wavelength/frequency.
- Using an optical fibre repeater to boost/regenerate signals can help to reduce signal degradation over long transmission distances.

Page 91 — Superposition and Coherence

1 a) The frequencies and wavelengths of the two sources must be equal ***[1 mark]*** and the phase difference must be constant ***[1 mark]***.
b) Interference will only be noticeable if the amplitudes of the two waves are approximately equal ***[1 mark]***.

2 B ***[1 mark]***

Answers

Page 93 — Two-Source Interference

1 a) Distance between sources (slit width) ***[1 mark]***, spacing between two consecutive maxima or minima (fringe spacing) ***[1 mark]*** and the distance from the sources to the screen ***[1 mark]***.

b) Any two from: Don't look into the laser beam / don't point the beam at a person / don't point the laser beam at a reflective surface / display a laser warning sign / wear laser safety goggles / turn the laser off when not in use ***[1 mark for two correct suggestions]***
Laser light can permanently damage your eyes/retinas ***[1 mark]***.

2 a) $\lambda = c / f = 330 / 1320 =$ **0.25 m** ***[1 mark]***

b) Separation $= w = \lambda D / s = (0.25 \times 7.3) / 1.5 = 1.21666...$
$=$ **1.2 m (to 2 s.f.)** ***[1 mark]***

Page 95 — Diffraction Gratings

1 a) Use $d \sin\theta = n\lambda$.
For the first order, $n = 1$, so, $\sin\theta = \lambda / d$ ***[1 mark]***
No need to actually work out d. The number of lines per metre is $1 / d$. So you can simply multiply the wavelength by that.
$\sin\theta = 6.0 \times 10^{-7} \times 4.0 \times 10^5 = 0.24$
$\theta = \sin^{-1}(0.24) = 13.886... =$ **14° (to 2 s.f.)** ***[1 mark]***
For the second order, $n = 2$ and $\sin\theta = 2\lambda / d$.
You already have a value for λ / d. Just double it to get $\sin\theta$ for the second order.
$\sin\theta = 0.48$, $\theta = \sin^{-1}(0.48) = 28.685... =$ **29° (to 2 s.f.)** ***[1 mark]***

b) No. Putting $n = 5$ into the equation gives a value of $\sin\theta$ of 1.2, which is impossible ***[1 mark]***.

2 $d \sin\theta = n\lambda$, so for the 1st order maximum, $d \sin\theta = \lambda$
$\sin 14.2° = \lambda \times 3.7 \times 10^5$ ***[1 mark]***
$\lambda = 6.629... \times 10^{-7} =$ **6.6×10^{-7} m (or 660 nm) (to 2 s.f.)** ***[1 mark]***

Page 98 — Stationary (Standing) Waves

1

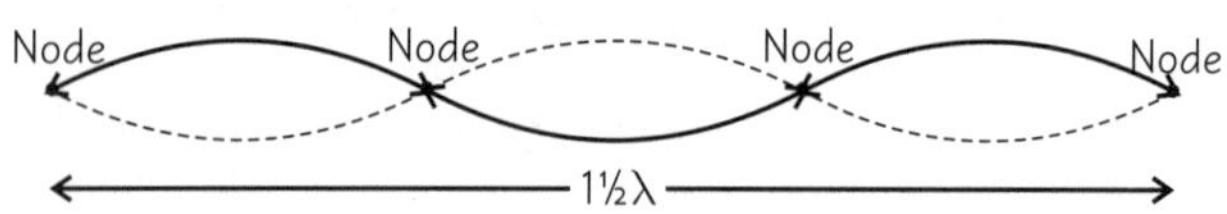

[1 mark for the correct shape, 1 mark for labelling the length]

2 Rearrange $v = \sqrt{\frac{T}{\mu}}$ for μ:
$\mu = \frac{T}{v^2} = \frac{3.2}{16^2} = 0.0125 \text{ kgm}^{-1}$ ***[1 mark]***
$\mu = \frac{M}{l}$, so $M = \mu l = 0.0125 \times 2.8 =$ **0.035 kg** (= 35 g) ***[1 mark]***

3 Distance between nodes $= \lambda/2$ ***[1 mark]***
$\lambda = 2 \times 15.0 \text{ cm} = 30.0 \text{ cm} = 0.30 \text{ m}$ ***[1 mark]***
$v = f\lambda$ so $v = 1.00 \times 10^9 \times 0.30$ ***[1 mark]***
$v =$ **3.0×10^8 ms^{-1}** ***[1 mark]***

Page 99 — Phasors

1

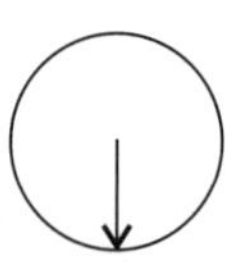

[1 mark for correct direction] The magnitude of the phasor is the same size as for the first sound wave ***[1 mark]***.

Topic 6 — Imaging and Signalling

Page 102 — Lenses

1 a) Rays meeting the lens parallel to the principal axis converge at the principal focus. / Waves parallel to the lens axis are given spherical curvature as they pass through the lens. This curvature is centred on the focal point. ***[1 mark]***

b) $\frac{1}{f} = \frac{1}{u} + \frac{1}{v}$ so $\frac{1}{v} = -\frac{1}{u} + \frac{1}{f} = -\frac{1}{0.2} + \frac{1}{0.15} = \frac{5}{3}$ ***[1 mark]***,
so $v = \frac{3}{5} =$ **0.60 m** ***[1 mark]***

2 a) $m = \frac{\text{size of image}}{\text{size of object}} = \frac{47.2}{12.5} = 3.776 =$ **3.78 (to 3 s.f.)** ***[1 mark]***

b) $m = \frac{v}{u}$, giving $v = m \times u$ ***[1 mark]***
$v = 3.776 \times 4.0 = 15.104 =$ **15 mm (to 2 s.f.)** ***[1 mark]***

c) $P = \frac{1}{f} = \frac{1}{u} + \frac{1}{v}$ ***[1 mark]***, so $P = \left(\frac{1}{0.0040}\right) + \frac{1}{0.015104} = 316.20...$
$=$ **320 D (to 2 s.f.)** ***[1 mark]***

Page 105 — Information in Images

1 a) Number of bits $= \log_2(65\,536) = 16$ ***[1 mark]***.
Number of bytes = number of bits ÷ 8 = 16 ÷ 8 = **2 bytes** ***[1 mark]***

b) total information = number of pixels × bits per pixel
$= (1920 \times 1080) \times 16 = 3.3177... \times 10^7$
$=$ **3.32×10^7 bits (to 3 s.f.)** ***[1 mark]***

c) width of square = 1920 × 0.25 = 480 pixels ***[1 mark]***
resolution = real size of object ÷ pixels representing object
$= 1.5 \div 480 = 3.125 \times 10^{-3}$
$=$ **3.1×10^{-3} m pixel^{-1} (to 2 s.f.)** ***[1 mark]***

2 a)

The diagram should show a fairly uniform mid-grey outside, with a lighter square in the centre. ***[1 mark]***

b) Noise can be removed by replacing each pixel with the median of itself and the eight pixels surrounding it. ***[1 mark]***

c) Write the values out in numerical order:
97, 98, 98, 99, 100, 100, 101, 101, 185
The value in the middle of the list (the 5th value) is 100.
So replace the 185 in the middle of the array with 100:

100	99	100
97	100	98
101	101	98

[1 mark]

Page 107 — Digital and Analogue Signals

1 a) time period of shortest repeating part $= (1.25 - 0.25) \times 10^{-3}$
$= 1.0 \times 10^{-3}$ s
$f = \frac{1}{T} = \frac{1}{1.0 \times 10^{-3}} =$ **1000 Hz** ***[1 mark]***

b)

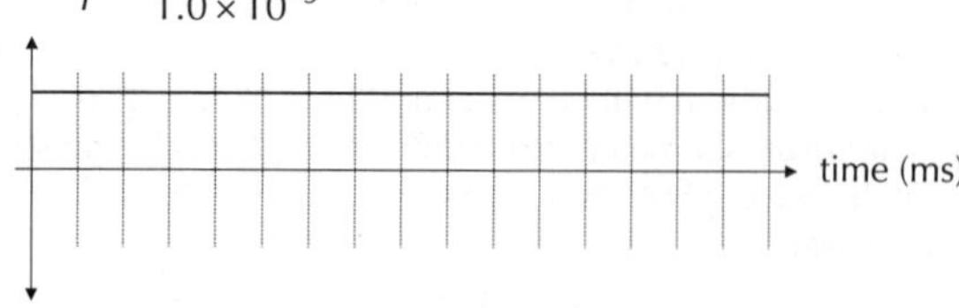

[1 mark for a straight line and 1 mark for the line being above the horizontal axis]
The graph is a straight line because the sample frequency is the same as the fundamental frequency.

Answers

Page 109 — Sampling and Transmitting Signals

1 $b = \log_2\left(\frac{V_{total}}{V_{noise}}\right)$ ***[1 mark]*** $= \log_2\left(\frac{160}{10}\right) = \log_2 16 =$ **4 bits** ***[1 mark]***

You don't need to convert the variations from mV to V because the units cancel in the fraction.

2 a) Rate of transmission = samples per second × bits per sample
= 8000 × 8
= **64 000 bits per second** ***[1 mark]***

b) 1 byte = 8 bits.
So, 64 000 bits per second = **8000 bytes per second** ***[1 mark]***

3 a) rate of transmission = samples per second × bits per sample, so

samples per second = $\frac{\text{rates of transmission}}{\text{bits per sample}} = \frac{128 \times 10^3}{16}$
= **8000 Hz** ***[1 mark]***

b) rate of transmission = number of bits to transmit ÷ time taken

number of bits = $(2.0 \times 10^6) \times 8 = 1.6 \times 10^7$ bits ***[1 mark]***

rate of transmission = $\frac{1.6 \times 10^7}{110} = 145\,454 = 150$ kbits^{-1} (to 2 s.f.)

[1 mark] **150 > 128, so yes** — the connection is sufficient to stream the radio station. ***[1 mark]***

Page 111 — Ultrasound Imaging

1 E.g. A transducer is used to send pulses of ultrasound waves into the patient's body ***[1 mark]***. Waves are partially reflected and partially transmitted when they meet an interface, with the proportion reflected increasing with the difference in density between the media ***[1 mark]***. When the ultrasound waves meet an interface in the patient's body, for example between different tissues, some of the ultrasound waves are reflected back, which are then detected by the transducer ***[1 mark]***. From the time taken for the signals to return, a computer attached to the transducer can work out the location of different tissue boundaries, and build up an image of inside the patient's body ***[1 mark]***.

2 a) Sound waves with short wavelengths diffract less than sounds waves with long wavelengths ***[1 mark]***. So decreasing the wavelength means that the waves spread out less, which improves the precision of the location of tissue surfaces ***[1 mark]***. This means the image produced is of a higher quality ***[1 mark]***.

b) The shorter the pulse length, the less likely it is that a reflected ultrasound wave will reach the transducer whilst it is still transmitting the pulse ***[1 mark]***. The transducer cannot receive the reflected waves whilst it is transmitting, so if this did happen, data would be lost ***[1 mark]***. A longer time between the pulses means a reflected ultrasound wave is less likely to reach the transducer after it has transmitted the next pulse ***[1 mark]***. If this did happen, the computer would not be able to distinguish which transmitted wave corresponded with which reflected wave, and so image quality would be lost ***[1 mark]***.

Topic 7 — Particles

Page 113 — Atomic Structure

1 Inside every atom there is a nucleus which contains protons and neutrons ***[1 mark]***. Orbiting this core are the electrons ***[1 mark]***.

2 Proton number is 8, so there are 8 protons and electrons (it's neutral) ***[1 mark]***. The nucleon number is 16. This is the total number of protons and neutrons. Subtract the 8 protons and that leaves 8 neutrons ***[1 mark]***.

3 a) Atoms with the same number of protons but different numbers of neutrons are called isotopes ***[1 mark]***.

b) Any two from: e.g. They have the same chemical properties. / Their nuclei have different stabilities. / They have different physical properties. ***[2 marks available — 1 mark for each correct answer.]***

4 A $^4_2$He nucleus has 2 protons and 2 neutrons (but no electrons) ***[1 mark]***.
Charge = $2 \times 1.60 \times 10^{-19} = 3.20 \times 10^{-19}$ C ***[1 mark]***
Mass = $(2 \times 1.673 \times 10^{-27}) + (2 \times 1.675 \times 10^{-27})$
$= 6.696 \times 10^{-27}$ kg ***[1 mark]***

Specific charge = $\frac{\text{charge}}{\text{mass}} = \frac{3.20 \times 10^{-19}}{6.696 \times 10^{-27}} = 4.778... \times 10^7$

= $\mathbf{4.78 \times 10^7}$ **C kg^{-1} (to 3 s.f.)** ***[1 mark]***

The actual mass of a helium nucleus is slightly less than this due to energy being released when the nucleus was made. However, you don't need to worry about this at this level.

Page 115 — Stable and Unstable Nuclei

1 a) The strong nuclear force must be repulsive at very small nucleon separations to prevent the nucleus being crushed to a point ***[1 mark]***.

b) The protons repel each other with an electrostatic force and attract each other with the strong nuclear force. The strong force is not large enough to overcome this repulsion ***[1 mark]***. When two neutrons are added to the nucleus, they attract each other and the protons via the strong force. The strong force is now able to balance out the force of repulsion between the protons ***[1 mark]***.

2 a) $^{226}_{88}Ra \rightarrow ^{222}_{86}Rn + ^{4}_{2}\alpha$ ***[3 marks available — 1 mark for the alpha particle, 1 mark for the proton number of radon and 1 mark for the nucleon number of radon.]***

b) $^{40}_{19}K \rightarrow ^{40}_{20}Ca + ^{0}_{-1}\beta + \bar{\nu}_e$ ***[4 marks available — 1 mark for the beta particle, 1 mark each for the proton number and nucleon number of calcium, 1 mark for the antineutrino.]***

Page 117 — Particles and Antiparticles

1 $e^+ + e^- \rightarrow \gamma + \gamma$ ***[1 mark]*** This is called annihilation ***[1 mark]***

2 Energy and mass are equivalent ***[1 mark]***. When two particles collide, there is a lot of energy at the point of impact. This energy is converted to mass ***[1 mark]***.

3 The creation of a particle of matter also requires the creation of its antiparticle. In this case no antineutron has been produced ***[1 mark]***.

4 Total energy = $2 \times 9.84 \times 10^{-14} = 1.968 \times 10^{-13}$ J ***[1 mark]***

$E = hf$, so $f = \frac{E}{h} = \frac{1.968 \times 10^{-13}}{6.63 \times 10^{-34}} = 2.968... \times 10^{20}$

= $\mathbf{2.97 \times 10^{20}}$ **Hz (to 3 s.f.)** ***[1 mark]***

Page 119 — Forces and Exchange Particles

1 The electrostatic force is due to the exchange of virtual photons that only exist for a very short time ***[1 mark]***. The force is due to the momentum transferred to or gained from the photons as they are emitted or absorbed by a proton ***[1 mark]***.

2

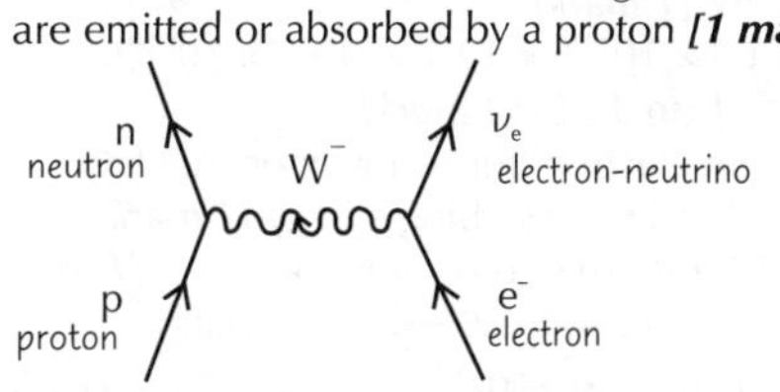

[1 mark for showing a W⁻ boson being exchanged from the electron to the proton, 1 mark for showing a neutron is produced.]

This is a weak interaction ***[1 mark]***.

Don't forget to put the arrows on your diagram.
Remember that the weak interaction uses W bosons, and is the only force that can change protons into neutrons or neutrons into protons.

Answers

Page 122 — Classification of Particles

1 Proton, electron and electron-antineutrino ***[1 mark]***. The electron and the electron-antineutrino are leptons ***[1 mark]***. Leptons are not affected by the strong interaction, so the decay can't be due to the strong interaction ***[1 mark]***.
Remember that it's really just the same as beta decay.
Some books might leave out the antineutrino, so don't be misled.

2 Mesons are hadrons but the muon is a lepton ***[1 mark]***. The muon is a fundamental particle but mesons are not. Mesons are built up from simpler particles ***[1 mark]***. Mesons interact via the strong interaction but the muon does not ***[1 mark]***.
You need to classify the muon correctly first and then say why it's different from a meson because of what it's like and what it does.

3 The correct answer is **B** ***[1 mark]***.
Protons and neutrons are both baryons, which are a type of hadron. Electrons are leptons.

Page 125 — Quarks

1 a) $\pi^- = d\bar{u}$ ***[1 mark]***
b) Charge of down quark = –1/3 unit.
Charge of anti-up quark = –2/3 unit.
Total charge = –1 unit ***[1 mark]***

2 The weak interaction converts a down quark into an up quark plus an electron and an electron-antineutrino ***[1 mark]***.
The neutron (udd) becomes a proton (uud) ***[1 mark]***.
The lepton number L_e is conserved in this reaction.

3 The Baryon number changes from 2 to 1, so baryon number is not conserved ***[1 mark]***. The strangeness changes from 0 to 1, so strangeness is not conserved ***[1 mark]***.

Topic 8 — Quantum Phenomena

Page 127 — The Photoelectric Effect

1 The plate becomes positively charged ***[1 mark]***.
Negative electrons in the metal absorb energy from the UV light and leave the surface ***[1 mark]***.
There's one mark for saying what happens, and a second mark for saying why.

2 An electron needs to gain a certain amount of energy (the work function) before it can leave the surface of the metal ***[1 mark]***.
$E = hf$, so if the frequency is too low, then the energy carried by each photon is less than this work function and so no electrons will be emitted ***[1 mark]***.

Page 130 — Electron Levels and Photon Emission

1 a) i) E (eV) = V = 12.1 eV ***[1 mark]***
ii) E (J) = E (eV) × (1.60 × 10^{-19}) = 12.1 × (1.60 × 10^{-19})
= **1.94 × 10^{-18} J (to 3 s.f.)** ***[1 mark]***
b) i) The movement of an electron from a lower energy level to a higher energy level by absorbing energy ***[1 mark]***.
ii) –13.6 + 12.1 = –1.5 eV. This corresponds to n = 3 ***[1 mark]***.
iii) n = 3 → n = 2: –1.5 – (–3.4) = **1.9 eV** ***[1 mark]***

$E = hf$, so $f = \frac{E}{h} = \frac{1.9 \times 1.60 \times 10^{-19}}{6.63 \times 10^{-34}} = 4.5852... \times 10^{14}$ Hz
= **4.6 × 10^{14} Hz (to 2 s.f.)** ***[1 mark]***

n = 2 → n = 1: –3.4 – (–13.6) = 10.2 eV
= **10 eV (to 2 s.f.)** ***[1 mark]***

$f = \frac{E}{h} = \frac{10.2 \times 1.60 \times 10^{-19}}{6.63 \times 10^{-34}}$
= 2.4615... × 10^{15} Hz
= **2.5 × 10^{15} Hz (to 2 s.f.)** ***[1 mark]***

n = 3 → n = 1: –1.5 – (–13.6) = 12.1
= **12 eV (to 2 s.f.)** ***[1 mark]***

$f = \frac{E}{h} = \frac{12.1 \times 1.60 \times 10^{-19}}{6.63 \times 10^{-34}}$
= 2.9200... × 10^{15} Hz
= **2.9 × 10^{15} Hz (to 2 s.f.)** ***[1 mark]***

Page 133 — The "Sum Over Paths" Theory

1 Probability ∝ (resultant phasor amplitude)^2
Probability of electron reaching point A ∝ $(6.3)^2$ = 39.69 ***[1 mark]***
Probability of electron reaching point B ∝ $(4.5)^2$ = 20.25 ***[1 mark]***
39.69 ÷ 20.25 = 1.96, so an electron is **2.0 (to 2 s.f.) times** as likely to reach point A as point B ***[1 mark]***.

2 The frequency of phasor rotation = 6.0 × 10^{14} rotations per second.
Use time (s) = distance (m) ÷ speed (ms^{-1}) to find the time taken for the photon to reach the detector.
t = 0.12 ÷ c = 0.12 ÷ 3.00 × 10^8 = 4.0 × 10^{-10} s ***[1 mark]***
So the number of phasor rotations along this path is:
$f \times t$ = 6.0 × 10^{14} × 4.0 × 10^{-10} = **2.4 × 10^5 rotations** ***[1 mark]***

3 Photons take every possible path. To find out the probability of a photon taking a particular path, the resultant phasors describing each individual path are summed ***[1 mark]***. Usually, these cancel each other out, with the quickest path (a straight line) having the largest resultant amplitude. The probability of a photon arriving at a certain point is proportional to the resultant phasor amplitude squared ***[1 mark]***. This means that light is most likely to take the quickest path, a straight line, as that is the path with the largest contribution to the resultant vector. As the detector is around the corner from the source, instead of in a straight line view of the source, the photons have a very low probability of reaching the detector ***[1 mark]***.

4 All of the phasors are pointing in the same direction, so the focal point is described by point B ***[1 mark]***.
The lens will focus the photons by ensuring the paths take the same time to reach the focal point.

Page 135 — Wave-Particle Duality

1 a) Electromagnetic radiation can show characteristics of both a particle and a wave. ***[1 mark]***
b) $\lambda = \frac{h}{p}$ so $p = \frac{h}{\lambda} = \frac{6.63 \times 10^{-34}}{590 \times 10^{-9}}$ ***[1 mark]***
= **1.1 × 10^{-27} kgms^{-1} (to 2 s.f.)** ***[1 mark]***

2 a) $\lambda = \frac{h}{mv} = \frac{6.63 \times 10^{-34}}{9.11 \times 10^{-31} \times 3.50 \times 10^6}$ ***[1 mark]***
= 2.079... × 10^{-10} = **2.08 × 10^{-10} m (to 3 s.f.)** ***[1 mark]***
b) Either $v = \frac{h}{m\lambda} = \frac{6.63 \times 10^{-34}}{1.67 \times 10^{-27} \times 2.079... \times 10^{-10}}$ ***[1 mark]***
= 1909.28... = **1910 ms^{-1} (to 3 s.f.)** ***[1 mark]***
Or momentum of protons = momentum of electrons
so $m_p \times v_p = m_e \times v_e$
$v_p = v_e \times \frac{m_e}{m_p}$ = 3.50 × 10^6 × $\frac{9.11 \times 10^{-31}}{1.67 \times 10^{-27}}$ ***[1 mark]***
= 1909.28... = **1910 ms^{-1} (to 3 s.f.)** ***[1 mark]***
c) The proton has a larger mass, so it will have a smaller speed, since the two have the same kinetic energy ***[1 mark]***.
Kinetic energy is proportional to the square of the speed, while momentum is proportional to the speed, so they will have different momenta ***[1 mark]***. Wavelength depends on the momentum, so the wavelengths will be different ***[1 mark]***.
This is a really hard question. If you didn't get it right, make sure you understand the answer fully. Do the algebra if it helps.

Index

Index

Index

Index

PHR54